Acta Numerica 2000

Acta
Numerica

Volume 9   2000

CAMBRIDGE
UNIVERSITY PRESS

CAMBRIDGE UNIVERSITY PRESS
Cambridge, New York, Melbourne, Madrid, Cape Town, Singapore,
São Paulo, Delhi, Dubai, Tokyo, Mexico City

Cambridge University Press
The Edinburgh Building, Cambridge CB2 8RU, UK

Published in the United States of America by Cambridge University Press, New York

www.cambridge.org
Information on this title: www.cambridge.org/9780521157674

© Cambridge University Press 2000

First published 2000
First paperback edition 2010

A catalogue record for this publication is available from the British Library

ISBN 978-0-521-78037-7 Hardback
ISBN 978-0-521-15767-4 Paperback

Contents

Acta Numerica (2000), pp. 1–38

Radial basis functions

M. D. Buhmann

Mathematical Institute,

Justus Liebig University,

35392 Giessen, Germany

E-mail: `Martin.Buhmann@math.uni-giessen.de`

Radial basis function methods are modern ways to approximate multivariate functions, especially in the absence of grid data. They have been known, tested and analysed for several years now and many positive properties have been identified. This paper gives a selective but up-to-date survey of several recent developments that explains their usefulness from the theoretical point of view and contributes useful new classes of radial basis function. We consider particularly the new results on convergence rates of interpolation with radial basis functions, as well as some of the various achievements on approximation on spheres, and the efficient numerical computation of interpolants for very large sets of data. Several examples of useful applications are stated at the end of the paper.

CONTENTS

1. Introduction

There is a multitude of ways to approximate a function of many variables: multivariate polynomials, splines, tensor product methods, local methods and global methods. All of these approaches have many advantages and some disadvantages, but if the dimensionality of the problem (the number of variables) is large, which is often the case in many applications from statistics to neural networks, our choice of methods is greatly reduced, unless

we resort solely to tensor product methods. In fact, sometimes we are given
scattered data, immediately excluding the use of tensor product methods.
Also, tensor product methods in high dimensions always require many, often too many data. In this situation, the method of choice is often a radial
basis function approach which is, incidentally, also highly useful in lower-
dimensional problems and as an alternative to (piecewise) polynomials, because of its excellent approximation properties; at any rate it is universally
applicable independent of dimension.

In order to formulate the problem as we will see it in this review, let Ξ
be a finite set of distinct points in $\mathbb{R}^n$, which are traditionally called *centres*
in radial basis function jargon, because our basis functions will be radially
symmetric about these points. The goal of our work is to approximate an
unknown function that is only given at those centres via a set of real numbers
f_ξ, $\xi \in \Xi$. These are almost always interpreted as function evaluations of
some smooth function $f : \mathbb{R}^n \supset \Omega \to \mathbb{R}$, so that $f_\xi = f(\xi)$. Here, Ω is a
domain in $\mathbb{R}^n$. This point of view will allow us to measure conveniently the
uniform approximation error between f and its approximant s. This error
depends on the choice of the approximant, on Ξ, on f, and in particular on
its smoothness.

In order to approximate with s, which is usually by interpolation, we
take a univariate continuous function ϕ that is radialized by composition
with the Euclidean norm on $\mathbb{R}^n$, or a suitable replacement thereof when
we are working on a sphere in n-dimensional Euclidean space, for instance.
This $\phi : \mathbb{R}_+ \to \mathbb{R}$ is the *radial basis function*. Additionally, we take the
given centres ξ from the given finite set Ξ of distinct points and use them
simultaneously for shifting the radial basis function and as interpolation
(collocation) points. Therefore, our standard radial function approximants
now have the form

$$s(x) = \sum_{\xi \in \Xi} \lambda_\xi \phi(\|x - \xi\|), \qquad x \in \mathbb{R}^n, \qquad (1.1)$$

suitable adjustments being made when x is not from the whole space, and
the coefficient vector $\lambda = (\lambda_\xi)_{\xi \in \Xi}$ is an element of $\mathbb{R}^\Xi$. In many instances,
particularly those that will interest us in Section 3, the interpolation re-
quirements

$$s \mid_\Xi = f \mid_\Xi \qquad (1.2)$$

for given data $f \mid_\Xi$ lead to a *positive definite* interpolation matrix $A = \{\phi(\|\xi - \zeta\|)\}_{\xi,\zeta \in \Xi}$. In that case, we call the radial basis function 'positive
definite' as well. If it is, the linear system of equations that comes from (1.1)
and (1.2) and uses precisely that matrix A yields a unique coefficient vector
$\lambda \in \mathbb{R}^\Xi$ for the interpolant (1.1).

All radial basis functions of Section 3 have this property of positive definiteness, as does for instance the Gaussian radial basis function $\phi(r) = e^{-c^2 r^2}$ for all positive parameters c and the inverse multiquadric function $\phi(r) = 1/\sqrt{r^2 + c^2}$.

However, in some instances such as the so-called thin-plate spline radial basis function, the radial function ϕ is only *conditionally positive definite* of some order k on $\mathbb{R}^n$, say, a notion that we shall explain and use in the subsequent section. Now, in this event, polynomials $p(x) \in \mathbb{P}_n^{k-1}(x)$ of degree $k-1$ in n unknowns are augmented to the right-hand side of (1.1) so as to render the interpolation problem again uniquely solvable. Consequently we have as approximant

$$s(x) = \sum_{\xi \in \Xi} \lambda_\xi \phi(\|x - \xi\|) + p(x), \qquad x \in \mathbb{R}^n. \tag{1.3}$$

Then the extra degrees of freedom are taken up by requiring that the coefficient vector $\lambda \in \mathbb{R}^\Xi$ is orthogonal to the polynomial space $\mathbb{P}_n^{k-1}(\Xi)$, that is, all polynomials of total degree less than k in n variables restricted to Ξ:

$$\mathbb{R}^\Xi \ni \lambda \perp \mathbb{P}_n^{k-1}(\Xi) \iff \sum_{\xi \in \Xi} \lambda_\xi q(\xi) = 0, \quad \forall\, q \in \mathbb{P}_n^{k-1}. \tag{1.4}$$

In order to retain uniqueness, Ξ has to contain a $\mathbb{P}_n^{k-1}$-unisolvent subset in this case. When $k = 2$, for instance, this means that the centre-set must not be a subset of a straight line. This is a requirement that can easily be met in most cases.

The two probably best-known and most often applied radial basis functions are called multiquadrics and thin-plate splines, respectively. The former is, for a positive parameter c, $\phi(r) = \sqrt{r^2 + c^2}$ and the latter is $\phi(r) = r^2 \log r$, where in the second case (1.3) and (1.4) with $k = 2$ are applied. The multiquadric is, subject to a sign change, conditionally positive definite of order $k = 1$, but it turns out that the original interpolation problem without augmentation by constants is also nonsingular because of special properties of the multiquadric function (Micchelli 1986).

One more well-known example comes up when we set $c = 0$ in the multiquadric example: then we have the so-called linear radial basis function $\phi(r) = r$ which also gives a nonsingular interpolation problem without augmentation by constants. All these examples are useful for various forms of interpolation and approximation and they all allow this interpolation procedure in all dimensions n and all sets of distinct centres, independently of the geometry of the points. In contrast to spline approximation in more than one dimension, for instance, there is no triangulation or tessellation of the data points required, nor is there any restriction on the dimensionality of the problem. We note, however, that these nonsingularity properties are strictly linked to the fact that we use Euclidean norms; for p-norms with

$p > 2$, including $p = \infty$, or $p = 1$, singularity can occur in any dimension for unfortunate choices of data points ξ (Baxter 1991).

While it has been known for a long time (see Hardy (1990) for references and the history of the approach) that approximants of the above form exist and approximate well if the centres are sufficiently close together and the function f is smooth enough, it took quite a while to underpin these empirical results theoretically, that is, get existence, uniqueness and convergence results, and extend the known classes of useful radial basis functions to further examples. Now, however, research into radial basis functions is a very active and fruitful area and it is timely to stand back and summarize its new developments in this article. Among the plethora of new papers that are published every year on radial basis functions we have made out and selected five major directions which have had important new developments and which we will review in this work.

One major feature, for example, that has been looked at recently (again) and that we address, is the accuracy of approximation with radial basis functions when the centres are scattered points in a domain, and, ultimately, form a dense subset of the domain – a subject, incidentally, well fitted to begin this review paper in the next section, because the whole theoretical development started some 20 years ago in France with Duchon's contributions (1976, 1978, 1979) to exactly this question in a special context.

Apart from the fundamental question of unique solvability of the ensuing linear system – which has by now been completely answered for large classes of radial basis functions that are conditionally positive definite, mostly following Micchelli (1986) – an important question is that of convergence and convergence rates of those interpolants to the function f that is being approximated by collocation to $f|_\Xi$ if f is in a suitable smoothness space. Duchon (1976, 1978, 1979) gave answers to this question that address the important special case of thin-plate splines $\phi(r) = r^2 \log r$ and its siblings, for instance the odd powers of r ('pseudo-cubics' $\phi(r) = r^3$, etc.), and recent research has improved some of his 20 year-old results in various directions, including inverse theorems, and theorems about optimality of convergence orders, the sets of centres still always being finite. Several of these relevant results we address in the next section. We do not comment further, however, on the nonsingularity results, as they have been reviewed often before (Powell (1992a), for instance).

Next, some theorems are given on the new classes of radial functions with compact support that are currently under investigation. They must, in the author's opinion, be seen as an alternative to the standard radial functions of global support, like thin-plate splines or the famous and extremely useful multiquadric $\phi(r) = \sqrt{r^2 + c^2}$ (Hardy 1990), but not as an excluding alternative, because the approximation orders they give are much less impressive than the dimension-dependent orders of the familiar radial functions. Some possible applications, for instance, for numerical solutions of partial differential equations come to mind when those radial functions are used because they can act as finite elements, and they are being tested at the moment by several colleagues for this very purpose. Results are mildly encouraging, although the mathematical analysis is still lacking (see, e.g., Fasshauer (1999)); some mathematical underpinning is given in Franke and Schaback (1998) and Pollandt (1997). We will explain further and give examples in Section 6.

In Section 4, several recent results about the efficient implementation of the radial basis function interpolation are given, especially iterative methods for the computation of interpolants when the number $|\Xi|$ of centres is very large. This work is called for when the radial functions are of global support and increasing with increasing argument, as they often are, because no direct or simple iterative methods will then work satisfactorily, the matrices being ill-conditioned, and sometimes highly so, with exponentially increasing condition numbers. This is particularly unfortunate because, in applications, large numbers of data occur frequently.

Section 5 is devoted to radial basis functions on spheres. Indeed, several radial basis functions were created to interpolate data given on the Earth's surface, for instance potential or temperature data (Hardy 1990). This has inspired many researchers to consider the question of how to approximate efficiently when the data are from a sphere and when the whole idea of distance defining the above approximants is adjusted properly to distances on spheres, that is, geodesic distances. Several groups are currently working on these approximations, which need not be interpolants, although the main focus is, as always with radial basis functions, on interpolation, and we give a brief review of some important new results.

The final section is devoted to applications, mostly to some initial attempts at the numerical solution of differential equations with radial basis function methods.

At the end of this introduction we remind the reader that everything said here is the result of a selection, not comprehensive, and explained, of course, from the author's personal point of view. It is certain that several relevant theorems have been omitted as a consequence. On the other hand, there will be a fairly comprehensive book by the author that gives more details and proofs of many results that are merely stated here.

2. Convergence rates

As usual in the study of methods for the approximation of functions, one of the central themes in the analysis of radial basis functions is their convergence behaviour when the centres become dense in a domain or in the whole underlying Euclidean space. This is highly relevant because it shows how

well we can approximate smooth functions even in the practical case when
the centres become close together but do not cover a whole domain. Several
of the results were initiated by the work of Duchon and this explains the
title of the following subsection.

2.1. Improvements and extensions to Duchon's convergence theorems

Many of the theorems about convergence rates are related to the so-called
variational approach and founded on ideas of Duchon (1976, 1978, 1979)
who considered the – even today – important special cases of radial basis
functions of thin-plate spline type in n dimensions, namely

$$\phi(r) = \begin{cases} r^{2k-n} \log r, & \text{if } 2k - n \text{ is an even integer,} \\ r^{2k-n}, & \text{if } 2k - n \text{ is not an even integer,} \end{cases} \qquad (2.1)$$

where we admit nonintegral k but always demand $k > \frac{1}{2}n$. The typical case
we always think of is the thin-plate spline in two dimensions, namely $n = k =
2$ and therefore $\phi(r) = r^2 \log r$. The aforementioned odd powers – linear or
cubic, for instance – also belong to this class. Several further important cases
such as multiquadrics or shifted logarithms are in fact derived from the above
by composition with $\sqrt{r^2 + c^2}$: namely, $\phi(r) = r$ altered in this fashion gives
multiquadrics and $\phi(r) = \log r$ provides the shifted logarithm $\log \sqrt{r^2 + c^2}$
(although in this case $2k = n$). The reason for this transformation is to gain
infinite smoothness when c is positive (recall that all of the above are not
smooth at the origin when composed with the Euclidean norm).

In this context, the approximants are usually taken from the 'native
spaces' X of distributions (Jones (1982), for instance, for generalized functions or distributions) in n unknowns whose total kth degree partial derivatives are square-integrable: we call the space that depends on the choice of ϕ
the space $X := D^{-k}L^2(\mathbb{R}^n)$. The Sobolev embedding theorem tells us that
this space consists of continuous functions as long as $k > \frac{n}{2}$. This is the first
and perhaps most important example of the general notion of native spaces,
namely semi-Hilbert spaces X that 'belong' to the radial function and are
defined by all distributions $f : \mathbb{R}^n \to \mathbb{R}$ that render a certain ϕ-dependent
seminorm $\|f\|_\phi$ finite. In the present case, the seminorm is the homogeneous
Sobolev norm of order k. We will return to this concept soon in somewhat
more generality.

A beautiful convergence result that holds on a very general domain Ω
and does not yet require explicitly native space seminorms (although they
are used implicitly in the proof, as we shall see) is the following one of
Powell (1994). So long as all domains as general as those in the statement
of the theorem are admitted, this is the best possible bound achievable for
this general class of domains, that is, the constant C on its right-hand side

cannot, for $h \to 0$, be replaced by $o(1)$. In order to state the result we let

$$h := \sup_{x \in \Omega} \inf_{\xi \in \Xi} \|x - \xi\|. \tag{2.2}$$

This notion is also used in the rest of this section, as is the notation $\|\cdot\|_{\infty,\Omega}$ for the Chebyshev norm restricted to Ω.

Theorem 1. Let ϕ be from the class (2.1) with $k = n = 2$ and Ω be bounded and not contained in a straight line. Let s be the radial basis function interpolant to $f|_\Xi$ satisfying (1.2) for $\Xi \subset \Omega$ where k keeps the same meaning as in the introduction, that is, linear polynomials are added to s and the appropriate side conditions (1.4) demanded. Then there is an h- and Ξ-independent C such that

$$\|s - f\|_{\infty,\Omega} \leq C h \sqrt{\log(h^{-1})}, \qquad 0 < h < 1.$$

In fact we can be even more specific about the constant C in the above error estimate. It is, of course, independent of h and Ξ, but its dependence on f can be expressed by $C = \tilde{c}\|f\|_\phi$, where $\tilde{c}$ depends only on Ω and $\|f\|_\phi$ is the homogeneous Sobolev seminorm of order 2 of f, depending on the aforementioned native space. The general approach to convergence estimates of this form is always to bound the error $|s(x) - f(x)|$ by a fixed constant multiple of

$$\|f\|_\phi \sqrt{\Phi(\alpha)} \tag{2.3}$$

that depends on the radial basis function (2.1) and the dimension n, where $\alpha = (\alpha_\xi) \in \mathbb{R}^\Xi$ are the coefficients of the representation

$$x = \sum_{\xi \in \Xi} \alpha_\xi \xi, \qquad x \in \Omega,$$

and Φ is the so-called power functional

$$\Phi(\alpha) = \phi(0) - 2 \sum_{\xi \in \Xi} \alpha_\xi \phi(\|x - \xi\|) + \sum_{\xi \in \Xi} \alpha_\xi \sum_{\zeta \in \Xi} \alpha_\zeta \phi(\|\xi - \zeta\|).$$

Consequently, the main work lies in bounding this power functional from above. If this is done judiciously we can obtain optimal bounds, because the bound of the error function by a suitable constant multiple of (2.3) is best possible (Wu and Schaback 1993).

Bejancu (1997) has generalized this result to arbitrary k and n, and his theorem includes the above result (see also Matveev (1997)). There are no further restrictions on Ω except, in general, its $\mathbb{P}_n^{k-1}$-unisolvency which we demand for the following theorem.

Theorem 2. Let ϕ be from the class (2.1). Let Ω be bounded and contain a $\mathbb{P}_n^{k-1}$-unisolvent subset. Let s be the radial basis function interpolant (1.3) to $f|_\Xi$ for $\Xi \subset \Omega$ as in the introduction where k keeps the same meaning, that is, $\mathbb{P}_n^{k-1}$ polynomials p are added to s with the appropriate side conditions (1.4) on the $\lambda \in \mathbb{R}^{\Xi}$. Then there is an h-independent constant C such that

$$\|s - f\|_{\infty,\Omega} \leq C \begin{cases} h\sqrt{\log(1/h)}, & \text{if } 2k - n = 2, \\ \sqrt{h}, & \text{if } 2k - n = 1, \quad \text{and} \quad 0 < h < 1. \\ h, & \text{in all other cases,} \end{cases}$$

Johnson (1998b) has the following improved convergence orders that do, however, require Ω to be a domain with Lipschitz continuous boundary and satisfying an interior cone condition (see any of the cited papers by Duchon for the standard definition of this concept). Now, (2.1) for general n and k are admitted and H^{2k} denotes the usual Sobolev space.

Theorem 3. Let $\tilde{\Omega} \subset \Omega$ be compact and $f \in H^{2k}(\Omega)$ be supported in $\tilde{\Omega}$. Then we have for any $\Xi \subset \Omega$ that contains a $\mathbb{P}_n^{k-1}$-unisolvent subset and an h-independent constant C and the interpolant s on Ξ the error estimate in the p-norm

$$\|s - f\|_{p,\Omega} \leq Ch^{2k+\min[n/p-n/2,0]}, \qquad 0 < h < 1, \qquad (2.4)$$

where $1 \leq p \leq \infty$.

Note that the best rate occurs in (2.4) when $p = 2$ but $p = \infty$ gives the inferior rate $2k - \frac{1}{2}n$.

2.2. Upper bounds on approximation orders and inverse theorems; saturation orders

It is a remarkable new development to have *upper bounds* on the obtainable convergence rate rather than the lower bounds thereon as above. Johnson (1998a) shows that the rates of Theorem 3 are almost optimal. (As is well known, the optimal ones in case of $\Xi = h\mathbb{Z}^n$ and $p = \infty$ are $2k$ (Buhmann 1990a, 1990b) – we will come back to this soon.) We still let the radial basis function be from the above class (2.1), although many of the results cited include multiquadric interpolation, for example.

Theorem 4. Let $1 \leq p \leq \infty$ and let Ω be the unit ball. Suppose $\{\Xi = \Xi_h \subset \Omega\}_{h>0}$ is a sequence of finite sets of centres with distance (2.2) for each. Then there exists an infinitely smooth f such that, for the best $L^p(\Omega)$-approximation s to f of the form (1.1) with the appropriate polynomials added, the error on the left-hand side of (2.4) is *not* $o(h^{2k+1/p})$ as h tends to zero.

Therefore, in Theorem 3 we are not far from the optimal result. We do get the optimal results of $\mathcal{O}(h^{2k})$ for uniform convergence ($p = \infty$) on grids $\Xi = h\mathbb{Z}^n$ as mentioned already, but also, as Bejancu (1999) shows us, on finite grids $\Xi = \Omega \cap h\mathbb{Z}^n$, where Ω is a cube with sides parallel to the axes.

This extension of the results on $\Xi = h\mathbb{Z}^n$ to $\Xi = \Omega \cap h\mathbb{Z}^n$ relies on the *locality* of the cardinal interpolants on $h\mathbb{Z}^n$ which can be expressed conveniently in Lagrange form

$$s(x) = \sum_{\xi \in \Xi} f(\xi) L\left(h^{-1}(x - \xi)\right) = \sum_{i \in \mathbb{Z}^n} f(hi) L\left(h^{-1}x - i\right), \qquad x \in \mathbb{R}^n.$$

In this expression, the Lagrange function L that satisfies $L(k) = \delta_{0k}$ for all multi-integers k, and may be expanded as

$$L(x) = \sum_{i \in \mathbb{Z}^n} \lambda_i \phi(\|x - i\|), \qquad x \in \mathbb{R}^n,$$

enjoys the remarkable property of the fast (exponential) decay of $|L(x)|$ (Madych and Nelson 1990), which renders the approximant local with respect to the data. This, in turn, gives rise to the same convergence orders when the domain of approximation is restricted from the whole of the Euclidean space to a cube with sides parallel to the axes.

Theorem 5. Let Ω, $\Xi = \Omega \cap h\mathbb{Z}^n$ and the interpolant s be as above, and let $\tilde{\Omega}$ be a compact subset of the interior of Ω. Then, for $f \in \mathrm{Lip}^{2k+1}(\Omega)$, and for an h-independent constant C,

$$\|s - f\|_{\infty, \tilde{\Omega}} \leq C h^{2k}, \qquad 0 < h < 1.$$

Precise upper bounds on the approximation order that can be identified for $\Xi = (h\mathbb{Z})^n$ may be stated in a very general context even for h-dependent radial basis functions ϕ_h that are from a sequence of radial functions $\{\phi_h\}_{h>0}$. Therefore we study approximants from spaces

$$S_h(\phi_h) = \mathrm{span}\left\{\phi_h\left(\frac{\cdot}{h} - j\right) \mid j \in \mathbb{Z}^n\right\}.$$

A special case is, of course, $\phi_h \equiv \phi$ for all h, perhaps taken from one of our (2.1). For the statement of the next theorem we let Ω be the unit ball and $\sigma : \mathbb{R}^n \to \mathbb{R}$ a smooth (C^∞) cut-off function that is supported in Ω and satisfies $\sigma\mid_{\frac{1}{2}\Omega} = 1$. We recall that the notation $\check{f}$ stands for the inverse Fourier transform

$$\check{f}(x) = \frac{1}{(2\pi)^n} \int e^{ix \cdot t} f(t) \, dt,$$

integrals being over $\mathbb{R}^n$ unless stated otherwise.

Theorem 6. Let $1 \leq p \leq \infty$. Let $\hat{\phi}_h : \mathbb{R}^n \to \mathbb{R}$ be measurable and such that, for some $j_0 \in \mathbb{Z}^n \backslash \{0\}$ and some $\varepsilon \in (0,1)$:

(1) $\hat{\phi}_h(x) \neq 0$ for almost all $x \in \varepsilon\Omega$;

(2) the *inverse Fourier transform* $\left(\sigma\left(\frac{\cdot}{\varepsilon}\right) \frac{\hat{\phi}_h(\cdot + 2\pi j_0)}{\hat{\phi}_h} \right)^{\vee}$ is absolutely integrable;

(3) for a measurable function $\tilde{\rho}$ which is locally bounded and for all positive m,

$$\left\| h^{2k}\, \tilde{\rho} - \frac{\hat{\phi}_h(h \cdot + 2\pi j_0)}{\hat{\phi}_h(h\cdot)} \right\|_{\infty, m\Omega} = o(h^{2k}), \qquad h \to 0.$$

Then the L^p-approximation order from $S_h(\phi_h)$ cannot be more than $2k$, that is, the distance in the L^p-norm between the L^p-closure of $S_h(\phi_h)$ and the class of band-limited f whose Fourier transform is infinitely differentiable cannot be $o(h^{2k})$ as h tends to zero.

We note that the class of all band-limited f whose Fourier transform is infinitely differentiable is a class of very smooth local functions and if the L^p-approximation order to such smooth functions cannot be more than h^{2k}, it cannot be more than h^{2k} to any general nontrivial larger set of less smooth functions.

A typical example occurs when we again use radial basis functions of the form (2.1) where $\hat{\phi}(r)$ is a constant multiple of r^{-2k}. Then condition (1) of the above theorem is certainly true for $\hat{\phi}_h \equiv \hat{\phi}$. Moreover, condition (2) is true because the smoothness of the cut-off function, the smoothness of the function $\hat{\phi}(\| \cdot + 2\pi j_0 \|)$ in a neighbourhood of the origin for nonzero j_0 and the fact that $2k > n$ imply by Lemma 2.7 of Buhmann and Micchelli (1992) that condition (2) holds. Finally, we can take $\tilde{\rho} = \hat{\phi}^{-1}$ and get the required result from condition (3), namely that $\mathcal{O}(h^{2k})$ is best possible.

This is the obtainable (saturation) order and an inverse theorem of Schaback and Wendland (1998) tells us that all functions for which a better order is obtainable must be trivial in the sense of polyharmonic functions. For its statement we recall the standard notation Δ for the Laplace operator.

Theorem 7. Let Ω be as in Theorem 2, $\Xi \subset \Omega$ a finite centre-set with distance h as in (2.2) and ϕ as in (2.1). If, for any $f \in C^{2k}(\Omega)$, and all compact $\tilde{\Omega} \subset \Omega$,

$$\|s - f\|_{\infty, \tilde{\Omega}} = o(h^{2k}), \qquad 0 < h < 1,$$

then $\Delta^k f = 0$ on Ω.

A very similar, also inverse, but more general theorem from the same paper is the following one with which we close this section. In order to state it we come back to the notion of native spaces and recall that for a

radial basis function with positive distributional Fourier transform $\hat{\phi}(\|\cdot\|)$:
$\mathbb{R}^n \setminus \{0\} \to \mathbb{R}$, the square of the native space norm is

$$\|f\|_\phi^2 := \frac{1}{(2\pi)^n} \int \frac{1}{\hat{\phi}(\|t\|)} |\hat{f}(t)|^2 \, dt \qquad (2.5)$$

and the native space X is the space of all distributions f on $\mathbb{R}^n$ for which
(2.5) is finite. In the case of the thin-plate splines or, more generally, (2.1),
X agrees with $D^{-k}L^2(\mathbb{R}^n)$, because $\hat{\phi}(\|t\|)^{-1}$ is a constant multiple of $\|t\|^{2k}$,
and by the Parseval–Plancherel theorem (Stein and Weiss 1971); this being
a special case, other positive $\hat{\phi}$ are admitted as well. For the statement of
the following theorem we recall that the radial basis function is *conditionally
positive definite* of order k in $\mathbb{R}^n$ if the matrix A with centres from $\Xi \subset \mathbb{R}^n$
is nonnegative definite on the subspace of coefficient vectors $\lambda \in \mathbb{R}^\Xi$ that
are orthogonal to $\mathbb{P}_n^{k-1}(\Xi)$. The functions (2.1) are all conditionally positive
definite of order k subject to a possibly needed sign change. We use the
notation $\tau(x) \sim t(x)$ if both $\tau(x)/t(x)$ and $t(x)/\tau(x)$ are uniformly bounded
for the appropriate range of x.

Theorem 8. Let Ω be a bounded domain containing a $\mathbb{P}_n^{k-1}$-unisolvent
subset. Let ϕ be conditionally positive definite of order k and satisfy

$$\hat{\phi}(r) \sim r^{-2k}, \qquad r > 0.$$

If $f \in C(\Omega)$ and there exists $\mu > k$ such that

$$\|s - f\|_{\infty,\Omega} \leq Ch^\mu, \qquad h \to 0,$$

for the radial basis function interpolants s defined on all finite sets of centres
$\Xi \subset \Omega$ with h given by (2.2), then f is an element of X. (The constant C
depends on f but not on h.)

An example for the application of this result is the radial basis function
(2.1) with k there and in Theorem 8 being the same.

3. Compact support

This section deals with radial basis functions that are compactly supported,
quite in contrast to everything else we have seen before in this article. All of
the radial basis functions that we have considered so far have global support,
and in fact many of them, such as multiquadrics, do not even have isolated
zeros (thin-plate splines do, by contrast). Moreover, the radial basis functions $\phi(r)$ are usually increasing with growing argument $r \to \infty$, so that
square-integrability, for example, and especially absolute integrability are
immediately ruled out. In most cases, this poses no severe restrictions; we
can, in particular, interpolate with these functions and get good convergence
rates nonetheless, as we have seen in the previous section. There are, however, practical applications that demand local support of the basis functions

such as finite element applications (but note one of the approaches to partial
differential equations in the last section which works specifically with globally supported ϕ) or applications with very quickly growing data, or where
frequent evaluations with substantial amounts of centres are required, *etc.*
Therefore there is now a theory of radial basis functions of compact support where the entire class of ϕs gives rise to positive definite interpolation
matrices for distinct centres.

We note that, however, in the approximation theory community there
is an ongoing discussion about the usefulness of radial basis functions of
compact support because of their inferior convergence properties (unless we
wish to again forego the advantages of compact support – see our discussion
about scaling in the text below), and in comparison to standard, piecewise
polynomial finite elements. The latter, however, are harder to use in grid-
free environments and $n > 3$, because they require triangulations to be found
first.

As we shall see in this section, there exist at present essentially two approaches to constructing univariate, compactly supported $\phi : \mathbb{R}_+ \to \mathbb{R}$ such
that the interpolation problem is uniquely solvable with a positive definite
collocation matrix $A = \{\phi(\|\xi - \zeta\|)\}_{\xi,\zeta\in\Xi}$. There will be no restriction on
the geometry of the set Ξ of centres, but there are – in fact, there must be
– bounds on the maximal spatial dimension n which is admitted for each
radial function ϕ so that positive definiteness is retained. So they are still
useful in grid-free and high-dimensional applications.

In contrast to the radial basis functions of the previous section, the approximation orders we state, unfortunately, are not nearly as good as the
maximal ones available of the well-known globally supported radial functions such as (2.1). This indeed puts a stricter limit to the usefulness of
compactly supported radial basis functions than we are used to for the globally supported ones. Be that as it may, it is nonetheless interesting to study
the question of when compactly supported radial functions give nonsingular
and convergent interpolation schemes.

3.1. Wendland's functions

Initially, there were the approaches by Wu and by Wendland, where the radial basis functions consist of only one polynomial piece on the unit interval
$[0, 1]$ and are otherwise zero. Although by this means they are piecewise
polynomial seen as a univariate function, the resulting approximants are, of
course, not. The whole idea is based on the earlier work by Askey (1973)
who observed by considering Fourier transforms that the truncated power
function $\phi(r) = \phi_0(r) = (1 - r)_+^\ell$ gives rise to positive definite interpolation matrices A for $\ell \geq [n/2] + 1$. Already here we see, incidentally, an

upper bound on the spatial dimension n if we fix the degree of the piecewise polynomial.

In order to derive a large class of compactly supported radial basis functions starting from Askey's result, Schaback and Wu (1995) introduced the two so-called operators on radial functions

$$Df(x) := -\frac{1}{x}f'(x), \qquad x > 0,$$

and

$$If(x) := \int_x^\infty rf(r)\, dr, \qquad x > 0,$$

that are defined for suitably differentiable or asymptotically decaying f, respectively, and are inverse to each other. Additive polynomial terms do not arise on integration since we always restrict ourselves to compactly supported functions. Next, Wendland (1995) and Wu (1995) use the fact that the said interpolation matrix A for the truncated power ϕ_0 remains positive definite if the basis function

$$\phi(r) = \phi_{n,k}(r) = I^k\phi_0(r), \qquad r \geq 0, \tag{3.1}$$

is used when $\ell = k + [n/2] + 1$. The way to establish that fact is by considering the Fourier transform of the n-variate radially symmetric function $\phi(\|\cdot\|)$, which is also radially symmetric and computed by the *univariate* Hankel transform

$$\hat{\phi}(r) = (2\pi)^{n/2}r^{1-n/2}\int_0^\infty s^{n/2}J_{n/2-1}(rs)\phi(s)\, ds, \qquad r = \|x\| \geq 0, \tag{3.2}$$

where $J_{n/2-1}$ is a Bessel function. This is the radial part of the radially symmetric Fourier transform of $\phi(\|\cdot\|)$. Hence we have to show that positivity of the Fourier transform, which is necessary and sufficient for the positive definiteness of the matrix A by Bochner's theorem, prevails when the operator I above is applied, as long as the restrictions on dimension are observed. This proof is carried out by studying the action of I on the Hankel transform (3.2) and by direct computation and use of identities of Bessel functions.

Starting from this, Wendland (1995, 1998), in particular, developed an entire theory of the radial basis functions of compact support which are piecewise polynomial and are positive definite. This theory encompasses recursions for their coefficients when they are expanded in linear combinations of powers and truncated powers of lower order, convergence results, and minimality of polynomial degree for given dimension and smoothness. Two results below serve as examples for the whole theory. The first states the minimality of degree for given smoothness and dimension and the second is a convergence result.

Theorem 9. The radial function (3.1) gives rise to a positive definite interpolation matrix A with radial basis function $\phi = \phi_{n,k}$, and with distinct centres Ξ in $\mathbb{R}^n$. Further, among all such functions for dimension n and smoothness C^{2k}, it is of minimal polynomial degree.

Theorem 10. Let ϕ be defined by (3.1), let f be a function in the Sobolev space $H^{k+(n+1)/2}(\mathbb{R}^n)$, and let k be at least 1 when $n = 1$ or $n = 2$. Then, for a compact domain Ω with centres $\Xi \subset \Omega$, the interpolant (1.1) satisfies

$$\|s - f\|_{\infty,\Omega} = \mathcal{O}\left(h^{k+1/2}\right), \qquad h \to 0,$$

where h is given by (2.2).

Examples are, for $\ell \geq \lceil n/2 \rceil + 1$,

$$\phi(r) = (1 - r)_+^{\ell+1}((\ell + 1)r + 1),$$
$$\phi(r) = (1 - r)_+^{\ell+2}((\ell^2 + 4\ell + 3)r^2 + (3\ell + 6)r + 3),$$

in a form proposed by Fasshauer.

3.2. Further contributions

The second class of radial basis functions of compact support (Buhmann 1998, 2000) are reminiscent of the famous thin-plate splines, albeit truncated in a suitable way, and of a certain convolution form. It contains, for instance, the following cases. (We state the value of the function only on the unit interval; elsewhere it is zero. It can, of course, be scaled suitably depending on the distances between the centres.)

Namely, two examples that give twice and three-times continuously differentiable functions, respectively, in three dimensions and two dimensions are as follows. We state them first because they are useful to illustrate the goal of our later result. The parameter choices for the theorem below $\alpha = \delta = \frac{1}{2}$, $\rho = 1$, and $\lambda = 2$, give, for $n = 3$,

$$\phi(r) = 2r^4 \log r - \frac{7}{2}r^4 + \frac{16}{3}r^3 - 2r^2 + \frac{1}{6}, \qquad 0 \leq r \leq 1,$$

while the choices $\alpha = \frac{3}{4}$, $\delta = \frac{1}{2}$, $\rho = 1$, and $\lambda = 2$, provide, for $n = 2$,

$$\phi(r) = \frac{112}{45}r^{\frac{9}{2}} + \frac{16}{3}r^{\frac{7}{2}} - 7r^4 - \frac{14}{15}r^2 + \frac{1}{9}, \qquad 0 \leq r \leq 1.$$

Another two-dimensional ($n = 2$) example which is twice times continuously differentiable is

$$\phi(r) = \frac{1}{18} - r^2 + \frac{4}{9}r^3 + \frac{1}{2}r^4 - \frac{4}{3}r^3 \log r, \qquad 0 \leq r \leq 1.$$

Theorem 11. Let $0 < \delta \leq \frac{1}{2}$, $\rho \geq 1$ be reals, and suppose $\lambda \neq 0$ and $\alpha > -1$ are also real quantities with

$$
\lambda \in \begin{cases}
(-\frac{1}{2}, \infty), & \alpha \leq \min[\frac{1}{2}, \lambda - \frac{1}{2}], & \text{if } n = 1, \quad \text{or} \\
[1, \infty), & -\frac{1}{2} < \alpha \leq \frac{1}{2}\lambda, & \text{if } n = 1, \quad \text{and} \\
(-\frac{1}{2}, \infty), & \alpha \leq \min[\frac{1}{2}(\lambda - \frac{1}{2}), \lambda - \frac{1}{2}], & \text{if } n = 2, \quad \text{and} \\
(0, \infty), & \alpha \leq \frac{1}{2}(\lambda - 1), & \text{if } n = 3, \quad \text{and} \\
(\frac{1}{2}(n - 5), \infty), & \alpha \leq \frac{1}{2}(\lambda - \frac{1}{2}(n - 1)), & \text{if } n > 3.
\end{cases}
$$

Then the radial basis function

$$
\phi(r) = \int_0^\infty \left(1 - r^2/\beta\right)_+^\lambda \beta^\alpha (1 - \beta^\delta)_+^\rho \, \mathrm{d}\beta, \qquad r \geq 0, \tag{3.3}
$$

has a positive Fourier transform and therefore, by Bochner's theorem, gives rise to positive definite interpolation matrices A with centres Ξ from $\mathbb{R}^n$. Moreover, $\phi(\| \cdot \|) \in C^{1 + \lceil 2\alpha \rceil}(\mathbb{R}^n)$.

There is also a convergence estimate for the above radial functions available which includes scaling of the radial basis function, because, as the distances between the centres become smaller, we wish to decrease the support of the radial function as well. Otherwise we would lose the advantages of compact support since the support relative to the distance between the centres would increase. Note that, in the statement of the following theorem, the approximand f is continuous by the Sobolev embedding theorem precisely as long as $1 + \alpha$ is positive by the conditions in the previous theorem (this is therefore a necessary condition).

Theorem 12. Let ϕ be as in the previous theorem and suppose additionally $\rho > 1$, $2\alpha \leq \lambda - n/2 - 3 + \lfloor \rho \rfloor$. Let Ξ be a finite set in a compact domain Ω. Let s be the scaled interpolant (1.1)

$$
s(x) = \sum_{\xi \in \Xi} \lambda_\xi \phi(\eta^{-1} \| x - \xi \|), \qquad x \in \mathbb{R}^n,
$$

to $f \in L^2(\mathbb{R}^n) \cap D^{-n/2 - 1 - \alpha} L^2(\mathbb{R}^n)$, with the interpolation conditions $s|_\Xi = f|_\Xi$ satisfied. Then the uniform convergence estimate

$$
\|f - s\|_{\infty, \Omega} \leq C h^{1 + \alpha} \eta^{-n/2 - 1 - \alpha}
$$

holds for $h \to 0$ and positive bounded η, the positive constant C being independent of both h and η.

It is interesting to observe that the function classes of the first subsection of this section can be integrated into the more general class discussed in this subsection. Namely, when the operator D is applied to our radial functions (3.3), it gives

$$D^\lambda \phi(r) = \Gamma(\lambda + 1)2^\lambda \int_{r^2}^1 \beta^{\alpha - \lambda}(1 - \sqrt{\beta})^\rho \, \mathrm{d}\beta, \qquad 0 \leq r \leq 1,$$

for $\delta = \frac{1}{2}$ and any integral nonnegative λ. Therefore, by one further application of the differentiation operator and explicit evaluation,

$$D^{\lambda+1}\phi(r) = \Gamma(\lambda + 1)2^{\lambda+1} r^{2\alpha - 2\lambda}(1 - r)_+^\rho = \Gamma(\lambda + 1)2^{\lambda+1}(1 - r)_+^\rho, \quad r \geq 0,$$

for $\alpha = \lambda$. Now let $\lambda = k - 1$, $k \geq 1$, $\rho = [\frac{1}{2}n] + k + 1$ and recall that, on the other hand, we know that the radial functions $\phi_{n,k}$ of Wendland are such that

$$D^k \phi_{n,k}(r) = (1 - r)_+^{[\frac{1}{2}n] + k + 1}, \qquad r \geq 0.$$

Therefore, the functions generated in this fashion can also be derived from (3.3), albeit with a parameter α that does not fulfil the conditions of Theorem 11, so it is a type of continuation. The functions from Wu (1995) also belong to that class as they were shown by Wendland to be special cases of his functions. Therefore, since the class of Wendland functions described here contains those introduced by Wu (1995), our class covers both. Finally, because we know that all those functions have *positive Fourier transforms*, there follow immediately new results about the positivity of Hankel transforms (3.2) which extend the work by Misiewicz and Richards (1994), and this is in spite of the fact that the parameter α here does not yield the conditions of Theorem 11.

4. Iterative methods for implementation

Among the most useful radial basis functions which provide good, accurate approximations are the thin-plate splines, for instance in two dimensions, and multiquadrics. However, these two, like many other examples of radial basis functions with *global support*, lead to linear interpolation systems that require special considerations if $|\Xi|$ is more than a few hundred when we calculate the coefficients λ_ξ. Otherwise the computational cost of $\mathcal{O}(|\Xi|^3)$ is prohibitively expensive for direct methods, storage being another major obstacle if the size of the set of centres is too large – even with the fastest and biggest workstations currently available.

There are several approaches currently under intense investigation, mainly in Cambridge and in Christchurch, New Zealand, for dealing with this problem, two of which we shall explain here to exemplify what can be done today when we have 50000 centres, say. This is the BGP (Beatson–Goodsell–Powell) method of local Lagrange functions, another highly relevant and successful class of methods being the fast multipole approaches (Beatson and Newsam (1992); see also Beatson and Light (1997) and Beatson, Cherrie and Mouat (1999)). Like the fast multipole methods, the BGP algorithm

(Beatson, Goodsell and Powell 1995) is iterative and depends on an initial structuring of the centres Ξ before the start of the iteration, although that one is less complicated than the fairly sophisticated hierarchical structure demanded for the multipole schemes, especially when $n > 2$.

4.1. The BGP method

We explain the method for $n = 2$, but we remark that the case $n = 3$ is under investigation and will probably be admissible and have suitable software soon. The special further complication for $n = 3$ is the important three-dimensional structuring of the data; conceptually there are no changes in the algorithm we describe below. We also take thin-plate splines only, although the method has actually been already tested successfully on multi-quadrics as well (Faul and Powell 1999), and once again there are almost no changes except for a different k which plays an important rôle below. This is because the method makes, as we shall see, extensive use of the variational properties explained in Section 2 of this paper, and those can be extended to multiquadrics, for instance, although they were first found in connection with thin-plate splines.

The interpolant we wish to compute still has the same form as in the introduction for $k = n = 2$ and $\phi(r) = r^2 \log r$ but we denote the actual, sought interpolant by s^*, whereas s denotes in this section only the active approximation to s^* at each stage of the algorithm. This is useful for describing our iterative scheme. The basic idea of the algorithm is derived from a Lagrange formulation of the interpolant

$$s^*(x) = \sum_{\xi \in \Xi} f(\xi) L_\xi(x), \qquad x \in \mathbb{R}^n, \tag{4.1}$$

instead of (1.1), where each Lagrange function L_ξ satisfies the Lagrange conditions

$$L_\xi(\zeta) = \delta_{\zeta\xi}, \qquad \xi, \zeta \in \Xi, \tag{4.2}$$

and is of the form

$$L_\xi(x) = \sum_{\zeta \in \Xi} \lambda_{\zeta\xi} \phi(\|x - \zeta\|) + p_\xi(x), \qquad x \in \mathbb{R}^n. \tag{4.3}$$

Here, $p_\xi \in \mathbb{P}_2^1$ and $\lambda_{.,\xi} \perp \mathbb{P}_2^1(\Xi)$ for each ξ as usual. Clearly, the computation of such full Lagrange functions would be just as expensive as solving the full usual linear interpolation system of equations. Therefore the idea is to replace (4.2) by local Lagrange conditions which require for each ξ only that the identity holds for some $q = 30$ points, say, ζ that are nearby ξ. Hence we take Lagrange functions that are still of the form (4.3) but with at most q nonzero coefficients. We end up with an approach to our interpolation problem that resembles a domain decomposition approach, because we divide the set of centres into subsets and regard the interpolation method as a local method on those subsets.

So that we can associate with each ξ the Lagrange function L_ξ and its 'active' point set of centres, we call the latter $\mathcal{L}_\xi \subset \Xi$, $|\mathcal{L}_\xi| = q$. In addition to having q elements, $\mathcal{L}_\xi$ must contain a unisolvent subset with respect to $\mathbb{P}_n^{k-1}$ as required before. We also extract another set of centres with a unisolvent subset Σ of approximately the same size q from Ξ for which no local Lagrange functions are computed because that set is sufficiently small to allow direct solution of the interpolation problem.

In order that an iterative method can be applied, we order the centres $\Xi \setminus \Sigma = \{\xi_i\}_{i=1}^m$ for which local Lagrange functions are computed and have the final requirement that, for all $i = 1, 2, \ldots, m$,

$$\xi_i \in \mathcal{L}_{\xi_i} \subset \Xi \setminus \{\xi_1, \xi_2, \ldots, \xi_{i-1}\}.$$

We require that the q points in the set above are those among the centres that follow ξ_i which are the closest ones to ξ_i. There may be ties in the necessary ordering procedure that can be broken randomly. The Lagrange conditions that must be satisfied by the local Lagrange functions are now the locally restricted conditions

$$L_\xi(\zeta) = \delta_{\zeta\xi}, \qquad \zeta \in \mathcal{L}_\xi. \tag{4.4}$$

We employ the same notation for the local Lagrange functions as for the full ones, as the latter will no longer occur in this section. Using these local Lagrange functions, (4.1) is naturally no longer a representation of the exact approximant s^*, but only an approximation thereof, and here is where the iteration and iterative correction come in. Of course the Lagrange functions, that is, their coefficients, are computed in advance once and for all and stored before the beginning of the iterations. (This is an $\mathcal{O}(|\Xi|)$ process.) A useful comparison of this approach with the Newton formulation of univariate polynomial interpolants is made by Powell (1999).

In those iterations, we make an iterative refinement of the approximation at each step of the algorithm by correcting its residual through updates

$$s(x) \longrightarrow s(x) + L_\xi(x) \times c_\xi(x),$$

where

$$c_\xi(x) = \frac{1}{\lambda_{\xi\xi}} \sum_{\zeta \in \mathcal{L}_\xi} \lambda_{\zeta\xi}(f(\zeta) - s(\zeta)). \tag{4.5}$$

We shall see in the proof of the next theorem why $\lambda_{\xi\xi}$ is positive and we may therefore divide by $\lambda_{\xi\xi}$. The correction (4.5) is added for all $\xi \in \Xi \setminus \Sigma$ for each sweep of the algorithm. The final stage of each sweep consists of

the correction
$$s(x) \longrightarrow s(x) + \sigma(x),$$
where σ is the full standard thin-plate spline solution of the interpolation problem with centres Σ computed with a direct method
$$\sigma(\xi) = f(\xi) - s(\xi), \qquad \xi \in \Sigma.$$

Here also, s denotes the current approximation to s^* after all intermediate steps. This finishes the sweep of the algorithm, and, if we started with a trivial approximation $s = s_0 = 0$ to s^*, the sweep replaces s_m by s_{m+1} that goes into the next sweep. The stopping criterion can be, for instance, that we terminate the algorithm if all the residuals are sufficiently small. The main work is clearly the computation of the residuals, the coefficients of the local Lagrange made available before the start. For that, a method such as the one described in the next subsection about multipole methods can be used.

It is quite a novelty that there is a convergence proof of this algorithm, although it had been known for some time that the method performs exceptionally well in practical computations even if Ξ contains as many as 50000 points. It turns out that it is not unusual to have an increase of accuracy of one digit per sweep of the algorithm, that is, each sweep reduces $\max |s(\xi) - f(\xi)|$ by a factor of ten, which indicates very fast convergence.

We provide the essentials of the proof of this theorem (Faul and Powell 1999) – which is in fact not very long – because it is highly instructive about the working of the algorithm and involves many concepts that are typical for the use and analysis of radial basis functions, such as native spaces and their seminorms and semi-inner products.

Theorem 13. Let $\{s_j\}_{j=0}^{\infty}$ be a sequence of approximations to s^* generated by $s_0 = 0$ and the above algorithm. Let the radial function be the thin-plate spline function and $n = 2$. Then $\lim_{j \to \infty} s_j = s^*$.

Proof. We show first that the native space norm $\|s^* - s_j\|_\phi$ is monotonically decreasing with increasing j. Let $(\cdot, \cdot)$ be the semi-inner product that corresponds to the native space norm, namely

$$(u, v) = \frac{1}{(2\pi)^n} \int \frac{1}{\hat{\phi}(\|t\|)} \, \hat{u}(t)\overline{\hat{v}(t)} \, \mathrm{d}t. \qquad (4.6)$$

Here, the requirements $\|u\|_\phi < \infty$, $\|v\|_\phi < \infty$ are sufficient for the above integral to exist, by the Cauchy–Schwarz inequality. In the thin-plate spline case and in two dimensions, this is by Parseval–Plancherel a multiple of

$$\int \frac{\partial^2 u(x,y)}{\partial x^2} \frac{\partial^2 v(x,y)}{\partial x^2} + 2\frac{\partial^2 u(x,y)}{\partial x \partial y} \frac{\partial^2 v(x,y)}{\partial x \partial y} + \frac{\partial^2 u(x,y)}{\partial y^2} \frac{\partial^2 v(x,y)}{\partial y^2} \, \mathrm{d}x \, \mathrm{d}y$$

and has as kernel the space of linear polynomials.

We start with $s = s_0 = 0$. Each full sweep of the algorithm replaces s_j by s_{j+1}, but within each sweep there are further $m + 1$ updates . By those updates, $s_{j,0} = s_j$ is replaced by $s_{j,1}$, and so on until $s_{j,m}$ is replaced by $s_{j+1} = s_{j,m+1} = s_{j,m} + \sigma$ where σ is defined above. At the $(j - 1)$st stage, for each index i, $\|s^* - s_{j-1,i-1}\|_\phi$ is replaced by

$$\|s^* - s_{j-1,i-1} - \theta_{\xi_i} L_{\xi_i}\|_\phi,$$

where $\theta_{\xi_i} = c_{\xi_i}$. It is elementary that we get

$$\|s^* - s_{j-1,i-1} - \theta L_{\xi_i}\|_\phi^2 = \min!$$

when the linear parameter θ assumes the value

$$\theta = \frac{(s^* - s_{j-1,i-1}, L_{\xi_i})}{\|L_{\xi_i}\|_\phi^2}. \tag{4.7}$$

We claim that (4.7) is the same as θ_{ξ_i} and in the proof of this fact we shall also show as a by-product that the denominator in (4.7) is positive. Indeed, recalling that the additional polynomials used are in the kernel of the inner product (4.6), we get from the reproducing kernel identity

$$
\begin{aligned}
\|L_{\xi_i}\|_\phi^2 &= \left(\sum_{\zeta \in \mathcal{L}_{\xi_i}} \lambda_{\zeta \xi_i} \phi(\| \cdot - \zeta \|), \sum_{\tau \in \mathcal{L}_{\xi_i}} \lambda_{\tau \xi_i} \phi(\| \cdot - \tau \|) \right) \\
&= \sum_{\zeta \in \mathcal{L}_{\xi_i}} \lambda_{\zeta \xi_i} \sum_{\tau \in \mathcal{L}_{\xi_i}} \lambda_{\tau \xi_i} \phi(\|\zeta - \tau\|) \\
&= \lambda_{\xi_i \xi_i},
\end{aligned}
$$

by (4.4) and because p_ζ is annihilated by the side conditions on the coefficients $\lambda_{\zeta \xi_i}$. This also shows, incidentally, that the important inequality

$$\lambda_{\xi_i \xi_i} > 0 \tag{4.8}$$

holds because $\|L_{\xi_i}\|_\phi^2$ is not zero. Otherwise L_{ξ_i} would be in the kernel of our semi-inner product and not able to satisfy the cardinality conditions. Moreover, by the same token, we get from the reproducing kernel properties

$$
\begin{aligned}
(s^* - s_{j-1,i-1}, L_{\xi_i}) &= \sum_{\zeta \in \mathcal{L}_{\xi_i}} \lambda_{\zeta \xi_i} (s^* - s_{j-1,i-1}, \phi(\| \cdot - \zeta \|)) \\
&= \sum_{\zeta \in \mathcal{L}_{\xi_i}} \lambda_{\zeta \xi_i} (s^*(\zeta) - s_{j-1,i-1}(\zeta)) \\
&= \sum_{\zeta \in \mathcal{L}_{\xi_i}} \lambda_{\zeta \xi_i} (f(\zeta) - s_{j-1,i-1}(\zeta)).
\end{aligned}
$$

Next, we have to consider the alteration to $\|s^* - s_{j-1,m}\|_\phi$ as soon as σ is added to $s_{j-1,m}$. In order to prove the monotonic decrease of $\|s^* - s_{j-1,m}\|_\phi$

when this happens, we need to prove that the following inner product vanishes

$$\left(s^* - s_{j-1,m} - \sigma, \sum_{\tau \in \Sigma} \hat{\lambda}_\tau \phi(\|\cdot - \tau\|) + \hat{p} \right)$$

$$= \left(s^* - s_{j-1,m} - \sigma, \sum_{\tau \in \Sigma} \hat{\lambda}_\tau \phi(\|\cdot - \tau\|) \right)$$

$$= 0.$$

In the above, the coefficients $\hat{\lambda}_\tau$ are real and $\hat{p} \in \mathbb{P}_n^{k-1}$. This orthogonality relation is established using the same facts as those we required for showing $\|L_{\xi_i}\|_\phi^2 = \lambda_{\xi_i \xi_i}$ and therefore we do not repeat the arguments.

In summary, we now know that $\|s_{j,i} - s^*\|_\phi$ tends to a limit for *all* fixed i, and j increasing, because it is bounded below by zero and monotonically decreasing, as it is a subsequence of the monotonically decreasing doubly indexed sequence $\{\|s_{j,i} - s^*\|_\phi\}_{j=0, 0 \leq i \leq m}^\infty$. Moreover, by the definition of our semi-inner product

$$\|s^* - s_{j-1,i-1} - \theta_{\xi_i} L_{\xi_i}\|_\phi^2 = \|s_{j-1,i-1} - s^*\|_\phi^2 - \frac{(s^* - s_{j-1,i-1}, L_{\xi_i})^2}{\|L_{\xi_i}\|_\phi^2}.$$

Therefore, for all centres

$$(s^* - s_{j-1,i-1}, L_{\xi_i}) \to 0, \qquad j \to \infty, \ \xi_i \in \Xi \setminus \Sigma, \tag{4.9}$$

because $\|s_{j,i} - s^*\|_\phi$ converges. In particular, it follows that

$$(s^* - s_j, L_{\xi_1}) \to 0, \qquad j \to \infty. \tag{4.10}$$

Moreover, $s_{j,1} = s_j + (s^* - s_j, L_{\xi_1}) L_{\xi_1} / \|L_{\xi_1}\|_\phi^2$, so that we also get

$$\left(s^* - s_j - \frac{(s^* - s_j, L_{\xi_1}) L_{\xi_1}}{\|L_{\xi_1}\|_\phi^2}, L_{\xi_2} \right) \to 0, \qquad j \to \infty.$$

Therefore

$$(s^* - s_j, L_{\xi_2}) \to 0, \qquad j \to \infty,$$

and indeed

$$(s^* - s_j, L_\xi) \to 0, \qquad j \to \infty, \ \xi \in \Xi \setminus \Sigma. \tag{4.11}$$

Finally, we observe

$$(s^* - s_j, L_{\xi_i}) = \sum_{\zeta \in \mathcal{L}_{\xi_i}} \lambda_{\zeta \xi_i} (s^*(\zeta) - s_j(\zeta)) \to 0, \qquad j \to \infty. \tag{4.12}$$

Recalling that $s^* - s_j$ restricted to Σ vanishes anyway, and recalling that $\lambda_{\xi_i \xi_i}$ is positive, we now go backwards and start from $i = m$ with an induction argument, whereupon (4.8), (4.9) and (4.12) imply $s_j(\zeta) \to s^*(\zeta)$ as $j \to \infty$

for all $\zeta \in \Xi$. This implies $s_j \to s^*$ for $j \to \infty$, as demanded, because the space spanned by the translates $\phi(\| \cdot -\xi\|)$ plus the polynomial space is finite-dimensional. □

4.2. Fast multipole methods

Another approach for approximation and iterative refinement to the radial basis function interpolants is that of *fast multipole methods*. These algorithms are based on analytic expansions of the underlying radial functions for large argument. We see that this is possible for all non-compactly supported radial basis functions that have been mentioned, because they are smooth away from their centres. For the compactly supported ones, the approach for fast evaluation and solution of the linear system is, of course, not required. The salient ideas date back to a paper of Greengard and Rokhlin (1987) where the methods are used to solve numerically integral equations – this is related to our radial basis function approximations because the sums and coefficients ('weights') can be viewed as discretizations of integrals.

The methods require structuring the data in a hierarchical way before the onset of the iteration, and computing so-called far-field expansions (Laurent series expansions for large arguments, as already alluded to) in advance. The goal is to reduce the cost of a single approximate evaluation of the radial function sum down to $\mathcal{O}(1)$ and storage to $\mathcal{O}(|\Xi|)$. The far-field expansions exploit the fact that radial basis functions of the form (2.1) are analytic except at the origin, even when made multivariate through composition with Euclidean norms. Therefore, they can be approximated well away from the origin by a truncated Laurent expansion. The accuracy (that is, the length of the truncated expansion) to this can be preset and made to match any chosen accuracy ε, for instance a small positive multiple of the machine precision of the computer in use, but of course the cost of the method, that is, the multiplier in the operation count, rises with larger accuracy. In tandem with the hierarchical structure of the centre-set, which we shall explain in some detail below, this allows approximative evaluation of many radial basis function terms whose centres are close to each other simultaneously by a single finite Laurent expansion that is inexpensive, as long as the argument x is far from the said cloud of centres. By contrast, all radial functions whose translates are close to x are computed exactly and explicitly. We shall call those centres the near field below.

The hierarchical structure of the centres is fundamental to the algorithm because it decides and orders what is far and what is near any given x where we wish to evaluate. It is built up as follows. We assume for the sake of an easy exposition that $\Xi \subset [0,1]^2$ and that the ξ are fairly uniformly distributed in that unit square so that uniform subdivisions are acceptable. We form a *quad-tree* of centres which contains as a root Ξ and as the next children the intersections of the four quarter squares of the unit square with Ξ. They are in turn divided into four grandchildren each in the same fashion. We stop this process at a predetermined level that in part determines the accuracy of the calculation in the end. Each member of this family is called a panel. We point out that there are implementations where the divisions are only by two and not by four (Powell 1993). There is no principal reason for making this choice that forms a binary tree instead of a quad-tree and both approaches have been tried successfully.

Now we must define the far field and the near field for each x where $s(x)$ is computed. Given x, all centres ξ that are in the near field give rise to explicit evaluations of $\phi(\|x - \xi\|)$. The near field consists simply of contributions from all points which are not 'far', according to the following definition: we say x is far away from a panel T and therefore from all centres in that panel, if there is at least one more panel between x and T, and if this panel is on the same level of parenthood as T itself.

Next, we have to decide how to group the far points. All panels Q are in the *evaluation list* of a panel T if Q is either at the same or a coarser (higher) level than T and every point in T is far away from Q, and if, finally, T contains a point that is *not* far away from the parent of Q. Thus the far field of an x whose closest centre is in a panel T is the sum of all

$$s_Q(x) = \sum_{\xi \in Q} \lambda_\xi \phi(\|x - \xi\|) \qquad (4.13)$$

such that Q is in the evaluation list of T. For each (4.13), a common Laurent series is computed. Since we do not know the value of x that is to be inserted into (4.13), we compute the *coefficients* of the Laurent series and insert x later on.

In a set-up process, we compute these expansions of the radial function for large argument, that is, their coefficients, and store them; when x is provided at the evaluation stage, we combine those expansions for all centres from each evaluation list, and in an additional, final step we can simplify further by approximating the whole far field by one Taylor series.

A typical Laurent series expansion to thin-plate spline terms, which we still use as a paradigm for algorithms with more general classes of radial basis functions, is as follows. To this end, we work in two dimensions $n = 2$ and identify two-dimensional real space with one-dimensional complex space.

Lemma 1. Let z and t be complex numbers, and let

$$\phi_t(z) := \|t - z\|^2 \log \|t - z\|.$$

Then, for all $\|z\| > \|t\|$, and denoting the real part by Re,

$$\phi_t(z) = \mathrm{Re}\left\{ (\bar{z} - \bar{t})(z - t)\left(\log z - \sum_{k=1}^{\infty} \frac{1}{k}\left(\frac{t}{z}\right)^k \right) \right\},$$

which is the same as

$$(\|z\|^2 - 2\mathrm{Re}\,(\bar{t}z) + \|t\|^2)\log\|z\| + \mathrm{Re}\left\{\sum_{k=0}^{\infty}(a_k\bar{z} + b_k)z^{-k}\right\},$$

where $b_k = -\bar{t}a_k$ and $a_0 = -t$ and $a_k = t^{k+1}/[k(k+1)]$ for positive k. Moreover, if the above series is truncated after $p+1$ terms, the remainder is bounded above by

$$\frac{\|t\|^2}{(p+1)(p+2)}\frac{c+1}{c-1}\left(\frac{1}{c}\right)^p$$

with $c = \|z/t\|$.

In summary, the principal steps of the whole algorithm are as follows.

Set-up

(1) Perform the repeated subdivision of the square down to $\log|\Xi|$ levels and sort the elements of Ξ into the finest level panels.
(2) Form the Laurent series expansions (*i.e.*, compute their coefficients) for all fine level panels R.
(3) Translate centres of expansions so that the expansions can be re-used for other centres and, by working up the tree towards coarser levels, form analogous expansions for all less refined levels.
(4) Working down the tree from the coarsest level, compute Taylor expansions of the *whole* far field for each panel Q.

Evaluation at x

(1) Locate the finest level panel Q containing the centre closest to x.
(2) Evaluate $s(x)$ by computing near field contributions explicitly and by using the Taylor approximation of the entire far field. For the far field, we use all panels that are far away from x and are not subsets of any coarser panels already considered.

The computational cost without set-up is $\mathcal{O}(|\Xi|)$ because of the hierarchical structure. The set-up cost is $\mathcal{O}(\log|\Xi|)$, because of the tree structure, but the constant contained in this estimate may be large because of the computation of the various expansions and the design of the tree. This is so although each expansion is an $\mathcal{O}(1)$ procedure. In practice it turns out that the method is superior to direct computation if $|\Xi|$ is at least of the order of 200 points.

In which way is this algorithm now related to the computation of interpolation coefficients? It is related in one way because the efficient evaluation of the linear combination of thin-plate spline translates is required in the first algorithm presented in this section. There the residuals $f_\xi - s(\xi)$ played an important rôle, $s(\xi)$ being the current linear combination of thin-plate splines at a centre and the current approximation to the solution we require. Therefore, to make the BGP algorithm efficient, fast evaluation of s is needed, partly because our radial basis functions are conditionally positive definite (the preconditioning making the matrices positive (semi-)definite matrices) and partly because the condition numbers severely influence the convergence properties of conjugate gradient methods.

However, the importance of fast availability of residuals is not only restricted to the BGP method. Other iterative methods for computing radial basis function interpolants such as conjugate gradient methods are also in need of these residuals. In order to apply those, however, a preconditioning method is usually needed.

There are various possible improvements to the algorithm of this subsection. For instance, we can use an adaptive method for subdividing into the hierarchical structure of panels, so that the panels may be of different sizes, but always contain about the same number of centres. This is particularly advantageous when the data are highly non-uniformly distributed.

5. Interpolation on spheres

Because of the many applications that suit radial basis functions in geodesy, there is already a host of papers that specialize radial basis function approximation and interpolation to spheres. Freeden and co-workers (1981, 1986, 1995) have made a very large number of contributions to this aspect of approximation theory of radial basis functions. There are excellent and long review papers available from the work of this group (see the cited references) and we will therefore be relatively brief in this section. Of course, we no longer use the conventional Euclidean norm in connection with a univariate radial function when we approximate on the $(n-1)$ sphere S^{n-1} within $\mathbb{R}^n$ but apply so-called geodesic distances. Therefore the standard notions of positive definite functions and conditional positive definiteness no longer apply, and one has to study new concepts of (conditionally) positive definite functions on the $(n-1)$ sphere. This started with Schoenberg (1942), who characterized positive definite functions on spheres as those ones whose expansions in series of Gegenbauer polynomials always have nonnegative coefficients. Xu and Cheney (1992) studied strict positive definiteness on spheres and gave necessary and sufficient conditions. This was further generalized by Ron and Sun in 1996

Recent papers by Jetter, Stöckler and Ward (1999) and Levesley, Light, Ragozin and Sun (1999) use native spaces, (semi-)inner products and reproducing kernels (cf. Saitoh (1988) for a treatise on the theory of reproducing kernels) to derive approximation orders in a very similar fashion to the work summarized in Section 2. They all apply the *spherical harmonics* $\{Y_k^{(\ell)}\}_{k=1}^{d_\ell}$ that form an orthonormal basis of the d_ℓ-dimensional space of polynomials on the sphere in $\mathbb{P}_n^\ell(S^{n-1}) \cap \mathbb{P}_n^{\ell-1}(S^{n-1})^\perp$. They are called spherical harmonics because they are the restrictions of polynomials of total degree ℓ to the sphere and are in the kernel of the Laplace operator Δ. The dimension d_ℓ is computable, and this is not difficult, but it is not important to us at this stage.

Then, a *native space* X is defined for all expansions of functions on the sphere in spherical harmonics: namely, the native space's elements are

$$f(x) = \sum_{\ell=0}^{\infty} \sum_{k=1}^{d_\ell} \hat{f}_{\ell k} Y_k^{(\ell)}(x), \qquad x \in S^{n-1}, \tag{5.1}$$

whose coefficients $\hat{f}_{\ell k}$ satisfy certain square summability conditions with prescribed positive real weights $a_{\ell k}$. In other words, the native space is defined by

$$X = \left\{ f \,\middle|\, \sum_{\ell=0}^{\infty} \sum_{k=1}^{d_\ell} \frac{|\hat{f}_{\ell k}|^2}{a_{\ell k}} < \infty \right\}. \tag{5.2}$$

The native space X given in (5.2) and functions (5.1) will give rise to a reproducing kernel that is positive definite, but if we enlarge the space by starting the first sum in (5.2) only at $\ell = \kappa$ and thereby weakening conditions, we can also get conditionally positive definite (reproducing) kernels (Levesley et al. 1999) for the ensuing spaces X_κ. Then the native space will be a semi-Hilbert space, that is, the inner product that we shall describe shortly has a nullspace

$$K = \left\{ f \,\middle|\, f = \sum_{\ell=0}^{\kappa-1} \sum_{k=1}^{d_\ell} \hat{f}_{\ell k} Y_k^{(\ell)} \right\}.$$

Now, a standard choice for the positive weights for defining the space X which are often independent of k is $a_{\ell k} = (1 + \lambda_\ell)^{-s}$. This gives rise to the Sobolev space $H^s(S^{n-1})$. Here the $\lambda_\ell = \ell(\ell + n - 2)$ are eigenvalues of the Laplace–Beltrami operator.

The inner product that the native space is equipped with can be described by

$$\langle f, g \rangle = \sum_{\ell=0}^{\infty} \sum_{k=1}^{d_\ell} \frac{1}{a_{\ell k}} \hat{f}_{\ell k} \hat{g}_{\ell k},$$

where the coefficients are defined through (5.1); they are still assumed to be positive. The reproducing kernel that results from this Hilbert space X with the above inner product and that corresponds to the function of our previous radial basis functions in the native space is, when x and y are on the sphere,

$$\phi(x,y) = \sum_{\ell=0}^{\infty} \sum_{k=1}^{d_\ell} a_{\ell k} Y_k^{(\ell)}(x) Y_k^{(\ell)}(y), \qquad x, y \in S^{n-1}. \tag{5.3}$$

This can be simplified by the famous addition theorem (Stein and Weiss 1971) to $\phi(x,y) = \phi(x^T y)$, where

$$\phi(t) = \frac{1}{\omega_{n-1}} \sum_{\ell=0}^{\infty} d_\ell a_{\ell k} P_\ell(t), \tag{5.4}$$

ω_{n-1} being the measure of the unit sphere, if the coefficients are constant with respect to k. Here, the d_ℓ are as above and P_ℓ is a Gegenbauer polynomial (Abramowitz and Stegun 1972) normalized by $P_\ell(1) = 1$. Therefore, we now use (5.3) or (5.4) for interpolation on the sphere, in the same place and with the same centres Ξ as before, but they are from the sphere themselves of course. Convergence estimates are available from all three sources mentioned above that vary in approaching the convergence question. Using the mesh norm

$$h = \sup_{x \in S^{n-1}} \inf_{\xi \in \Xi} \arccos(x^T \xi),$$

Jetter et al. prove the following theorem. The notation $|\Xi|$ is for the cardinality of the set Ξ as before.

Theorem 14. Let X and Ξ be as above with the given mesh norm h. Let κ be a positive integer such that $h \le 1/(2\kappa)$. Then, for any $f \in X$, there is a unique interpolant s in

$$\text{span } \{\phi(\xi, \cdot) \mid \xi \in \Xi\}$$

that interpolates f on Ξ and satisfies the error estimate

$$\|s - f\|_\infty^2 \le \frac{5(|\Xi| + 1)}{\omega_{n-1}} \|f\|_\phi^2 \sum_{\ell=\kappa+1}^{\infty} \left(d_\ell \max_{k=1,\dots,d_\ell} a_{\ell k} \right). \tag{5.5}$$

Corollary 1. Let the assumptions of the previous theorem hold and suppose further that $|\Xi| + 1 \le C_1 \kappa^{n-1}$ and

$$\frac{C_2}{1 + \kappa} \le h \le \frac{1}{2\kappa}.$$

Then the said interpolant s provides

$$\|s - f\|_\infty = \mathcal{O}\left(\left(\frac{h}{C_2} \right)^{(\alpha-n)/2} \right)$$

or

$$\|s - f\|_\infty = \mathcal{O}\left(\frac{\exp(-\alpha C_2/2h)}{h^{(n-1)/2}} \right),$$

respectively, if $d_\ell \times \max_{k=1}^{d_\ell} a_{\ell k}$ is bounded by a constant multiple of $(1+\ell)^{-\alpha}$ for an $\alpha > n$ or by a constant multiple of $\exp(-\alpha(1+\ell))$ for a positive α, respectively.

An error estimate of Levesley et al., which includes conditionally positive definite kernels for $\kappa = 1$ or $\kappa = 2$, is as follows, for $n = 2$ (see also Freeden and Hermann (1986) for a similar, albeit slightly weaker result).

Theorem 15. Let X, Ξ, h and κ be as above, let s be the minimal norm interpolants to $f \in X_\kappa$ on Ξ. When ϕ is twice continuously differentiable on $[1 - \varepsilon, \varepsilon]$ for some $\varepsilon \in (0, 1)$, then

$$\|s - f\|_\infty \le Ch^2 \|f\|_\phi,$$

so that, in particular, a polynomial $p \in \mathbb{P}_3^{\kappa-1}(S^2)$ is added to the s used in Theorem 14.

6. Applications

6.1. Numerical solution of partial differential equations

Given that radial basis functions are known to be useful to approximate multivariate functions efficiently, it is suitable to apply them to approximate solutions of partial differential equations numerically. Three approaches have been tried and tested in this direction, namely collocation techniques, variational formulations and boundary element methods, all in order to solve elliptic partial differential equations with boundary values given. There are various reasons why radial basis functions are useful for these three approaches. The first of these is useful because we know much about existence and accuracy of radial basis function interpolants, especially when the data are scattered, which is useful for non-grid collocation. The second resembles typical finite element applications, where usually radial basis functions of compact support are used to mimic the standard finite element approach with multivariate piecewise polynomials. Finally, boundary element methods are suitable in several cases when radial basis functions are known to be fundamental solutions (Green's functions) of elliptic partial differential operators, most notably powers of the Laplace operator. An example is the thin-plate spline radial basis function and the bi-Laplacian operator. After all, in boundary element methods, explicit solutions of the associated homogeneous problem are required in advance, for which it is immensely helpful to have Green's functions to work with.

Naturally, an important decision is the choice of radial basis function, especially whether globally or locally supported ones should be used. In a Galerkin approach, locally supported elements are almost always employed. Further, the use of radial basis functions becomes particularly interesting when nonlinear partial differential equations are solved or non-grid approaches are used, for instance because of non-smooth domain boundaries, where non-uniform knot placement is always important to modelling the solution to good accuracy.

We begin with a description of the collocation approach, which is the first approach one is tempted to try because of our knowledge of interpolation properties of the radial basis functions. The first important decision is whether to use the well-known, standard globally supported radial functions such as multiquadrics or the new compactly supported ones that are described earlier in this review. Since the approximation properties of the latter are not as good as the former ones, we have a trade-off between accuracy on one hand and sparsity of the collocation matrix on the other hand. Compactly supported ones, if scaled suitably, give banded collocation matrices, while the globally supported ones give no sparsity to speak of. When we use the compactly supported radial functions we have, in fact, another trade-off which we have not mentioned so far, because even their scaling pits accuracy against sparseness of the matrix. If we impose no scaling to the radial basis function as in Theorem 12, we do have satisfactory convergence, as shown in Theorem 10, but basically the radial basis function behaves as a globally supported one, with essentially full matrices, since the centres become dense in each of the radial functions' supports. In the other extreme case, when we scale so that there is always a uniformly bounded number of centres inside each support, we run the risk of losing convergence altogether – but the interpolation matrix may be diagonal (and nonsingular, of course). There are several approaches to fixing this problem, and we will mention two of them while describing algorithms.

One typical linear partial differential equation problem suitable for collocation techniques reads

$$Lu(x) \ = \ f(x), \quad x \in \Omega \subset \mathbb{R}^n, \tag{6.1}$$
$$Bu|_{\partial\Omega} \ = \ q, \tag{6.2}$$

where Ω is a domain with suitably smooth – at least Lipschitz-continuous – boundary $\partial\Omega$ and f, q are prescribed functions. Here L is a linear differential operator and B a boundary operator. We will soon come to some specific nonlinear examples in the context of boundary element techniques.

The usual approach to collocation is then for centres Ξ that are partitioned in two disjoint sets Ξ_1 and Ξ_2, the former from the domain, the latter from its boundary, to solve the Hermite–Birkhoff interpolation system

$$\Lambda_\xi u_\Xi \ = \ f(\xi), \quad \xi \in \Xi_1,$$
$$\Lambda_\zeta u_\Xi \ = \ q(\zeta), \quad \zeta \in \Xi_2. \tag{6.3}$$

The approximants u_Ξ are defined by the sums

$$u_\Xi(x) = \sum_{\xi \in \Xi_1} c_\xi \Lambda_\xi \phi(\|x - \xi\|) + \sum_{\zeta \in \Xi_2} d_\zeta \Lambda_\zeta \phi(\|x - \zeta\|).$$

The Λ_ξ and Λ_ζ are suitable discrete functionals to describe our operators L and B on the discrete set of centres. In the above display the operators are applied with respect to the variable x. Examples for such discrete functionals to replace the operators L and B are obtained, for example, by replacing derivatives by symmetric differences or one-sided differences for the boundary in case of Neumann problems. Thus we end up with a square symmetric system of linear equations whose collocation matrix is nonsingular if, for instance, the radial basis function is positive definite, smooth enough for application of the operators Λ, and the discrete linear functionals are linearly independent functionals in the dual space of the native space of the radial basis functions (see Section 2 and Wu (1992) for the details). An error estimate is given in Franke and Schaback (1998). For those error estimates, it has been noted that more smoothness of the radial basis function is required than for a comparable finite element setting in order to get the same approximation orders, but clearly, the radial basis function setting has the distinct advantage of availability in any dimension and the absence of grids or triangulations.

If a compactly supported radial basis function is used, the necessary scaling leads to the aforementioned trade-off between accuracy and bandwidth of the matrix. In fact, the *conditioning* of the collocation matrix is also affected, becoming worse with smaller η, with $\phi(\cdot/\eta)$ being used, although the matrix is positive definite. A Jacobi preconditioning by the diagonal values helps here, so the matrix A is replaced by $P^{-1}AP^{-1}$ where $P = \sqrt{\text{diag}(A)}$, the diagonal elements of the matrix being positive. Moreover, one can use a *multilevel method* (Narcowich, Schaback and Ward (1999), Fasshauer (1999)) where numerical approximations $\{u_k\}_{k=0}^N$ are computed on nested sets of centres $\Xi_k \supset \Xi_{k-1}$, $k = 1, 2, \ldots, N$, $\Xi_N = \Xi$, and, within each sweep of the algorithm, a new approximation to the desired solution is computed as follows. For instance, to solve $Lu = f$ on the domain Ω with Dirichlet boundary conditions only, at each step k of one sweep one computes, starting with $u_0 = 0$,

$$L\tilde{u}_k = (f - Lu_{k-1})$$

and sets $u_k = \tilde{u}_k + u_{k-1}$. Unfortunately, little is known about the convergence behaviour of such a multilevel method.

In the event that a Galerkin method is applied, for instance, to the Helmholtz equation with Neumann conditions when $L = -\Delta + I$ and $B = \frac{\partial}{\partial n}$, we end up with a square system of linear equations, the stiffness equations for u_Ξ,

$$a(u_\Xi, \phi(\| \cdot - \xi\|)) = (f, \phi(\| \cdot - \xi\|))_{L^2(\Omega)}, \qquad \xi \in \Xi,$$

with

$$a(u, v) = \int_\Omega (\nabla u)^T (\nabla v) + uv.$$

If ϕ is a radial basis function of compact support such that its Fourier transform satisfies the decay estimate $|\hat\phi(r)| = \mathcal{O}(r^{-2k})$, then Franke and Schaback (1998) establish the convergence estimate

$$\|u - u_\Xi\|_{H^1(\Omega)} \le Ch^{\sigma-1}\|u\|_{H^\sigma(\Omega)},$$

where h is given by (2.2) and $k \ge \sigma > n/2 + 1$.

We now outline the third method, that is, a boundary element (BEM) method, following Pollandt (1997). The dual reciprocity method uses the second Green's formula and a fundamental solution $\phi(\| \cdot \|)$ of the Laplace operator Δ, in order to reformulate a boundary value problem as a boundary integral problem over a space of one dimension lower. This will then lead through discretization to a linear system with a full matrix for collocation by radial basis functions, in the way familiar from other applications of radial basis function interpolation. The radial basis function that occurs in that context is this fundamental solution, and, naturally, it is highly relevant in this case that the Laplace operator is rotationally invariant and has radial functions as Green's functions. We give a concrete example. Namely, for a nonlinear problem on a domain $\Omega \subset \mathbb{R}^n$ with Dirichlet boundary conditions such as the following one with a nonlinear right-hand side

$$\Delta u(x) = u^2(x), \quad x \in \Omega \subset \mathbb{R}^n, \tag{6.4}$$

$$u|_{\partial\Omega} = q, \tag{6.5}$$

the goal is to approximate the solution u of the elliptic partial differential equation on the domain by g plus a boundary term $\tilde r$ that satisfies $\Delta \tilde r \equiv 0$ on the domain. To this end, one gets after an application of Green's formula the equation on the boundary that $u(x)$ is the same as

$$\int_\Omega u(y)^2 \phi(\|x-y\|)\,\mathrm{d}y - \int_{\partial\Omega} \phi(\|x-y\|)\frac{\partial}{\partial n_y}u(y) - u(y)\frac{\partial}{\partial n_y}\phi(\|x-y\|)\,\mathrm{d}\Gamma_y$$

$$\tag{6.6}$$

for $x \in \Omega$, where $\frac{\partial}{\partial n_y}$ is the normal derivative with respect to y on $\Gamma = \partial\Omega$. The radially symmetric ϕ is still the fundamental solution of the Laplace operator used in the formulation of the differential equation above. Further, one gets after two applications of Green's formula the equation

$$\frac{1}{2}\left(u(x) - g(x)\right) + \int_{\partial\Omega} \phi(\|x-y\|) \times \frac{\partial}{\partial n_y}\left(u(y) - g(y)\right) -$$

$$(q(y) - g(y))\,\frac{\partial}{\partial n_y}\,\phi(\|x-y\|)\,\mathrm{d}\Gamma_y = 0, \quad x \in \partial\Omega.$$

We will later use this equation to approximate the boundary part of the solution, that is, the part satisfying the boundary conditions. Now we *assume* that there are real coefficients λ_ξ such that the infinite expansion (which will be truncated later on)

$$u^2(y) = \sum_\xi \lambda_\xi \, \widetilde{\phi}\,(\|y - \xi\|), \qquad y \in \Omega,$$

holds, and set

$$g(y) = \sum_\xi \lambda_\xi \, \widetilde{\Phi}\,(\|y - \xi\|), \qquad y \in \Omega;$$

so that $\Delta g = u^2$ everywhere with no boundary conditions. Therefore we are first solving a homogeneous problem. Here $\widetilde{\phi}$ is a suitable radial basis function, which is to be distinguished from the fundamental solution ϕ, with $\Delta \widetilde{\Phi}\,(\|\cdot\|) = \widetilde{\phi}(\|\cdot\|)$, and the centres ξ are from Ω. We now replace the infinite sums by finite ones (*i.e.*, we approximate the infinite expansions by finite sums for a suitable Ξ), so that exchanging summation and the Laplace operator and term-by-term differentiation in the sum is admitted:

$$u^2(y) = \sum_{\xi \in \Xi} \lambda_\xi \, \widetilde{\phi}\,(\|y - \xi\|), \qquad y \in \Omega, \tag{6.7}$$

and

$$g(y) = \sum_{\xi \in \Xi} \lambda_\xi \, \widetilde{\Phi}\,(\|y - \xi\|), \qquad y \in \Omega.$$

We require that the equation in the display after (6.6) holds for finitely many points $x = \zeta_j \in \partial\Omega$, $j = 1, 2, \ldots, t$, only. Then we solve for the coefficients λ_ξ by requiring that (6.7) holds for $y = \xi$, for all $\xi \in \Xi$. This fixes the λ_ξ by interpolation (collocation in the language of differential equations), whereas the equation after (6.6) determines the normal derivative $\frac{\partial}{\partial n_y}\, u(y)$ on $\partial\Omega$, where we are replacing $\frac{\partial}{\partial n_y}\, u(y)$ by another approximant, a polynomial spline $\tau(y)$, for instance. Thus the spline is found by requiring the above equation for all $x = \zeta_j \in \partial\Omega$, $j = 1, 2, \ldots, t$, and choosing a suitable t. Finally, an approximation $\widetilde{u}(x)$ to $u(x)$ is determined on Ω by the identity

$$\widetilde{u}(x) := g(x) + \int_{\partial\Omega} (q(y) - g(y)) \, \frac{\partial}{\partial n_y}\, \phi(\|x - y\|) \, \mathrm{d}\Gamma_y -$$

$$\int_{\partial\Omega} \phi(\|x - y\|) \left(\tau(y) - \frac{\partial g(y)}{\partial n_y} \right) \mathrm{d}\Gamma_y, \quad x \in \Omega, \tag{6.8}$$

where the boundary term $\widetilde{r}$ corresponds to the second term on the right-hand side of the display.

Now, all expressions on the right-hand side are known. This is an outline of the approach but we have skipped several important details. Nonetheless, one can clearly see how radial basis functions appear in this algorithm;

indeed, it is most natural to use them here, since many of them are fundamental solutions of Laplace operators or their iterates in certain dimensions. In the above example and $n = 2$, $\phi(r) = (2\pi)^{-1} \log r$, $\widetilde{\phi}(r) = r^2 \log r$ (thin-plate splines) and $\widetilde{\Phi}(r) = \frac{1}{16} r^4 \log r - \frac{1}{32} r^4$ are the correct choices. As the collocation matrices that appear in the boundary integral equation we are left with in the method are dense and as interpolation is not absolutely necessary in this approach, it can be substituted by quasi-interpolation. Quasi-interpolation has asymptotically essentially the same approximation behaviour (i.e., convergence speed) as interpolation but does not require any data-dependent linear systems to be solved; see also Buhmann (1990a). Convergence theorems for the method are available in the paper by Pollandt.

6.2. Other applications

We outline a few further practical applications of radial basis functions. They include, for example, mappings of two- or three-dimensional images such as portraits or underwater sonar scans into other images for comparison. Here interpolation comes into play because some special features of an image may have to be preserved while others need not be mapped exactly, thus enabling a comparison of some features that may differ while at the same time retaining others. Such so-called 'markers' can be, for example, certain points of the skeleton in an x-ray that has to be compared to another one, taken at another time. The same structure appears if we wish to compare sonar scans of a harbour at different times, the rocks being suitable as markers this time. Thin-plate splines turned out to be excellent for such very practical applications (Barrodale and Zala 1997, 1999). Work of this kind led to the invention of the methods for fast evaluation of thin-plate splines and other radial basis functions discussed above, because, after the interpolation, the computed interpolant had to be evaluated on a very fine square grid in two dimensions for analysis or display (Powell 1993).

Measurements of gravitational potential or temperature on the Earth's surface at 'scattered' meteorological stations or measurements on other multidimensional objects, may give rise to interpolation problems that require scattered data. Multiquadric approximations are performing well for this type of use (Hardy 1990). Much work on radial basis functions on spheres which we have only touched upon in this article originates from those applications of geophysics.

Many applications involve high-dimensional interpolation or approximation problems when data are coming through many 'channels', for instance electrodes measuring brain activity from nerve cells. Typical applications from neural physics produce 50–100 dimensional data taken from those electrodes recording measurements from the brain and require post-processing for smoothing, for example. Radial basis functions have been used extensively for this (*e.g.*, Eckhorn (1999), Hochreiter and Schmidhuber (1999), Anderson, Das and Keller (1998)).

The approximation to so-called learning situations by neural networks usually leads to very high-dimensional interpolation problems with scattered data. Girosi (1992) mentions radial basis functions as a very suitable approach to this, partly because of their availability in arbitrary dimensions, and their smoothness. A typical application is in fire detectors. An advanced type of fire detector has to consider several measured parameters, such as colour, spectrum, intensity, movement of an observed object from which it must decide whether, for instance, it is looking at a fire in the room or not, because the apparent fire is reflected sunlight. There is a learning procedure before the implementation of the device, where several prescribed situations (these are the data) are tested and the values zero (no fire) and one (fire) are interpolated, so that the device can 'learn' to interpolate between these standard situations for general cases later when it is used in real life. Radial basis function methods have been tried very successfully for this application because they are excellent tools for high-dimensional problems that will undoubtedly find many more applications in real life, such as polynomial splines have done in at least the last 30 years and still do now.

REFERENCES

M. Abramowitz and I. Stegun, eds (1972), *Handbook of Mathematical Functions*, National Bureau of Standards, Washington.

R. W. Anderson, S. Das and E. L. Keller (1998), 'Estimation of spatiotemporal neural activity using radial basis function networks', *J. Comput. Neuroscience* **5**, 421–441.

N. Arad, N. Dyn and D. Reisfeld (1994), 'Image warping by radial basis functions: applications to facial expressions', *Graphical Models and Image Processing* **56**, 161–172.

R. Askey (1973), 'Radial characteristic functions', MRC Report 1262, University of Wisconsin, Madison.

I. Barrodale and C. A. Zala (1997), 'MJDP–BCS industrial liaison: applications to defence science', in *Approximation and Optimization: Tributes to M. J. D. Powell* (M. D. Buhmann and A. Iserles, eds), Cambridge University Press, Cambridge, pp. 31–46.

I. Barrodale and C. A. Zala (1999), 'Warping aerial photographs to orthomaps using thin plate splines', *Adv. Comput. Math.* **11**, 211–227.

B. J. C. Baxter (1991), 'Conditionally positive functions and p-norm distance matrices', *Constr. Approx.* **7**, 427–440.

R. K. Beatson and W. A. Light (1992), 'Quasi-interpolation in the absence of polynomial reproduction', in *Numerical Methods of Approximation Theory* (D. Braess and L. L. Schumaker, eds), Birkhäuser, Basel, pp. 21–39.

R. K. Beatson and W. A. Light (1997), 'Fast evaluation of radial basis functions: methods for 2-dimensional polyharmonic splines', *IMA J. Numer. Anal.* **17**, 343–372.

R. K. Beatson and G. N. Newsam (1992), 'Fast evaluation of radial basis functions: I', *Comput. Math. Appl.* **24**, 7–19.

R. K. Beatson, G. Goodsell and M. J. D. Powell (1995), 'On multigrid techniques for thin plate spline interpolation in two dimensions', in Vol. 32 of *Lectures in Applied Mathematics*, pp. 77–97.

R. K. Beatson, J. B. Cherrie and C. T. Mouat (1999), 'Fast fitting of radial basis functions: methods based on preconditioned GMRES iteration', *Adv. Comput. Math.* **11**, 253–270.

A. Bejancu (1997), 'The uniform convergence of multivariate natural splines', DAMTP Technical Report, University of Cambridge.

A. Bejancu (1999), 'Local accuracy for radial basis function interpolation on finite uniform grids', *J. Approx. Theory* **99** 242–257.

M. D. Buhmann (1990a), 'Multivariate interpolation in odd-dimensional Euclidean spaces using multiquadrics', *Constr. Approx.* **6**, 21–34.

M. D. Buhmann (1990b), 'Multivariate cardinal-interpolation with radial-basis functions', *Constr. Approx.* **6**, 225–255.

M. D. Buhmann (1998), 'Radial functions on compact support', *Proceedings of the Edinburgh Mathematical Society* **41**, 33–46.

M. D. Buhmann (2000), 'A new class of radial basis functions with compact support', to appear in *Math. Comput.*

M. D. Buhmann and C. A. Micchelli (1991), 'Multiply monotone functions for cardinal interpolation', *Adv. Appl. Math.* **12**, 359–386.

M. D. Buhmann and C. A. Micchelli (1992), 'On radial basis approximations on periodic grids', *Mathematical Proceedings of the Cambridge Philosophical Society* **112**, 317–334.

J. Duchon (1976), 'Interpolation des fonctions de deux variables suivant le principe de la flexion des plaques minces', *Rev. Française Automat. Informat. Rech. Opér. Anal. Numer.* **10**, 5–12.

J. Duchon (1978), 'Sur l'erreur d'interpolation des fonctions de plusieurs variables pars les D^m-splines', *Rev. Française Automat. Informat. Rech. Opér. Anal. Numer.* **12**, 325–334.

J. Duchon (1979), 'Splines minimizing rotation-invariant semi-norms in Sobolev spaces', in *Constructive Theory of Functions of Several Variables* (W. Schempp and K. Zeller, eds), Springer, Berlin/Heidelberg, pp. 85–100.

N. Dyn, F. J. Narcowich and J. D. Ward (1997), 'A framework for interpolation and approximation on Riemannian manifolds', in *Approximation and Optimization: Tributes to M. J. D. Powell* (M. D. Buhmann and A. Iserles, eds), Cambridge University Press, Cambridge, pp. 133–144.

R. Eckhorn (1999), 'Neural mechanisms of scene segmentation: recordings from the visual cortex suggest basic circuits for linking field models', *IEEE Trans. Neural Net.* **10**, 1–16.

R. Estrada (1998), 'Regularization of distributions', *Int. J. Math. Math. Sci.* **21**, 625–636.

G. Fasshauer (1999), 'Solving differential equations with radial basis functions: multilevel methods and smoothing', *Adv. Comput. Math.* **11**, 139–159.

A. C. Faul and M. J. D. Powell (1999), 'Proof of convergence of an iterative technique for thin plate spline interpolation in two dimensions', *Adv. Comput. Math.* **11**, 183–192.

M. Floater and A. Iske (1996), 'Multistep scattered data interpolation using compactly supported radial basis functions', *J. Comput. Appl. Math.* **73**, 65–78.

C. Franke and R. Schaback (1998), 'Solving partial differential equations by collocation using radial basis functions', *Comput. Math. Appl.* **93**, 72–83.

W. Freeden (1981), 'On spherical spline interpolation and approximation', *Math. Meth. Appl. Sci.* **3**, 551–575.

W. Freeden and P. Hermann (1986), 'Uniform approximation by harmonic splines', *Math. Z.* **193**, 265–275.

W. Freeden, M. Schreiner and R. Franke (1995), 'A survey on spherical spline approximation', Technical Report 95–157, University of Kaiserslautern.

F. Girosi (1992), 'Some extensions of radial basis functions and their applications in artificial intelligence', *Comput. Math. Appl.* **24**, 61–80.

M. von Golitschek and W. A. Light (2000), 'Interpolation by polynomials and radial basis functions on spheres', to appear in *Constr. Approx.*

L. Greengard and V. Rokhlin (1987), 'A fast algorithm for particle simulations', *J. Comput. Phys.* **73**, 325–348.

R. L. Hardy (1990), 'Theory and applications of the multiquadric-biharmonic method', *Comput. Math. Appl.* **19**, 163–208.

S. Hochreiter and J. Schmidhuber (1999), 'Feature extraction through Lococode', *Neural Computation* **11**, 679–714.

K. Jetter, J. Stöckler and J. D. Ward (1999), 'Error estimates for scattered data interpolation on spheres', *Math. Comput.* **68**, 733–747.

M. J. Johnson (1997), 'An upper bound on the approximation power of principal shift-invariant spaces', *Constr. Approx.* **13**, 155–176.

M. J. Johnson (1998a), 'A bound on the approximation order of surface splines', *Constr. Approx.* **14**, 429–438.

M. J. Johnson (1998b), 'On the error in surface spline interpolation of a compactly supported function', manuscript, University of Kuwait.

M. J. Johnson (1999), 'Approximation in $L^p(\mathbb{R}^d)$ from spaces spanned by the perturbed integer translates of a radial function', manuscript, University of Kuwait.

D. S. Jones (1982), *The Theory of Generalised Functions*, Cambridge University Press, Cambridge.

J. Levesley, W. A. Light, D. Ragozin and X. Sun (1999), 'A simple approach to the variational theory for interpolation on spheres', in *New Developments in Approximation Theory* (M. D. Buhmann, M. Felten, D. Mache and M. W. Müller, eds), Birkhäuser, Basel, pp. 117–143.

W. R. Madych and S. A. Nelson (1990), 'Polyharmonic cardinal splines', *J. Approx. Theory* **60**, 141–156.

O. V. Matveev (1997), 'On a method for interpolating functions on chaotic nets', *Math. Notes* **62**, 339–349. Translated from *Mat. Zametki* (1997) **62**, 404–417.

C. A. Micchelli (1986), 'Interpolation of scattered data: distance matrices and conditionally positive definite functions', *Constr. Approx.* **1**, 11–22.

J. K. Misiewicz and D. St. P. Richards (1994), 'Positivity of integrals of Bessel functions', *SIAM J. Math. Anal.* **25**, 596–601.

F. J. Narcowich, R. Schaback and J. D. Ward (1999), Multilevel interpolation and approximation, *Appl. Comput. Harm. Analysis* **7**, 243–261.

R. Pollandt (1997), 'Solving nonlinear equations of mechanics with the boundary element method and radial basis functions', *Internat. J. Numer. Methods Engrg* **40**, 61–73.

M. J. D. Powell (1992a), 'The theory of radial basis function approximation in 1990', in *Advances in Numerical Analysis II: Wavelets, Subdivision, and Radial Functions* (W. A. Light, ed.), Oxford University Press, Oxford, pp. 105–210.

M. J. D. Powell (1992b), 'Tabulation of thin plate splines on a very fine two-dimensional grid', in *Numerical Methods of Approximation Theory* (D. Braess and L. L. Schumaker, eds), Birkhäuser, Basel, pp. 221–244.

M. J. D. Powell (1993), 'Truncated Laurent expansions for the fast evaluation of thin plate splines', *Numer. Alg.* **5**, 99–120.

M. J. D. Powell (1994), 'The uniform convergence of thin-plate spline interpolation in two dimensions', *Numer. Math.* **67**, 107–128.

M. J. D. Powell (1999), 'Recent research at Cambridge on radial basis functions', in *New Developments in Approximation Theory* (M. D. Buhmann, M. Felten, D. Mache and M. W. Müller, eds), Birkhäuser, Basel, pp. 215–232.

A. Ron and X. Sun (1996), 'Strictly positive definite functions on spheres in Euclidean spaces', *Math. Comput.* **65**, 1513–1530.

S. Saitoh (1988), *Theory of Reproducing Kernels and its Applications*, Longman, Harlow.

R. Schaback and H. Wendland (1998), 'Inverse and saturation theorems for radial basis function interpolation', Technical Report, University of Göttingen.

R. Schaback and Z. Wu (1995), 'Operators on radial basis functions', *J. Comput. Appl. Math.* **73**, 257–270.

I. J. Schoenberg (1942), 'Positive definite functions on spheres', *Duke Math. J.* **9**, 96–108.

R. Sibson and G. Stone (1991), 'Computation of thin plate splines', *SIAM J. Sci. Statist. Comput.* **12**, 1304–1313.

E. M. Stein and G. Weiss (1971), *Introduction to Fourier Analysis on Euclidean Spaces*, Princeton University Press, Princeton.

H. Wendland (1995), 'Piecewise polynomial, positive definite and compactly supported radial functions of minimal degree', *Adv. Comput. Math.* **4**, 389–396.

H. Wendland (1998), 'Error estimates for interpolation by radial basis functions of minimal degree', *J. Approx. Theory* **93**, 258–272.

H. Wendland (1997), 'Sobolev-type error estimates for interpolation by radial basis functions', in *Surface Fitting and Multiresolution Methods* (A. LeMéhauté and L. L. Schumaker, eds), Vanderbilt University Press, Nashville, pp. 337–344.

Z. Wu (1992), 'Hermite–Birkhoff interpolation of scattered data by radial basis functions', *Approx. Theory Appl.* **8**, 1–10.

Z. Wu (1995), 'Multivariate compactly supported positive definite radial functions', *Adv. Comput. Math.* **4**, 283–292.

Z. Wu and R. Schaback (1993), 'Local error estimates for radial basis function interpolation of scattered data', *IMA J. Numer. Anal.* **13**, 13–27.

Y. Xu and E. W. Cheney (1992), 'Strictly positive definite functions on spheres', *Proc. Amer. Math. Soc.* **116**, 977–981.

Acta Numerica (2000), pp. 39–131

The numerical analysis of bifurcation problems with application to fluid mechanics

K. A. Cliffe

AEA Technology, Harwell Laboratory,
Didcot, Oxfordshire OX11 0RA, England
E-mail: andrew.cliffe@aeat.co.uk

A. Spence

School of Mathematics, University of Bath,
Claverton Down, Bath BA2 7AY, England
E-mail: A.Spence@bath.ac.uk

S. J. Tavener

Department of Mathematics,
The Pennsylvania State University,
University Park, PA 16802, USA
E-mail: tavener@math.psu.edu

In this review we discuss bifurcation theory in a Banach space setting using the singularity theory developed by Golubitsky and Schaeffer to classify bifurcation points. The numerical analysis of bifurcation problems is discussed and the convergence theory for several important bifurcations is described for both projection and finite difference methods. These results are used to provide a convergence theory for the mixed finite element method applied to the steady incompressible Navier–Stokes equations. Numerical methods for the calculation of several common bifurcations are described and the performance of these methods is illustrated by application to several problems in fluid mechanics. A detailed description of the Taylor–Couette problem is given, and extensive numerical and experimental results are provided for comparison and discussion.

CONTENTS

1. Introduction

The numerical analysis of bifurcation problems is concerned with the stable, reliable and efficient computation of solutions to multiparameter nonlinear problems. We shall consider numerical methods for solving nonlinear equations of the form

$$F(x, \lambda) = 0, \tag{1.1}$$

where F is a smooth operator in an appropriate Banach space setting, x is a state variable and λ represents one or more parameters. In applications the main interest is often the determination of qualitative changes in x as λ varies. Problems like (1.1) arise in the consideration of steady states of the dynamical system

$$\frac{dx}{dt} + F(x, \lambda) = 0, \tag{1.2}$$

and indeed the study of the solution set of (1.1) is usually the first step in an analysis of the behaviour of solutions to (1.2).

   The material in this review is applicable to a wide range of problems although we shall concentrate on problems arising in fluid dynamics, and so for us (1.2) represents the dynamical Navier–Stokes equations. The nonlinear character of the Navier–Stokes equations gives rise to multiple solutions and possibly complicated dynamics and this nonlinear behaviour is central to problems in fluid dynamics, where the idea of dynamical similarity introduces various nondimensional groups, for instance the Reynolds number and Rayleigh number, plus geometric parameters: for example, in the Taylor–Couette problem discussed in Section 8 there are the aspect ratio and the radius ratio.

In fluid mechanics we are therefore confronted with nonlinear partial differential equations that depend on a number of parameters. This is precisely the domain of bifurcation theory. The overall goal, when studying a fluid mechanics problem, is to understand the complete behaviour of the system as a 'function' of the parameters. Relevant questions are: How many steady states are there? Are they stable or unstable? (It is important to have the ability to compute unstable steady states as well as stable ones, since solutions arising from bifurcations along unstable branches often interact with stable solutions producing otherwise inexplicable phenomena.) How does the structure of the steady state solution set change as the parameters are varied? Do solutions always respect the symmetry of the domain or is there symmetry breaking? How do time-dependent solutions arise? We shall address some aspects of these questions in this review. Other very important questions about which we have nothing to say here include: How do the initial conditions affect the evolution of the system? What types of long-term dynamical behaviour are possible? How does fluid turbulence arise?

In fluid mechanics the nonlinearity of the governing equations combined with the nontrivial geometry of the domain means that there are many problems where limited progress can be made with analytical techniques and one needs to use numerical methods.

There are two main numerical approaches to help answer some of the above questions for the Navier–Stokes equations. Either the time-dependent problem is discretized in space and the resulting system of ordinary equations is evolved forwards in time for various fixed values of the parameters. This approach is called 'simulation', and is the main technique used in the computational fluids community. The alternative approach is to discretize the steady problem to obtain a system of nonlinear equations, and then use methods from nonlinear analysis (*e.g.*, the implicit function theorem, singularity theory) to compute paths of steady solutions and provide stability assignment using numerical continuation methods and eigenvalue information. We shall concentrate on the latter approach here.

The numerical analysis of continuation methods was developed in the late 1970s by Keller (1977), Rheinboldt (1978) and Menzel and Schwetlick (1978), though many of the key ideas appear earlier in applications, especially buckling problems, for example Anselone and Moore (1966), Ricks (1972) and Abbot (1978). Several codes were then developed for numerical continuation and bifurcation analysis, the earliest being PITCON (see Rheinboldt (1986)) and AUTO developed by Doedel (see Doedel and Kernevez (1986)) but with recent extensions by Doedel, Champneys, Fairgrieve, Kuznetsov, Sandstede and Wang (1997). AUTO can treat steady state and time-dependent problems and discretized boundary value problems. We refer to the article by Allgower and Georg (1993) for a detailed discussion on numerical continuation.

Once reliable algorithms for numerical path following and simple bifurcation phenomena were devised then attention naturally shifted to multiparameter problems and the construction of numerical approaches based on the use of singularity theory (for example Beyn (1984), Jepson and Spence (1984), Jepson and Spence (1985b)). At the same time the convergence theory for discretization methods was concerned with the obvious questions: If a continuous problem has a particular singularity, under what conditions can it be guaranteed that the discretized problem has a singularity of the same type? Does the numerical method converge with the same rate of convergence as at nonsingular points? Do we observe superconvergence when using projection methods? In an important series of papers Brezzi, Rappaz and Raviart (1980, 1981a, 1981b) answered many of these questions, though again some key ideas and results were provided independently (see, for example, Kikuchi (1977), Fujii and Yamaguti (1980) and Moore and Spence (1981)).

There are many books on bifurcation theory: for example, Chow and Hale (1982) give an all-round treatment, Vanderbauwhede (1982) gives an early account of bifurcation in the presence of symmetries, and the important books by Golubitsky and Schaeffer (1985), and Golubitsky, Stewart and Schaeffer (1988) look at multiparameter bifurcation problems using singularity theory. Early conference proceedings are Rabinowitz (1977), Mittelmann and Weber (1980), Küpper, Mittelmann and Weber (1984), Küpper, Seydel and Troger (1987), Roose, Dier and Spence (1990) and Seydel, Küpper, Schneider and Troger (1991). H. B. Keller's book *Numerical Methods in Bifurcation Problems* (Keller 1987) is a published version of lectures delivered at the Indian Institute of Science, Bangalore. W. C. Rheinboldt's book (Rheinboldt 1986) is a collection of his papers and also gives information and listing of the code PITCON for numerical continuation of parameter-dependent nonlinear problems. The books by Kubíček and Marek (1983) and Seydel (1994) contain discussion of numerical methods and many interesting examples. A comprehensive treatment, including a full discussion of numerical methods using singularity theory, is to appear in the forthcoming book by Govaerts (2000).

One of the successes of numerical bifurcation techniques has been the ability to reproduce and help understand experimental results of the Taylor–Couette flow of a fluid confined between two concentric cylinders. Because this flow may be controlled quite precisely in the laboratory it provides an opportunity for rigorous experimental and numerical comparison. Of course the numerical techniques have been applied in a wide variety of other problems in fluid mechanics and have contributed significantly to the theoretical understanding of confined flows.

The detailed plan of this review is as follows. In Section 2 some of the main ideas in singularity theory are outlined first for scalar equations, then for multiparameter problems and problems with a simple reflectional symmetry. In Section 3 a review of bifurcation theory in Banach spaces is presented, covering the four main bifurcations to be expected in one-parameter problems. Section 4 discusses the convergence theory for numerical methods (both projection and finite difference methods), with special attention being paid to obtaining superconvergence results for bifurcation parameters when using Galerkin methods. In Section 5 mixed finite element methods for the Navier–Stokes equations are analysed. Section 6 contains a demonstration of superconvergence results using the $Q_2 - P_1$ finite element method applied to some classical problems in fluid mechanics. In Section 7 implementation details are provided for some of the main algorithms used to compute bifurcation points. Section 8 contains a detailed description of the Taylor–Couette problem, and presents extensive numerical and experimental results for comparison. In Section 9 other applications are discussed which use the numerical techniques described in this review. The review ends with a brief discussion about some important topics not covered here and suggests areas for future research.

2. Singularity theory

Golubitsky and Schaeffer (1979a, 1979b) pioneered the application of results from singularity theory to the study of bifurcation problems. Later, two books (Golubitsky and Schaeffer 1985, Golubitsky et al. 1988) provided a very careful explanation of the theory and techniques, as well as many illustrative examples and applications. Ideas from singularity theory were used in the numerical analysis of bifurcation problems by Brezzi and Fujii (1982) and Brezzi, Ushiki and Fujii (1984) to determine the effect of discretization errors, and by Beyn (1984), Jepson and Spence (1984, 1985b) to develop systematic numerical procedures for multiparameter nonlinear problems. Janovsky (1987) and Janovský and Plecháč (1992) further extended these ideas using minimally extended systems (see Section 7.5), and the forthcoming book by Govaerts (2000) gives a comprehensive account of numerical methods for bifurcation problems using singularity theory and minimally extended systems with bordered systems playing a key rôle in the linear algebra. There are many different aspects to singularity theory for bifurcation problems and we cannot hope to cover them all in this review: rather we concentrate on a few ideas to help motivate the material in later sections. However, we believe that a good understanding of the concepts and techniques in Golubitsky and Schaeffer (1985) and Golubitsky et al. (1988) is essential in order to develop reliable numerical techniques for multiparameter nonlinear problems.

The Lyapunov–Schmidt reduction procedure (see Section 3.2), is a process by which information about solutions near a singular point of a nonlinear

problem defined on a Banach space may be obtained by studying an equivalent *reduced* problem on a space of, typically, very small dimension. In
fact, if the singularity is such that the linearization of the problem evaluated
at the singularity has a one-dimensional kernel, then the reduced problem is
one-dimensional. Thus, it is appropriate to study nonlinear *scalar* problems
of the form

$$f(x, \lambda, \boldsymbol{\alpha}) = 0, \qquad f : \mathbb{R} \times \mathbb{R} \times \mathbb{R}^p \to \mathbb{R}, \qquad (2.1)$$

where x is a scalar state variable, λ a distinguished parameter, and $\boldsymbol{\alpha} \in \mathbb{R}^p$
a vector of control parameters. It is important to note that the view taken
in the singularity theory of Golubitsky and Schaeffer (1985) and Golubitsky
et al. (1988) is that in applications one will wish to plot the state variable
x against the special parameter λ for several fixed values of $\boldsymbol{\alpha}$. Thus we do
not interchange λ with one of the $\boldsymbol{\alpha}$s and λ plays a different rôle than the
other 'control' parameters. This approach leads to a different classification
of singularities than that obtained from standard singularity theory: see
Beyn (1984).

In Section 2.1, we first consider a simple problem with no control parameters. We consider multiparameter problems in Section 2.2, and in Section 2.3
give an example of the rôle played by symmetries. We draw some general
conclusions in Section 2.4.

2.1. Scalar problems

In this subsection we consider the numerical calculation of singular points
of the scalar problem

$$f(x, \lambda) = 0, \qquad x \in \mathbb{R}, \ \lambda \in \mathbb{R}, \qquad (2.2)$$

where $f(x, \lambda)$ is sufficiently smooth.

Analysis of this very simple case introduces some important ideas and
provides considerable insight into the behaviour of more complicated equa-
tions. First, note that it is convenient to write f^0 for $f(x_0, \lambda_0)$, f_λ^0 for
$f_\lambda(x_0, \lambda_0)$, etc. Now, if $f^0 = 0$ and $f_x^0 \neq 0$, then the Implicit Function
Theorem (IFT) ensures the existence of a smooth path, $x(\lambda)$, near (x_0, λ_0)
satisfying $f(x(\lambda), \lambda) = 0$. In this case we call (x_0, λ_0) a *regular* point. Of
more interest are singular points where $f_x^0 = 0$.

Consider the calculation of a singular point of (2.2). It is natural to form
the system

$$F(y) := \left[\begin{array}{c} f(x, \lambda) \\ f_x(x, \lambda) \end{array} \right] = 0 \in \mathbb{R}^2, \qquad y = \left(\begin{array}{c} x \\ \lambda \end{array} \right), \qquad (2.3)$$

and seek a zero of $F(y)$. A solution y_0 is regular if $F_y(y_0)$ is nonsingular,
which, as is easily checked, holds provided $f_\lambda^0 f_{xx}^0 \neq 0$, or, equivalently,

$$f_\lambda^0 \neq 0 \quad \text{and} \quad f_{xx}^0 \neq 0. \qquad (2.4)$$

If (2.3) and (2.4) hold then (x_0, λ_0) is a quadratic fold point. The reason for the name is clear when one sketches the solution curve near (x_0, λ_0), noting that near (x_0, λ_0), $\lambda = \lambda(x)$ with $\lambda(x_0) = \lambda_0$, and

$$\frac{\mathrm{d}\lambda}{\mathrm{d}x}(x_0) = 0, \quad \frac{\mathrm{d}^2\lambda}{\mathrm{d}x^2}(x_0) = -\frac{f_{xx}^0}{f_\lambda^0}. \tag{2.5}$$

We call (2.3) an *extended system*, and (2.4) provides two *side constraints*. Together, (2.3) and (2.4) provide the *defining conditions* for a quadratic fold point.

Quadratic fold points have several nice properties. First, Newton's method applied to (2.3) will converge quadratically for a sufficiently accurate initial guess. Second, a sensitivity analysis shows they are stable under perturbation. Assume $f(x, \lambda)$ is perturbed to $\hat{f}(x, \lambda, \epsilon) := f(x, \lambda) + \epsilon p(x, \lambda)$ and consider $\hat{F}(y, \epsilon) := (f + \epsilon p, f_x + \epsilon p_x) = 0$. Now $\hat{F}(y_0, 0) = 0$ and $\hat{F}_y(y_0, 0)$ is nonsingular and so the IFT shows that $y = y(\epsilon)$ near $\epsilon = 0$, with $y(\epsilon) = y_0 + \mathcal{O}(\epsilon)$, and $\hat{F}_y(y(\epsilon), \epsilon)$ nonsingular. Hence the perturbed problem $\hat{f}(x, \lambda, \epsilon) = 0$ has a quadratic fold point $(x(\epsilon), \lambda(\epsilon))$ satisfying $x(\epsilon) = x_0 + \mathcal{O}(\epsilon)$, $\lambda(\epsilon) = \lambda_0 + \mathcal{O}(\epsilon)$.

This type of sensitivity analysis is common in structural mechanics where the various physical imperfections in a system are 'lumped together' as a single artificial parameter. One might also consider $\epsilon = h^m$ where $\hat{f}$ is a discretization of f, h is a stepsize and m is the order of consistency. Clearly quadratic folds in f are preserved in $\hat{f}$ and it is not surprising that a similar result holds for more general problems under certain assumptions, as will be shown in Section 4.1.

2.2. Multiparameter problems

Let us change perspective now, and think of ϵ in the previous section as a control parameter to be varied rather than merely a perturbation parameter. The above analysis still applies, and provided $\hat{f}_\lambda(x(\epsilon), \lambda(\epsilon), \epsilon) \neq 0$ and $\hat{f}_{xx}(x(\epsilon), \lambda(\epsilon), \epsilon) \neq 0$ there is no requirement that ϵ remain small. Thus, we change notation by setting $\epsilon = \alpha$, and dropping the '^' symbol over the f, and consider the two-parameter problem

$$f(x, \lambda, \alpha) = 0, \quad x, \lambda, \alpha \in \mathbb{R}. \tag{2.6}$$

Provided the side constraints $f_\lambda \neq 0$ and $f_{xx} \neq 0$ continue to hold, then a path of quadratic fold points can be computed using Newton's method applied to

$$F(y, \alpha) = \begin{bmatrix} f(x, \lambda, \alpha) \\ f_x(x, \lambda, \alpha) \end{bmatrix} = 0, \quad y = \begin{pmatrix} x \\ \lambda \end{pmatrix}. \tag{2.7}$$

Since the side constraints appear in F_y, they can be easily monitored. If a zero occurs in a side constraint then a higher-order singularity has been detected.

In fact, a complete systematic procedure for multiparameter problems of the form (2.1) is given in Jepson and Spence (1985b) based on the singularity theory in Golubitsky and Schaeffer (1985). In Golubitsky and Schaeffer (1985) possible types of behaviour of solutions of (2.1) near a singular point are classified according to *contact equivalence*, namely, equivalence up to a smooth change of coordinates. This classification associates a number, the *codimension*, with each singularity, and if the codimension is finite then the singularity is equivalent to a *polynomial* canonical form. For example, the simplest singularity is the quadratic fold point, which has canonical form $f(x, \lambda) := x^2 - \lambda$ and has codimension zero. Clearly at $y_0 = (x_0, \lambda_0)^T = (0,0)^T$ then (2.3) and (2.4) are satisfied; conversely any f satisfying (2.3) and (2.4) is contact equivalent to $x^2 - \lambda$. In Jepson and Spence (1985b) the singularities of codimension less than 4 are arranged in a hierarchy (see also Table 2.4 of Golubitsky and Schaeffer (1985)), and this was used to provide an algorithm to obtain suitable extended systems and side constraints for the calculation of the singularities. For example, there are two codimension 1 singularities: a transcritical bifurcation ($\alpha = 0$ in Figure 1) that arises in a path of fold points when $f_\lambda = 0$; and a hysteresis bifurcation ($\alpha = 0$ in Figure 2) that arises in a path of fold points when $f_{xx} = 0$. To compute a transcritical bifurcation in a stable manner we need 2 parameters, namely λ and α, and the extended system is $F(y) := (f, f_x, f_\lambda)^T = 0$, $y = (x, \lambda, \alpha)^T$. A transcritical bifurcation point, $y_0 = (x_0, \lambda_0, \alpha_0)^T$ say, will be a regular solution if (a) $f_\alpha^0 \neq 0$, and (b) the side constraints $f_{xx}^0 \neq 0$ and $(f_{x\lambda}^0)^2 - f_{xx}^0 f_{\lambda\lambda}^0 \neq 0$ hold. The canonical form is $f(x, \lambda) := x^2 - \lambda^2$. The condition $f_\alpha^0 \neq 0$ is a *universal unfolding* condition that, roughly speaking, ensures that the control parameter α enters in f in such a way as to provide all qualitatively distinct solutions of $f(x, \lambda, \alpha) = 0$ as α varies near α_0. The transcritical bifurcation has codimension 1, since 1 control parameter is needed in the universal unfolding $f(x, \lambda, \alpha) = 0$. Figure 1 shows the unfoldings of a transcritical bifurcation, and Figure 2 shows the unfoldings of a hysteresis point (also of codimension 1) which has extended system $F(y) := (f, f_x, f_{xx})^T = 0$ and side constraints $f_\lambda \neq 0$, $f_{xxx} \neq 0$. (See p. 136 of Golubitsky and Schaeffer (1985) for the universal unfolding condition for a hysteresis point.)

It is important to note that one would not expect to see the codimension 1 singularities, that is, transcritical or hysteresis bifurcation points, in a one-parameter physical problem. Rather, two parameters are needed to observe them and to locate them numerically. Also, as we see in Figures 1 and 2, they are destroyed by perturbations. It is not surprising, then, that the con-

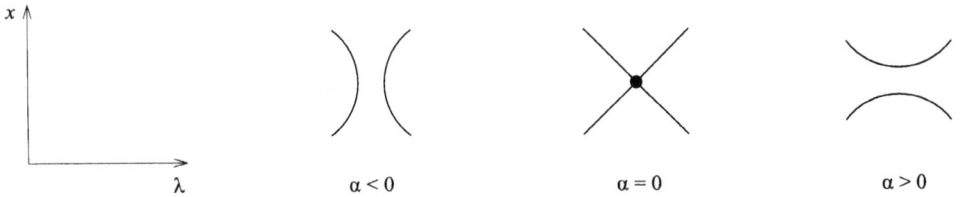

Fig. 1. Solution diagrams for $f(x, \lambda, \alpha) = x^2 - \lambda^2 + \alpha = 0$. The transcritical bifurcation point is destroyed for $\alpha \neq 0$

Fig. 2. Solution diagrams for $f(x, \lambda, \alpha) = x^3 + \alpha x - \lambda = 0$. The hysteresis point is destroyed for $\alpha \neq 0$ and there are no singular points for $\alpha > 0$

vergence theory of discretizations near bifurcation points in one-parameter problems proves very technical and is perhaps of limited usefulness.

A key result (Jepson and Spence 1985b, Theorem 3.10) is that a multiparameter problem is universally unfolded if and only if the extended systems produced from the hierarchy are nonsingular. This has important numerical implications, but also shows that one needs to consider singularities with the correct number of control parameters. If this is done then the effect of perturbations (and discretizations) can be readily understood.

We refer the reader to Golubitsky and Schaeffer (1985), Jepson and Spence (1985b), Golubitsky et al. (1988), Janovský and Plecháč (1992), Janovsky (1987) and Govaerts (2000) for more details about the use of singularity theory in the numerical analysis of bifurcations.

2.3. Problems with reflectional symmetry

A classification of singularities satisfying various symmetries can also be given. Simple reflectional symmetries are discussed in Golubitsky and Schaeffer (1985) and more complicated symmetries and mode interactions are discussed in Golubitsky et al. (1988). We content ourselves here with a few remarks about the simple Z_2 (*i.e.*, reflection) symmetry.

If $f(x, \lambda)$ satisfies the equivariance (symmetry) condition

$$f(-x, \lambda) = -f(x, \lambda), \tag{2.8}$$

then a classification of singularities arises that reflects the symmetry in the

48   K. A. CLIFFE, A. SPENCE AND S. J. TAVENER

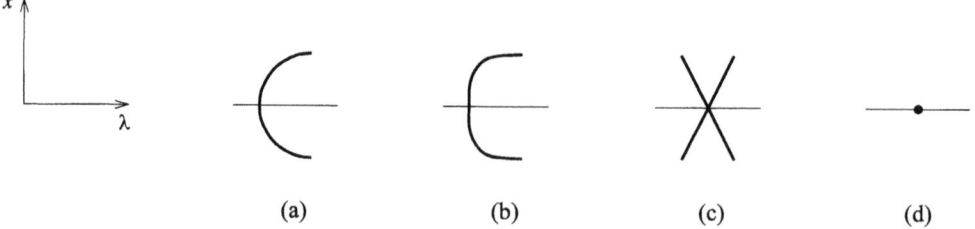

(a)   (b)   (c)   (d)

Fig. 3. Canonical solution diagrams for Z_2-symmetric singularities of
codimension ≤ 1: (a) $f(x, \lambda) = x^3 - \lambda x = 0$, (b) $f(x, \lambda) = x^5 - \lambda x = 0$,
a quadratic symmetry breaking bifurcation, (c) $f(x, \lambda) = x^3 - \lambda^2 x = 0$,
a C− coalescence point, and (d) $F(x, \lambda) = x^3 + \lambda^2 x = 0$, a C+ coalescence
point. Unfoldings of (b), (c) and (d) are given on p. 260 of
Golubitsky and Schaeffer (1985)

problem and is different from that for problems with no symmetry. First note
that if (2.8) is satisfied then $f(x, \lambda)$ is odd in x and so we may write $f(x, \lambda) = xa(x^2, \lambda)$ for some function $a(x^2, \lambda)$. Also, if (x, λ) satisfies $f(x, \lambda) = 0$ then
so does $(-x, \lambda)$. Thus the solution diagrams are symmetric about the λ
axis: see Figure 3. The simplest singularity (*i.e.*, codimension 0) has the
canonical form $f(x, \lambda) := x^3 - \lambda x = x(x^2 - \lambda)$ and gives rise to the common
symmetric pitchfork bifurcation diagram (see Figure 3(a)). The singularities
given in Figure 3(b), (c) and (d) have codimension 1 and typically will only
be observed in a two-parameter setting (see Cliffe and Spence (1984)).

In Chapter VIII of Golubitsky and Schaeffer (1985) it is shown that the
theory for Hopf bifurcation is intimately connected to that for Z_2-symmetric
problems. In particular, small amplitude periodic orbits of an autonomous
system of ODEs are in one-to-one correspondence with zeros of a nonlinear
problem that satisfies the Z_2-equivariance condition (2.8). The simplest
Hopf bifurcation corresponds to a codimension 0 singularity and hence is
likely to be observed in one-parameter problems.

2.4. Implications for bifurcation theory

As we noted at the beginning of this section, singularity theory is a very
important tool in our understanding of bifurcation problems. Specifically
for this review we make the following remarks.

(i) Singularity theory enables appropriate defining conditions for singular
points to be determined. Extended systems constructed incorporating
these defining conditions are regular at the singularity, and so by varying a control parameter, paths of singularities can be computed. Such
paths of singularities may in turn have singularities at more degenerate
(higher codimension) singularities. Reliable computational tools may
therefore be constructed on this basis and were used to perform the numerical computations described in Sections 6 and 8. A nice feature is

that the nondegeneracy conditions and unfolding conditions that need to be checked arise naturally in the numerical implementation: see, for example, Jepson and Spence (1985b), Janovský and Plecháč (1992).

(ii) In one-parameter problems one would expect to observe:

- quadratic fold points,
- Hopf bifurcations;

if a trivial solution exists,

- bifurcation from the trivial solution;

and, in the presence of reflectional symmetry,

- symmetric pitchfork bifurcations.

One would not expect to observe the more degenerate (higher codimension) singularities like transcritical bifurcations. Thus, in our discussion of bifurcation in Banach spaces we restrict attention to these four types of bifurcation. We note, however, that if there were a more complicated symmetry, for instance $O(2)$, $SO(2)$ or $O(3)$, then other bifurcations would arise (see, for example, Vanderbauwhede (1982), Cliffe, Spence and Tavener (2000)), but we do not discuss these cases here.

3. Bifurcation theory in Banach spaces

Consider nonlinear problems of the form

$$F(x, \lambda) = 0, \tag{3.1}$$

where F is a map from $V \times \mathbb{R} \to V$, for some Banach space V with norm $\| \cdot \|$. We assume F is smooth, that is,

$$F : V \times \mathbb{R} \to V \text{ is a } C^p \text{ mapping for } p \geq 3. \tag{3.2}$$

Denote the Frechet derivative of F at (x_0, λ_0) with respect to x (respectively λ) by F_x^0 or $D_x F^0$ (respectively F_λ^0 or $D_\lambda F^0$). We assume

$$F_x^0 : V \to V \text{ is a Fredholm operator of index 0 for all } (x, \lambda) \in V \times \mathbb{R}. \tag{3.3}$$

(Note: where convenient, we use the notation $F^0 = F(x_0, \lambda_0), \ldots$ etc.)
    Let us denote the set S by

$$S = \{(x, \lambda) \in V \times \mathbb{R} : F(x, \lambda) = 0\}.$$

It is often of interest in applications to compute paths or branches of solutions of (3.1), where λ is a distinguished parameter, for instance a flow rate or Reynolds number, and x is a state variable, for instance a temperature or velocity field. If $(x_0, \lambda_0) \in S$ with F_x^0 an isomorphism on V, then the Implicit Function Theorem (IFT) ensures the existence of a unique smooth path of solutions $x(\lambda) \in C^p$ satisfying $F(x(\lambda), \lambda) = 0$ for λ near λ_0, with $F_x(x(\lambda), \lambda)$ an isomorphism. A detailed account of this case appears in Brezzi, Rappaz and Raviart (1980, Section 2). It is not difficult to show (again using the IFT) that algebraically simple eigenvalues of $F_x(x(\lambda), \lambda)$ are also smooth functions of λ, and, as discussed in the introduction, our interest centres on cases when an eigenvalue crosses the imaginary axis, with a possible change in stability of steady solutions of $\dot{x} + F(x, \lambda) = 0$.

One of the simplest cases is that of *Hopf bifurcation*, where at $\lambda = \lambda_0$, say, a complex pair of eigenvalues crosses the imaginary axis as λ varies. In this case $F_x(x(\lambda), \lambda)$ is nonsingular for λ near λ_0. We defer discussion of this case till later, and for the moment we consider the case when a simple real eigenvalue crosses the imaginary axis.

Let (x_0, λ_0) be a simple singular point satisfying

$$F^0 = 0, \tag{3.4}$$

let $F_x^0 \in \mathcal{L}(V; V)$ be singular with algebraically simple zero eigenvalue, and

$$\begin{aligned} \dim \operatorname{Ker}(F_x^0) &= \operatorname{span}\{\phi_0 : \phi_0 \in V, \|\phi_0\| = 1\}, \\ \dim \operatorname{Ker}((F_x^0)') &= \operatorname{span}\{\psi_0 : \psi_0 \in V', \langle \phi_0, \psi_0 \rangle = 1\}, \end{aligned} \tag{3.5}$$

where V' denotes the dual of V with norm $\| \cdot \|'$, and $\langle \cdot, \cdot \rangle$ the duality pairing between V and V'. (Here, $\mathcal{L}(V; V)$ denotes the space of bounded linear operators on V.) Further, setting

$$V_1 := \operatorname{Ker}(F_x^0), \quad V_2 := \operatorname{Range}(F_x^0) = \{v \in V : \langle v, \psi_0 \rangle = 0\}, \tag{3.6}$$

we have

$$V = V_1 \oplus V_2, \tag{3.7}$$

and we may introduce the linear operator L, defined by

$$L := \left(F_x^0 \big|_{V_2} \right)^{-1}, \tag{3.8}$$

the inverse isomorphism of F_x^0 restricted to V_2, that is, the inverse isomorphism of $F_x^0|_{V_2}$.

In this review we shall only consider simple singularities, in the sense that $\dim \operatorname{Ker}(F_x^0) = \dim \operatorname{Ker}((F_x^0)^2) = 1$. Multiple zero eigenvalues arise, especially when there is a symmetry in the problem (*e.g.*, Bauer, Keller and Reiss (1975), Golubitsky et al. (1988)), and indeed we see a *double* singular point in the Taylor–Couette problem in Section 8.

The following lemma has many applications in bifurcation theory.

Lemma 3.1. ('ABCD Lemma'; Keller 1977) Let V be a Banach space and consider the linear operator $M : V \times \mathbb{R} \to V \times \mathbb{R}$ of the form

$$M := \begin{pmatrix} A & b \\ \langle \cdot, c \rangle & d \end{pmatrix},$$

where $A : V \to V$, $b \in V \setminus \{0\}$, $c \in V' \setminus \{0\}$, $d \in \mathbb{R}$. Then:

(i) if A is an isomorphism on V, then M is an isomorphism on $V \times \mathbb{R}$ if and only if $d - \langle A^{-1}b, c \rangle \neq 0$;

(ii) if $\dim \mathrm{Ker}(A) = \mathrm{codim}\,\mathrm{Range}(A) = 1$, then M is an isomorphism if and only if

    (a) $\langle b, \psi_0 \rangle \neq 0 \quad \forall\, \psi_0 \in \mathrm{Ker}(A') \setminus \{0\}$,
    (b) $\langle \phi_0, c \rangle \neq 0 \quad \forall\, \phi_0 \in \mathrm{Ker}(A) \setminus \{0\}$;

(iii) if $\dim \mathrm{Ker}(A) \geq 2$, then M is singular.

By analogy with the case when $V = \mathbb{R}^n$ and A is a matrix, one can think of M as being a 1-bordered extension of A. Keller (1977) considers ν-bordered extensions for $\nu \geq 1$, and these have application when $\dim \mathrm{Ker}(F_x^0) = \nu$.

3.1. Simple fold (limit or turning) points

In this section we consider the simplest singular point, namely a (quadratic) fold point in a Banach space setting (*cf.* Section 2.1). We assume

$$\langle F_\lambda^0, \psi_0 \rangle \neq 0, \tag{3.9}$$

where ψ_0 is defined in (3.5). Under (3.9) we have $\mathrm{Coker}\,[F_x^0, F_\lambda^0] = \{0\}$, and the behaviour of the solution set S near (x_0, λ_0) can be completely determined using the IFT. To do this, consider the system $H : (V \times \mathbb{R}) \times \mathbb{R} \to V \times \mathbb{R}$ introduced by Keller (1977),

$$H(y, t) := \begin{cases} F(x, \lambda), \\ \langle x - x_0, c \rangle + d(\lambda - \lambda_0) - t, \end{cases} \tag{3.10}$$

where $y = (x, \lambda)$ and $c \in V'$ satisfies

$$\langle \phi_0, c \rangle \neq 0 \tag{3.11}$$

(one possible choice is $c = \psi_0$). Then $H(y_0, 0) = 0$, and

$$H_y(y_0, 0) = \begin{bmatrix} F_x^0 & F_\lambda^0 \\ \langle \cdot, c \rangle & d \end{bmatrix}$$

is an isomorphism on $V \times \mathbb{R}$ using the ABCD Lemma. Hence near $t = 0$ there exists a smooth $y(t)$ satisfying $H(y(t), t) = 0$. It is a simple matter, by differentiating $F(x(t), \lambda(t)) = 0$ twice with respect to t, to show that

$$x_t(0) = \phi_0, \ \lambda_t(0) = 0, \ \lambda_{tt}(0) = -\langle F_{xx}^0 \phi_0 \phi_0, \psi_0 \rangle / \langle F_\lambda^0, \psi_0 \rangle, \tag{3.12}$$

(*cf.* (2.5)). Under (3.5) and (3.9), (x_0, λ_0) is called a *simple fold* (*limit* or *turning*) *point*. If, in addition,

$$\langle F_{xx}^0 \phi_0 \phi_0, \psi_0 \rangle \neq 0, \tag{3.13}$$

then (x_0, λ_0) is called a simple *quadratic* fold (Figure 4). As was indicated in Section 2.1, a quadratic fold point is the most typical singular point in a problem with no special features (*e.g.*, symmetry).

The accurate location of a quadratic fold may be accomplished in many ways, for example by finding a zero of λ_t, or, in finite dimensions, a point where $\det F_x = 0$ (see Section 7.2 and Griewank and Reddien (1984)), or by solving the extended system (Seydel 1979*a*, 1979*b*, Moore and Spence 1980)

$$T(y) := \begin{pmatrix} F(x, \lambda) \\ F_x(x, \lambda)\phi \\ \langle \phi, c \rangle - 1 \end{pmatrix}, \quad y = (x, \phi, \lambda) \in V \times V \times \mathbb{R}, \tag{3.14}$$

where c satisfies (3.11). We then have the following result.

Theorem 3.1. Assume (3.2), (3.3), (3.4), (3.5), (3.9) and (3.11). Then, near $\lambda = \lambda_0$, there exist smooth functions $x(t), \lambda(t), \mu(t), \phi(t)$, such that

- (i) $(x(t), \lambda(t))$ is the unique solution of $F(x, \lambda) = 0$ with $(x(0), \lambda(0)) = (x_0, \lambda_0)$;
- (ii) $x(t) = x_0 + t\phi_0 + \mathcal{O}(t^2)$; $\lambda(t) = \lambda_0 + \mathcal{O}(t^2)$;
- (iii) $F_x(x(t), \lambda(t))\phi(t) = \mu(t)\phi(t)$, $\phi(t) \in V$, $\mu(0) = 0$.

If, in addition, we assume (3.13) holds then

- (iv) $\lambda_{tt}(0) \neq 0$;
- (v) $\mu_t(0) \neq 0$;
- (vi) $T_y(y_0)$ is an isomorphism on $V \times V \times \mathbb{R}$, where T is given by (3.14), and $y_0 = (x_0, \phi_0, \lambda_0)$.

Proof. Part (i) follows directly from the IFT applied to (3.10). Part (ii) follows by examining the form of the tangent vector to S at (x_0, λ_0). Part (iii) follows by applying the IFT to the pair $F_x(x(t), \lambda(t))\phi - \mu\phi = 0$, $\langle \phi, c \rangle - 1 = 0$. Part (iv) is proved by differentiating $F(x(t), \lambda(t)) = 0$ twice, and part (v) by differentiating $F_x(x(t), \lambda(t))\phi(t) = \mu(t)\phi(t)$ once. The proof of part (vi) is in Moore and Spence (1980). □

Remarks.

(i) Note that the main tool used to prove these results is the Implicit Function Theorem applied to $H(y, t) = 0$. For this reason many authors refer to a fold point as a *regular* point. We prefer to use the term singular point because in many applications (x_0, λ_0) represents some critical phenomenon.

(ii) Condition (v) states that the eigenvalue $\mu(t)$ passes through zero with nonzero velocity. As we shall see, this is a very common type of non-degeneracy condition.

(iii) An immediate corollary of condition (v) is that a stable steady state of $\dot{x} + F(x, \lambda) = 0$ must lose (linearized) stability at a simple quadratic fold point since a real eigenvalue moves from the stable into the unstable half plane. A typical situation is shown in Figure 4 where the lower part of the branch is assumed to be stable. In the dynamical systems literature a fold point is commonly known as a 'saddle node'.

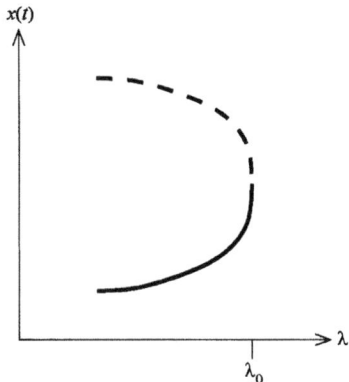

Fig. 4. Schematic illustrating the solution typical behaviour at a quadratic fold point. Here (━━) represents stable steady states of $\dot{x} + F(x, \lambda) = 0$, and (▪▪▪) represents unstable states

3.2. Lyapunov–Schmidt reduction

The Lyapunov–Schmidt reduction (see, for example Stackgold (1971) and Golubitsky and Schaeffer (1985)) plays a very important rôle in the theory of bifurcations. We follow the treatment in Brezzi, Rappaz and Raviart (1981a).

Consider the nonlinear problem (3.1)

$$F(x, \lambda) = 0,$$

subject to (3.2), (3.3), (3.4) and (3.5). Define the projection operator $Q : V \rightarrow V_2$ by

$$Qv = v - \langle v, \psi_0 \rangle \phi_0, \ v \in V,$$

induced by the direct sum decomposition (3.7). Then the equation $F(x, \lambda) = 0$ is equivalent to

$$QF(x, \lambda) = 0 \qquad (3.15)$$

and
$$(I - Q)F(x, \lambda) = 0. \tag{3.16}$$
For $x \in V$, there is a unique decomposition
$$x = x_0 + \alpha\phi_0 + v, \quad \alpha \in \mathbb{R}, \ v \in V_2.$$
Write
$$\lambda = \lambda_0 + \xi$$
and (3.15) becomes
$$\mathcal{F}(v, \alpha, \xi) := QF(x_0 + \alpha\phi_0 + v, \lambda_0 + \xi) = 0, \tag{3.17}$$
and hence, using the IFT on $\mathcal{F} = 0$ in V_2 and using (3.8) we obtain, for (α, ξ) small enough,
$$v = v(\alpha, \xi), \quad \text{with } v(0, 0) = 0. \tag{3.18}$$
Substituting into (3.16) we obtain the *bifurcation equation* (or *reduced problem*)
$$f(\alpha, \xi) := \langle F(x_0 + \alpha\phi_0 + v(\alpha, \xi), \lambda_0 + \xi), \psi_0 \rangle = 0. \tag{3.19}$$
Thus, for (α, ξ) small enough, solutions of (3.19) are in one-to-one correspondence with the solutions of (3.1). Note that this reduction process is a very powerful tool. Independently of the precise form of (3.1), provided the reduction process can be applied, the solution behaviour of (3.1) near a singular point can be analysed through a reduced problem of small dimension. The dimension of the reduced problem is usually (but need not be) equal to the dimension of $\mathrm{Ker}(F_x^0)$. Numerically convenient reduction procedures are discussed in Jepson and Spence (1984) and Janovský and Plecháč (1992).

It is easy to verify, using $\frac{\partial v}{\partial \alpha}(0, 0) = 0$, that

(i) $f(0, 0) = 0$,   (ii) $\frac{\partial f}{\partial \alpha}(0, 0) = 0$,

(iii) $\frac{\partial f}{\partial \xi}(0, 0) = \langle F_\lambda^0, \psi_0 \rangle$,   (iv) $\frac{\partial^2 f}{\partial \alpha^2}(0, 0) = \langle F_{xx}^0 \phi_0 \phi_0, \psi_0 \rangle$. $\quad$ (3.20)

Thus, using (3.9) and (3.20iii), the IFT ensures the existence of a unique path of solutions $\xi = \xi(\alpha)$ to $f(\alpha, \xi) = 0$. Proceeding in this way (see Brezzi et al. (1981 a) for details) one recovers the results of Theorem 3.1(i),(ii) where α is used to parametrize S near $\lambda = \lambda_0$ rather than t defined in (3.10). We shall see in Section 7.1 that t is a local approximate arclength.

Clearly
$$f(\alpha, \xi) = \langle F_\lambda^0, \psi_0 \rangle \xi + \langle F_{xx}^0 \phi_0 \phi_0, \psi_0 \rangle \alpha^2 + \text{h.o.t.}$$
In the language of singularity theory $f(\alpha, \xi)$ is 'contact equivalent to' (*i.e.*, can be smoothly transformed to) the form $\xi - \alpha^2$, which is the canonical form for a quadratic fold (see Section 2.1).

If (3.9) fails then $\frac{\partial f}{\partial \xi}(0,0) = 0$ and the analysis of the solutions of $f(\alpha, \xi) = 0$ near $(0,0)$ proceeds by considering the second derivatives of $f(\alpha, \xi)$. Following Brezzi et al. (1981a), let

$$A_0 := \frac{\partial^2 f}{\partial \alpha^2}(0,0), \ \ B_0 := \frac{\partial^2 f}{\partial \alpha \partial \xi}(0,0), \ \ C_0 := \frac{\partial^2 f}{\partial \xi^2}(0,0),$$

where

$$
\begin{aligned}
A_0 &= \langle F_{xx}^0 \phi_0 \phi_0, \psi_0 \rangle, \\
B_0 &= \langle F_{x\lambda}^0 \phi_0 + F_{xx}^0 \phi_0 w_0, \psi_0 \rangle, \\
C_0 &= \langle F_{\lambda\lambda}^0 + 2 F_{x\lambda}^0 w_0 + F_{xx}^0 w_0 w_0, \psi_0 \rangle,
\end{aligned}
\tag{3.21}
$$

with $w_0 \in V_2$ the unique solution in V_2 of

$$F_x^0 w_0 + F_\lambda^0 = 0. \tag{3.22}$$

The Morse Lemma (Nirenberg 1974, Chapter 3) shows that if

$$B_0^2 - A_0 C_0 > 0 \tag{3.23}$$

then near (x_0, λ_0) there are two C^{p-2} branches of solutions to (3.1) that intersect transversally. Local parametrizations of the two branches are given in Brezzi, Rappaz and Raviart (1981b, Section 2). We call such points *transcritical bifurcation points* since, under (3.23), $f(\alpha, \xi)$ is contact equivalent to $\alpha^2 - \xi^2$, the canonical form for a transcritical bifurcation given in Section 2.2. As discussed in Section 2.4, the singularity theory of Golubitsky and Schaeffer (1985) tells us that generically one would not expect to observe transcritical bifurcation points in one-parameter problems, but they would appear in two-parameter problems.

However, in two special cases of great practical importance, intersecting curves do generically arise in one-parameter problems. These are the case of 'bifurcation from the trivial solution' and 'bifurcation in the presence of symmetry'. We discuss these two cases in the following subsections.

3.3. Bifurcation from the trivial solution

Consider the nonlinear problem $F(x, \lambda) = 0$ with the additional property that

$$F(0, \lambda) = 0 \text{ for all } \lambda, \tag{3.24}$$

that is, the trivial solution, $x = 0$, is a solution for all λ. The important question is: 'For what values of λ do nontrivial solutions bifurcate from the trivial solution?' Such problems arise in many applications. One of the simplest, but very important, examples is the buckling of a slender elastic rod or column due to compression, a problem considered by Euler, Bernoulli and Lagrange (see, for example, Reiss (1969), Chow and Hale

(1982)). In fact buckling problems provide a rich source of such problems (see, for example, Keller and Antman (1969) and Rabinowitz (1977)), and the classic theoretical paper of Crandall and Rabinowitz (1971) is devoted to this case. A fluid mechanical example, which will be discussed in greater detail in Section 9, concerns a fluid layer subjected to a vertical temperature gradient, with cooler fluid lying over the top of warmer fluid. For appropriately chosen, physically reasonable boundary conditions, the non-convecting state is a solution of the governing equations for all values of the temperature gradient. In this so-called 'conducting' solution the buoyancy forces are balanced by the pressure gradient and heat is transferred by conduction alone. Above a critical temperature gradient this conducting solution becomes linearly unstable and a convecting state is observed.

We return to the mathematical analysis of (3.24). We see immediately that

$$F_\lambda(0, \lambda) = 0, \ F_{\lambda\lambda}(0, \lambda) = 0, \ \ldots \qquad (3.25)$$

and hence (3.22) gives that $w_0 = 0$. Hence $C_0 = 0$ and, in (3.21), B_0 reduces to $B_0 = \langle F_{x\lambda}^0 \phi_0, \psi_0 \rangle$. The nondegeneracy condition (3.23) becomes

$$\langle F_{x\lambda}^0 \phi_0, \psi_0 \rangle \neq 0. \qquad (3.26)$$

If γ denotes the eigenvalue of $F_x(0, \lambda)$ with $\gamma = 0$ at $\lambda = \lambda_0$, then it is readily shown using the IFT that $F_x(0, \lambda)$ has a simple eigenvalue $\gamma(\lambda)$ satisfying

$$F_x(0, \lambda)\phi(\lambda) = \gamma(\lambda)\phi(\lambda) \qquad (3.27)$$

with $\phi(\lambda_0) = \phi_0$, $\gamma(\lambda_0) = 0$, $\gamma_\lambda(\lambda) = \langle F_{x\lambda}^0 \phi_0, \psi_0 \rangle$. We now have the following theorem compiled from Crandall and Rabinowitz (1971) and Brezzi et al. (1981*b*).

Theorem 3.2. Assume (3.2), (3.3), (3.5), (3.24) and (3.26). Then near $(0, \lambda_0)$ there exists a nontrivial solution branch of $F(x, \lambda) = 0$ passing through $(0, \lambda_0)$.

  (i) If $A_0 \neq 0$ then

$$\lambda = \lambda_0 + \xi, \ x = \xi(-2B_0/A_0)\phi_0 + \mathcal{O}(\xi^2).$$

 (ii) If $A_0 = 0$ then

$$\lambda = \lambda_0 - \frac{1}{6}\frac{D_0}{B_0}\alpha^2 + \mathcal{O}(\alpha^3), \quad x = \alpha\phi_0 + \mathcal{O}(\alpha^2),$$

where

$$D_0 = \frac{\partial^3 f}{\partial \xi^3}(0, 0) := \langle F_{xxx}^0 \phi_0\phi_0\phi_0 - 3F_{xx}^0 \phi_0 z_0, \psi_0 \rangle, \qquad (3.28)$$

and z_0 is the unique solution in V_2 of

$$F_x^0 z_0 + F_{xx}^0 \phi_0\phi_0 = 0. \qquad (3.29)$$

Furthermore, with $\gamma(\lambda)$ an eigenvalue of $F_x(0, \lambda)$ defined by (3.27),

(iii) $\gamma_\lambda(\lambda_0) \neq 0$.

Condition (iii) is an eigenvalue crossing condition similar to (v) of Theorem 3.1. Thus we may deduce that a stable trivial solution loses stability at $\lambda = \lambda_0$ provided (3.26) holds. It is natural to ask about the stability of the bifurcating branches. Though stability assignment is possible by topological degree theory, Crandall and Rabinowitz (1973) present an elegant analysis using eigenvalues. Recall that $\gamma(\lambda)$ is the (smooth) eigenvalue of $F_x(0, \lambda)$ of smallest modulus near $\lambda = \lambda_0$. Along a branch of nontrivial solutions (given either by (i) or (ii) of Theorem 3.2), say $(x(t), \lambda(t))$, let the eigenvalue of smallest modulus be denoted $\mu(t)$, that is, $F_x(x(t), \lambda(t))\phi(t) = \mu(t)\phi(t)$, where $(x(0), \lambda(0)) = (x_0, \lambda_0)$, $\mu(0) = 0$, $\phi(0) = \phi_0$. The key result relating $\mu(t)$, $\lambda(t)$ and $\gamma(\lambda)$ is given in the next theorem.

Theorem 3.3. (Crandall and Rabinowitz 1973) Under the above assumptions,

 (i) $\mu(t)$ and $-t\lambda_t(t)\gamma_\lambda(\lambda_0)$ have the same zeros and, whenever $\mu(t) \neq 0$, the same sign;

(ii) $\lim_{t \to 0, \mu(t) \neq 0} \frac{-t\lambda_t(t)\lambda_\lambda(\lambda_0)}{\mu(t)} = 1$.

The first result of this theorem enables stability of bifurcating branches to be determined and, in particular, it is readily shown that, for the problem $\dot{x} + F(x, \lambda) = 0$, supercritical bifurcating branches are stable and subcritical branches are unstable. The exchange of stability at bifurcation from the trivial solution for three cases is illustrated in Figure 5.

The question of the computation of bifurcation points on a trivial solution branch reduces to a standard parameter-dependent eigenvalue problem, that is, find the zero eigenvalues of $F_x(0, \lambda)$. The question of computing the nontrivial bifurcating branches is discussed in Section 7.

3.4. Symmetry breaking bifurcation

Symmetries play an important rôle in many applications, for example in structural and fluid mechanics, some of which are described or referenced in Golubitsky et al. (1988). A group-theoretic approach is highly advantageous when studying linear or nonlinear problems in the presence of symmetry, and, in fact, a full understanding of the range of interactions and transitions that arise in applications is probably not possible without the use of group theory. A good introduction to the power of group-theoretic methods for linear problems is given in Bossavit (1986). In many applications the geometry of the domain imposes a natural symmetry. For example, the equations governing laminar flow in a circular pipe have the symmetries of the group $O(2)$ (comprising rotations through any angle between 0 and 2π,

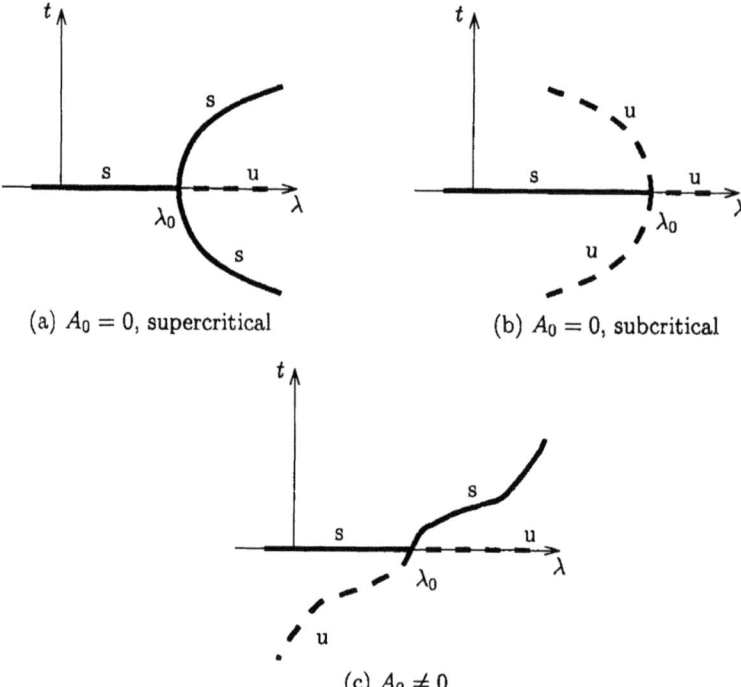

(a) $A_0 = 0$, supercritical　　　　　(b) $A_0 = 0$, subcritical

(c) $A_0 \neq 0$

Fig. 5. (a), (b), (c) are schematic diagrams illustrating the stability of bifurcating branches at bifurcation from the trivial solution. Here 's' denotes a stable branch, and 'u' an unstable branch

and a reflection (Vanderbauwhede 1982)), whereas the equations governing laminar flow in a (symmetrically) expanding two-dimensional channel (see Section 6), have a simple reflectional symmetry, that is, the symmetry of the group $Z_2 = \{1, -1\}$. This is the simplest nontrivial symmetry and in this review we shall restrict attention to this case.

We assume the following common equivariance condition (see Brezzi et al. (1981b), Werner and Spence (1984)):

> There exists a linear operator S on V with $S \neq I, S^2 = I$ such that $SF(x, \lambda) = F(Sx, \lambda)$. $\qquad$ (3.30)

Note that the two-element group $\{I, S\}$ is isomorphic to Z_2, and hence we call (3.30) a Z_2-equivariance condition. There is a natural decomposition of V into symmetric and anti-symmetric components, namely $V = V_s \oplus V_a$, where

$$V_s = \{x \in V : Sx = x\}, \quad V_a = \{x \in V : Sx = -x\}. \qquad (3.31)$$

For $x \in V_s$ and $\lambda \in \mathbb{R}$, the symmetric subspace is invariant under F, F_λ, etc. Assuming $x \in X_s$, then differentiating the equivariance condition (3.30) with respect to x gives

$$SF_x(x, \lambda)\phi = F_x(x, \lambda)S\phi, \quad \forall \phi \in V. \tag{3.32}$$

Clearly, $F_x : V_s \to V_s$ and $F_x : V_a \to V_a$, and so, provided $x \in X_s$, we may introduce $F_x|_{V_s}$ and $F_x|_{V_a}$. Differentiating (3.32) with respect to x gives

$$SF_{xx}(x, \lambda)uv = F_{xx}(x, \lambda)SuSv, \quad \forall u, v \in V; \tag{3.33}$$

hence for $x \in X_s$ and $v, w \in V_a$ or $v, w \in V_s$, we have $F_{xx}(x, \lambda)vw \in V_s$.

Let (x_0, λ_0) be a simple singular point with $x_0 \in X_s$ and let $\phi_0 \in \mathrm{Ker}(F_x^0)$. Substituting into (3.32) gives $SF_x^0\phi_0 = F_x^0S\phi_0 = 0$ and (3.5) requires $S\phi_0 = c\phi_0$ for some constant c. Multiplying by S gives $\phi_0 = cS\phi_0 = c^2\phi_0$, and hence $c = \pm 1$. The case $c = 1$ gives $\phi_0 \in V_s$, but this is less interesting since the symmetry is not broken and all nearby solutions lie in V_s. Instead we consider the case $c = -1$, the *symmetry breaking case*, that is,

$$\text{Assume } x_0 \in V_s \text{ and } \phi_0 \in V_a. \tag{3.34}$$

It is easy to show that

$$\psi_0 \in V_a', \text{ and } \langle x, \psi_0 \rangle = 0, \quad \forall x \in V_s. \tag{3.35}$$

Under (3.30) and (3.34) we immediately have the existence of a unique path of solutions in V_s to $F(x, \lambda) = 0$ near (x_0, λ_0); to show this simply apply the IFT to $F(x, \lambda) = 0$, $x \in V_s$. Denote this path by $x^s(\lambda)$ and introduce the new problem

$$\widetilde{F}(\tilde{x}, \lambda) := F\left(x^s(\lambda) + \tilde{x}, \lambda\right) = 0, \quad \tilde{x} \in V. \tag{3.36}$$

Clearly $\widetilde{F}(0, \lambda) = 0$ for all λ and all the results applying to bifurcation from the trivial solution apply. Note that (3.35) implies that $A_0 = 0$, and so Theorem 3.2(ii) applies for the symmetry breaking path. However, in addition, $S\widetilde{F}(\tilde{x}, \lambda) = \widetilde{F}(S\tilde{x}, \lambda)$ and so if $(\tilde{x}, \lambda)$ solves (3.36) so does $(S\tilde{x}, \lambda)$. This leads to the common *symmetric*-pitchfork bifurcation diagrams, which are symmetric with respect to the path $(x^s(\lambda), \lambda)$. Note that in problems with no symmetry pitchfork bifurcation arises in Theorem 3.2 if $A_0 = 0$ and λ and x depend quadratically on α. However, under (3.30) the bifurcating branches must be symmetric about the trivial solution. Importantly, the Lyapunov–Schmidt procedure goes through as one would expect, but also, as shown in Brezzi et al. (1981b), the reduced equation inherits an equivariance property from (3.30), namely

$$-f(\alpha, \xi) = f(-\alpha, \xi). \tag{3.37}$$

Hence $f(0, \xi) = 0$, $f_\xi(0, \xi) = 0$, $f_{\xi\xi}(0, \xi) = 0$, etc. ... for all ξ. In fact it is readily shown that the reduced equation in this case has the form

$$f(\alpha, \xi) = \frac{D_0}{6}\alpha^3 + B_0\alpha\xi + \text{h.o.t.}, \tag{3.38}$$

which we use later.

For completeness we state the analogous result to Theorem 3.2 for Z_2-symmetry breaking, that is, $F(x,\lambda) = 0$ with $SF(x,\lambda) = F(Sx,\lambda)$, $S^2 = I$. Note that, in (3.22) $w_0 \in V_s$ since $F_\lambda^0 \in V_s$; in (3.29), $z_0 \in V_s$ since $F_{xx}^0 \phi_0 \phi_0 \in V_s$; and $A_0 = 0$, $C_0 = 0$ using (3.35). Also, we may introduce $\gamma(\lambda)$, the eigenvalue of minimum modulus near λ_0 of $F_x(x^s(\lambda),\lambda)|_{V_a}$. The analogue of (3.14) has the same form but with a restricted domain, namely $T(y) = 0$ where $T : V_s \times V_a \times \mathbb{R} \to V_s \times V_a \times \mathbb{R}$ with

$$T(y) = \begin{pmatrix} F(x,\lambda) \\ F_x(x,\lambda)\phi \\ \langle \phi, c \rangle - 1 \end{pmatrix}, \quad y \in V_s \times V_a \times \mathbb{R}, \tag{3.39}$$

where $c \in V_a'$ satisfies (3.11).

Theorem 3.4. Assume (3.2), (3.3), (3.4), (3.5), (3.30) and (3.34). Assume also that

$$B_0 := \langle F_{x\lambda}^0 \phi_0 + F_{xx}^0 \phi_0 w_0, \psi_0 \rangle \neq 0,$$

where w_0 is given by (3.22). Then near (x_0, λ_0) there exist two nontrivial solution branches of $F(x,\lambda) = 0$ passing through (x_0, λ_0):

(i) a symmetric branch given by

$$\lambda = \lambda_0 + \xi, \ x^s(\lambda) = x_0 - \xi w_0 + \mathcal{O}(\xi^2) \in V_s;$$

(ii) an asymmetric branch given by

$$\lambda = \lambda_0 - \frac{1}{6}\frac{D_0}{B_0}\alpha^2 + \mathcal{O}(\alpha^4),$$

$$x = x_0 + \alpha\phi_0 + \alpha^2\left\{-\frac{1}{6}\frac{D_0}{B_0}w_0 + \frac{1}{2}z_0\right\} + \mathcal{O}(\alpha^3);$$

(iii) $\gamma_\lambda(\lambda_0) \neq 0$, where $\gamma(\lambda)$ is the eigenvalue of minimum modulus of $F_x(x^s(\lambda),\lambda)|_{V_a}$ near (x_0, λ_0);

(iv) $T_y(y_0)$ is an isomorphism on $V_s \times V_a \times \mathbb{R}$ where T is given by (3.39), and $y_0 = (x_0, \phi_0, \lambda_0) \in V_s \times V_a \times \mathbb{R}$.

Proof. Parts (i) and (ii) are proved in Brezzi et al. (1981b). Part (iii) is proved in Crandall and Rabinowitz (1971), but can easily be obtained by differentiating $F_x(x^s(\lambda),\lambda)\phi(\lambda) - \gamma(\lambda)\phi(\lambda) = 0$. Part (iv) is proved in Werner and Spence (1984). □

In this case we call (x_0, λ_0) a *simple Z_2-symmetry breaking bifurcation point* (or sometimes, a *symmetry breaking pitchfork* bifurcation point). The stability assignments for bifurcating branches are as in Figure 5(a) or (b).

Condition (iii) states that the eigenvalue $\gamma(\lambda)$ of $F_x(x^s(\lambda),\lambda)|_{V_a}$ changes sign with nonzero speed near $\lambda = \lambda_0$, and this fact can be used in the detection of symmetry breaking pitchfork bifurcations (see Section 7.3).

3.5. *Hopf bifurcation*

As mentioned at the beginning of this section, a Hopf bifurcation arises when a complex pair of eigenvalues of $F_x(x, \lambda)$ crosses the imaginary axis. This is one of the typical bifurcations one would expect to occur in a one-parameter problem of the form $\dot{x} + F(x, \lambda) = 0$ (see Hassard, Kazarinoff and Wan (1981), Sattinger (1973), Wiggins (1990), Crandall and Rabinowitz (1977)). This phenomenon has a long history with examples occurring in the work of Poincaré and Andronov (see Wiggins (1990) for a short account of the early history). A comprehensive account of both the theory and numerical analysis of Hopf bifurcations is given by Bernardi (1982). The account here is a summary of the treatment in Bernardi (1982).

Consider the nonlinear time-dependent problem

$$\frac{\mathrm{d}x}{\mathrm{d}t} + F(x, \lambda) = 0, \tag{3.40}$$

where F maps $V \times \mathbb{R}$ to V'. (This is the appropriate setting when $F(x, \lambda)$ involves spatial differentiation of x: see, for example, the theory for parabolic PDEs, Section 26 in Wloka (1987).) Assume:

(i) $F(x, \lambda) = Ax + G(x, \lambda)$ where A is an isomorphism from V (3.41) to V', satisfying $(Ax, x) \geq \alpha\|x\|^2$, $\alpha > 0$, and G is a C^p-mapping $(p \geq 3)$ from $\mathbb{R} \times V$ into V';

(ii) $$F(x_0, \lambda_0) = 0; \tag{3.42}$$

(iii) F_x^0 has two algebraically simple, purely imaginary eigen- (3.43) values $\pm i\omega_0$, $\omega_0 \neq 0$, with corresponding eigenvectors $\zeta_0, \bar{\zeta}_0$, and no other eigenvalues of the form $\pm in\omega_0 (n = 0, 2, 3, \ldots)$.

Since F_x^0 is an isomorphism from V to V', the IFT shows that, near (x_0, λ_0), there is a unique path of equilibrium solutions of (3.40), $x^e(\lambda) \in V$ say, such that $F(x^e(\lambda), \lambda)) = 0$, $x^e(\lambda_0) = x_0$. Also, for λ near to λ_0, the IFT ensures the existence of a unique pair of complex eigenvalues $\mu_\pm(\lambda) = \alpha(\lambda) \pm i\omega(\lambda)$ with $\alpha(\lambda_0) = 0$, $\omega(\lambda_0) = \omega_0$. The known path $x^e(\lambda)$ is 'subtracted out' by writing $x = x^e(\lambda) + v$ (*cf.* Section 4.3) and introducing $\hat{F}(v, \lambda) := F(x^e(\lambda) + v, \lambda)$. So

$$\hat{F}(v, \lambda) = Av + \hat{G}(v, \lambda)$$

for an appropriate $\hat{G}$.

Finally, we assume the complex pair of eigenvalues crosses the imaginary axis with nonzero speed, that is,

$$\frac{\mathrm{d}}{\mathrm{d}\lambda}(\alpha(\lambda))\Big|_{\lambda=\lambda_0} \neq 0. \tag{3.44}$$

A change of time variable, $s = \omega t$, in (3.40) produces

$$\mathcal{F}(v, \lambda, \omega) := \omega \frac{dv}{ds} + \hat{F}(v, \lambda) = 0 \qquad (3.45)$$

and we seek 2π-periodic solutions of (3.45) in a neighbourhood of $(0, \lambda_0, \omega_0)$ $\in \mathcal{X} \times \mathbb{R} \times \mathbb{R}$, where $\mathcal{X}$ is the closure of the space of 2π-periodic functions of $\mathcal{D}([0, 2\pi], V)$ for an appropriate norm (see Bernardi (1982, Section II)).

It is now possible to carry out a Lyapunov–Schmidt analysis on $\mathcal{F}(v, \lambda, \omega) = 0$. An important rôle in the theory is played by the operator

$$\mathcal{A}v := \omega_0 \frac{dv}{ds} + Av, \qquad (3.46)$$

which is shown to be an isomorphism from $\mathcal{X}$ to $\mathcal{X}'$, the dual of $\mathcal{X}$. Denote the inverse of $\mathcal{A}$ by $\mathcal{T}$. The results may be summarized in the following theorem (Bernardi 1982, p. 23).

Theorem 3.5. (Hopf bifurcation) Assume (3.41), (3.42), (3.43) and (3.44). Equation (3.40) has a unique branch of $T(\epsilon)$ periodic solutions in a neighbourhood of $(x_0, \lambda_0, \omega_0)$ which is of the form

$$\begin{aligned}
\lambda(\epsilon) &= \lambda_0 + \mathcal{O}(\epsilon^2), \\
\omega(\epsilon) &= \omega_0 + \mathcal{O}(\epsilon^2), \\
T(\epsilon) &= 2\pi/\omega_0 + \mathcal{O}(\epsilon^2), \\
x(\epsilon)(t) &= x^e(\lambda(\epsilon)) + v(\epsilon)(\omega(\epsilon)t),
\end{aligned}$$

where $v(\epsilon) = \epsilon(\zeta_0 + \bar{\zeta}_0) + \mathcal{O}(\epsilon^2)$. Furthermore, if a certain nondegeneracy condition holds (Bernardi 1982, (IV.20)), $\lambda(\epsilon) = \lambda_0 + \epsilon^2 \sigma_2(\epsilon)$, $\sigma_2(0) \neq 0$, ensuring a 'quadratic' (rather than quartic or higher-order) bifurcating branch.

As for fold and Z_2-symmetry breaking bifurcations, we can set up an extended system (cf. Jepson (1981), Griewank and Reddien (1983)). Consider $T(y) = 0$, where $T : V \times V^c \times \mathbb{R}^2 \to V \times V^c \times \mathbb{R}^2$ with

$$T(y) = \begin{pmatrix} F(x, \lambda) \\ F_x(x, \lambda)\zeta - i\omega\zeta \\ \langle \zeta, c \rangle - 1 \end{pmatrix}, \quad y = (x, \zeta, \lambda, \omega) \in V \times V^c \times \mathbb{R}^2 \qquad (3.47)$$

where $V^c = V + iV$, and $c \in (V^c)'$ satisfying $\langle \zeta_0, c \rangle \neq 0$. The following theorem is readily proved: see Jepson (1981), Griewank and Reddien (1983).

Theorem 3.6. Assume (3.41), (3.42), (3.43) and (3.44). Then $T_y(y_0)$ is an isomorphism on $V \times V^c \times \mathbb{R}^2$.

Remark. It was noted in Section 2.3 that the theory for Hopf bifurcation is closely connected to that for Z_2-symmetry breaking bifurcation. The same is true for stability assignment, and stability diagrams like those in Figure 5 can be shown to hold for Hopf bifurcation, where the bifurcating branches represent periodic solutions of $\dot{x} + F(x, \lambda) = 0$ (see Crandall and Rabinowitz (1977)).

4. Numerical approximation

The main theoretical results on the numerical approximation of bifurcation points were proved in the early to mid-1980s. In this review we shall concentrate on the work of Brezzi et al. (1980, 1981a, 1981b) and Descloux and Rappaz (1982), who considered numerical approximation in a projection method (especially a Galerkin method) framework, since we will use their results in the analysis of the mixed finite element method for the Navier–Stokes equations in Section 5. We shall also discuss the work of Moore and Spence (1981) and Moore, Spence and Werner (1986) who performed a 'Keller-like' analysis appropriate for finite difference methods. For completeness, we briefly mention other treatments at the end of this section. For the discussion on numerical approximation we shall denote discretizations of $F(x, \lambda) = 0$ by $F_h(x_h, \lambda_h) = 0$, approximations of (x_0, λ_0) by (x_h^0, λ_h^0), and the Jacobian of the approximate problem by $D_x F_h(x_h, \lambda_h)$. (Here we follow the notation in the Brezzi, Rappaz and Raviart papers in which x_h^0 denotes the numerical approximation of x_0, etc.)

Before we consider parameter-dependent problems, it is worth recalling the key ideas in the theory of the approximate solution of nonlinear operator equations (see, for example, Krasnosel'skii, Vainikko, Zabreiko, Rutitskii and Stetsenko (1972), Weiss (1974), Keller (1975) and López-Marcos and Sanz-Serna (1988)). Usually there are two main assumptions, namely a *consistency* condition on an approximating family, and a *stability* condition, which can take several forms. For example, following Keller's classic treatment (Keller 1975), consider the problem $F(x) = 0$, $F : V \to V$ with V a Banach space. The approximating family is written as $F_h(x_h) = 0$, $F_h : V_h \to V_h$ for some $h > 0$, with the spaces V and V_h linked by bounded restriction operators: for instance, we assume the existence of operators $r_h : V \to V_h$, such that $\|r_h x\| \to \|x\|$ as $h \to 0$ (for appropriate norms). The family $\{F_h\}$ is *consistent* with F at $x \in V$ to order p if

$$\|F_h(r_h x) - r_h F(x)\| \le Ch^p,$$

for h sufficiently small and C independent of h. (Throughout, C will denote a generic constant bounded independent of h.) The family $\{F_h\}$ is *stable* at $x \in V$ if $D_x F_h(r_h x)$ has a uniformly bounded inverse from V_h to V_h, that is,

$$\|D_x F_h(r_h x)^{-1}\| < C, \tag{4.1}$$

for h sufficiently small and C independent of h. We also assume that $D_x F_h$ satisfies a Lipschitz condition in an appropriate ball centred on $r_h x$. Existence and convergence (of order p) results are proved by Keller (1975,

Theorem 3.6). In particular we have the convergence result, that for h sufficiently small,

$$\begin{aligned} \|r_h x - x_h\| &\le C\|F_h(r_h x) - r_h F(x)\| \\ &= C\|F_h(r_h x)\|. \end{aligned}$$

The main practical difficulty is to prove the stability result. Only if x is a regular solution of $F(x, \lambda) = 0$, that is, if $D_x F(x, \lambda)$ is an isomorphism on V, should one expect to obtain a stability result like (4.1). A similar analysis for projection methods can also be given: see, for example, Section 19.3 of Krasnosel'skii et al. (1972), where again it is assumed that the exact solution is regular.

For parameter-dependent problems $F(x, \lambda) = 0$, we have seen that singular points are often regular points of certain extended systems. By analogy with the standard convergence theory for regular points, it is natural to expect that discretized parameter-dependent problems exhibit the same qualitative behaviour as the continuous problem, and that expected rates of convergence will be attained. This is indeed the case for the problems discussed here. We shall see that, under reasonable consistency and stability conditions, convergence results (of expected orders) are possible for branches of solutions, fold points, for bifurcations from a trivial solution, Z_2-symmetry breaking bifurcation points and Hopf bifurcation points. Further, when projection methods are used, superconvergence results for approximation of critical parameter values are obtained.

However, this is *not* the case for all bifurcation points, as is well known for discretizations of transcritical bifurcations: see, for instance, Descloux and Rappaz (1982, Section 4), and Brezzi et al. (1981b). This is not surprising from the standpoint of singularity theory, since we know from Figure 1 in Section 2.2 that, even in the scalar case, a transcritical bifurcation point is destroyed under perturbation when considered in the $x - \lambda$ plane.

In this section, we consider mainly projection methods as a means of generating numerical approximations, and, in particular, in the next three subsections we discuss the approximation of the bifurcation points described in Section 3. In Section 4.4 we discuss approximation using finite difference methods, and in Section 4.5 we briefly review the literature.

4.1. Fold points

We shall summarize in some detail the main results in Brezzi et al. (1980, 1981a, 1981b) and Descloux and Rappaz (1982). The setting is as follows. Consider the nonlinear problem (3.1), namely

$$F(x, \lambda) = 0, \quad F : V \times \mathbb{R} \to V$$

under assumptions (3.2) and (3.3). (We note that, in fact, Descloux and Rappaz (1982) consider a more general setting, but for this review we present their approach within the above framework.)

For each value of a real parameter $h > 0$ we introduce a finite-dimensional subspace V_h of V and consider the approximating problem

$$F_h(x_h, \lambda) = 0, \quad F_h : V_h \times \mathbb{R} \to V_h. \tag{4.2}$$

Near-regular points (x_0, λ_0) on S, (i.e., $F(x_0, \lambda_0) = 0$ with $D_x F(x_0, \lambda_0)$ an isomorphism), the solution set S contains a unique path parametrized by λ. Standard theory, under appropriate consistency conditions, shows that the discretized problem $F_h(x_h, \lambda_0) = 0$ has a unique solution (x_h^0, λ_0) with convergence of x_h^0 to x_0 as $h \to 0$. The approximation of the curve $(x(\lambda), \lambda)$ by the curve $(x_h(\lambda), \lambda)$ near $\lambda = \lambda_0$ also follows using a careful extension of the usual IFT (see Brezzi et al. (1980)).

Consider now the case when (x_0, λ_0) is a fold point, that is, in addition to assumptions (3.2), (3.3) we assume (x_0, λ_0) is a singular point of $F(x, \lambda) = 0$ satisfying (3.4), (3.5) and (3.9). A local analysis near a fold curve was carried out in Section 3.1 using equation (3.10). It was shown that, provided (3.11) is satisfied, $(y_0, 0)$ is a regular solution of $H(y, t) = 0$, where $y = (x, \lambda) \in V \times \mathbb{R}$. The obvious question is: 'How well is the location of the fold point approximated?'

We assume the discretization satisfies the following properties.

(i) $\{F_h\}$ is a family of C^p functions mapping $V_h \times \mathbb{R}$ into V_h.   (4.3)

(ii) $\{P_h\}$ is a family of projectors $P_h : V \to V_h$, with   (4.4)
      $\|(I - P_h)x\| \to 0$ as $h \to 0$, for all $x \in V$.

(iii) (Consistency.) For any fixed x, λ,

$$\|F(x, \lambda) - F_h(P_h x, \lambda)\| \to 0 \quad \text{as } h \to 0 \tag{4.5}$$

   with similar assumptions on the first $(p - 1)$ derivatives of $F(x, \lambda)$.

(iv) F and its first p derivatives are bounded independent of h (4.6)
      for all (x, λ) in a ball centred on $(P_h x_0, \lambda_0)$.

(v)   (Stability.) For some C independent of h,

$$\|D_x F_h(P_h x_0, \lambda_0)\xi_1 + D_\lambda F_h(P_h x_0, \lambda_0)\xi_2\| \geq C\|(\xi_1, \xi_2)\|, \tag{4.7}$$

   for $(\xi_1, \xi_2) \in (V_2 \cap X_h) \times \mathbb{R}$, where V_2 is given by (3.6).

The stability condition (4.7) is a condition on the total derivative of F_h on the space complementary to $\text{Ker}\{F_x^0\}$. The corresponding result for the continuous problem is easily shown to hold using (3.5) and (3.9).

Now define the numerical approximation of $H(y, t) = 0$ by

$$H_h(y_h, t) := \begin{cases} F_h(x_h, \lambda_h), \\ \langle x_h - x_0, c \rangle + d(\lambda_h - \lambda_0) - t, \end{cases} \tag{4.8}$$

where $y_h = (x_h, \lambda_h)$.

It is now not difficult to show (Descloux and Rappaz (1982, Lemma 3.1)) that, for h small enough, $D_y H_h(\pi_h y_0, 0)$ is an isomorphism on $V_h \times \mathbb{R}$ where $\pi_h y_0 = (P_h x_0, \lambda_0)$. One can then show the convergence of solution curves near the fold point. In fact we have the following theorem, which is a compilation of results from Descloux and Rappaz (1982).

Theorem 4.1. Let $y_0 = (x(0), \lambda(0))$ be a simple fold point on a solution branch $(x(t), \lambda(t))$ of $F(x, \lambda) = 0$. Under the above consistency and stability conditions, and for h sufficiently small, the following are true.

(i) There exists a locally unique path $(x_h(t), \lambda_h(t))$, for t near 0, satisfying $F_h(x_h(t), \lambda_h(t)) = 0$, with

$$|\lambda_h(t) - \lambda(t)| + \|x_h(t) - x(t)\|$$
$$\leq C\{\|F_h(P_h x(t), \lambda(t))\| + \|(I - P_h)x(t)\|\}$$

for some C independent of h, with similar bounds for $|\frac{d\lambda_h(t)}{dt} - \frac{d\lambda(t)}{dt}| + \|\frac{dx_h(t)}{dt} - \frac{dx(t)}{dt}\|$.

(ii) If the fold point is quadratic then $F_h(x_h(t), \lambda_h(t)) = 0$ has a quadratic fold point at $(x_h(t^0), \lambda_h(t^0))$ for some t^0 near 0.

(iii) Finally, if $F_h(x_h(t), \lambda_h(t)) = 0$ is a Galerkin approximation of $F(x(t), \lambda(t)) = 0$, then

$$|\lambda_h(t^0) - \lambda_0| \leq C\left\{\left\|\frac{dy_h(0)}{dt} - \frac{dy(0)}{dt}\right\|^2\right. \tag{4.9}$$

$$\left. + \|y_h(0) - y_0\|\left(\|y_h(0) - y_0\| + \inf_{\psi \in V_h \times \mathbb{R}} \|\psi - \bar{\psi}_0\|\right)\right\},$$

where $y_h(t) = (x_h(t), \lambda_h(t))$, *etc.*, and $\bar{\psi}_0$ is a known element of $V_h \times \mathbb{R}$ (see Section 3 in Descloux and Rappaz (1982)).

Remark. Theorem 4.1(i) indicates that the curve $(x_h(t), \lambda_h(t))$ approximates the exact curve to the expected order of convergence. The form of the bound on the right-hand side of (4.9) indicates the possibility of superconvergence when a Galerkin method is employed. The result is also in Brezzi et al. (1981a) (as we describe below) and in Griewank and Reddien (1989). We present numerical results illustrating superconvergence in Section 6.

Variationally posed nonlinear problems arise in a number of very important situations (see, for example, Brezzi et al. (1980, 1981a, 1981b) where the appropriate theoretical setting for a variety of problems is carefully laid out). With V and H Hilbert spaces, $V \subset H$, V dense and continuously embedded in H, $V \subset H \subset V'$, the scalar product in H may represent the duality pairing between V and V'. Let W be a reflexive Banach space such that $H \subset W \subset V'$ with continuous embeddings, and assume the canonical injection of W into V' is compact. Let $a : V \times V \to \mathbb{R}$ be a continuous bilinear V-elliptic form; and let $G : V \times \mathbb{R} \to W$ be a C^p mapping. Now consider the nonlinear problem: find $(x, \lambda) \in V \times \mathbb{R}$ such that

$$a(x, v) + (G(x, \lambda), v) = 0, \quad \forall v \in V. \tag{4.10}$$

We can now introduce the operators T, $T' \in \mathcal{L}(V'; V)$ defined by

$$a(Tf, v) = a(f, T'v) = (f, v), \quad \forall v \in V, \quad \forall f \in V',$$

and an equivalent problem to (4.10) is

$$F(x, \lambda) := x + TG(x, \lambda). \tag{4.11}$$

Note that in this setting $T : W \to V$ is compact. This is an appropriate formulation for many nonlinear problems in applications. The steady Navier–Stokes equations do not quite fit into this framework and they are considered in Section 5.

Define $T_h \in \mathcal{L}(V'; V_h)$ by

$$a(T_h f, v_h) = (f, v_h) \quad \forall v_h \in V_h, \forall f \in V'$$

and the projection $P_h \in \mathcal{L}(V; V_h)$ by

$$a(P_h x - x, v_h) = 0 \quad \forall v_h \in V_h, \ x \in V.$$

Then

$$T_h = P_h T$$

and the approximating problem has the form

$$F_h(x_h, \lambda) := x_h + T_h G(x_h, \lambda_h), \quad x_h \in V_h \tag{4.12}$$

where $T_h \in \mathcal{L}(W; V_h)$. The nice feature of this formulation is that the approximation of T by T_h (*i.e.*, approximation of a *linear* operator) determines the convergence rates for the solution of the nonlinear problem (4.11). In fact we shall assume the consistency condition

$$\lim_{h \to 0} \|T - T_h\| = 0 \tag{4.13}$$

where the norm is in $\mathcal{L}(W; V)$.

We shall make full use of the theorems in Brezzi et al. (1981a, 1981b) and so we discuss in detail how they obtain their convergence results. First it is appropriate to describe the Lyapunov–Schmidt reduction process applied to the approximate problem $F_h(x_h, \lambda_h) = 0$.

A key idea in Brezzi et al. (1981a) and (1981b) is to perform the Lyapunov-Schmidt reduction process on the discrete problem $F_h(x_h, \lambda_h) = 0$ about the *exact* solution (x_0, λ_0) using the *exact* ϕ_0, ψ_0 and Q (see Section 3.2). With

$$x_h = x_0 + \alpha\phi_0 + v, \qquad v \in V_2, \tag{4.14}$$
$$\lambda_h = \lambda_0 + \xi, \tag{4.15}$$

then $\mathcal{F}_h$ is defined by

$$\mathcal{F}_h(v, \alpha, \xi) := QF_h(x_0 + \alpha\phi_0 + v, \lambda_0 + \xi). \tag{4.16}$$

Assuming $\mathcal{F}_h$ is consistent with $\mathcal{F}$ given by (3.17), one obtains the existence of $v_h = v_h(\alpha, \xi)$ satisfying $\mathcal{F}_h(v_h, \alpha, \xi) = 0$. Thus the discrete reduced problem is given by

$$f_h(\alpha, \xi) := \langle F_h(x_0 + \alpha\phi_0 + v_h(\alpha, \xi), \lambda_0 + \xi), \psi_0 \rangle = 0 \tag{4.17}$$

and we have an equivalence between solutions of (4.17) and those of $F_h(x_h, \lambda_h) = 0$.

Note that, in this approach, consistency of $\{\mathcal{F}_h\}$ and $\{f_h\}$ as approximating families for $\mathcal{F}$ and f follows from (4.13). Stability is ensured by (3.5) and (3.9), since (3.5) ensures that $D_x\mathcal{F}$ is nonsingular. The induced nonsingularity of the approximating $D_x\mathcal{F}_h$ provides the equivalence of the approximation reduction, and (3.9) ensures $\frac{\partial f}{\partial \xi}(0,0) \neq 0$ and hence $\frac{\partial f_h}{\partial \xi}(\alpha, \xi) \neq 0$ for (α, ξ) near (0,0). The convergence theory is thus reduced to comparison of the solutions of $f(\alpha, \xi) = 0$ and $f_h(\alpha, \xi) = 0$. At a simple fold point $\frac{\partial f}{\partial \xi}(0,0) \neq 0$ and, if the fold is quadratic, then $\frac{\partial^2 f}{\partial \alpha^2}(0,0) \neq 0$ (see Section 3.1). Hence the existence of (α_h, ξ_h) such that $f_h(\alpha_h, \xi_h) = 0$, $\frac{\partial f_h}{\partial \xi}(\alpha_h, \xi_h) \neq 0$, $\frac{\partial^2 f_h}{\partial \xi^2}(\alpha_h, \xi_h) \neq 0$ is readily shown. For problems of the form $F(x, \lambda) = x + TG(x, \lambda) = 0$, we have the following theorem.

Theorem 4.2. (Brezzi et al. 1981a)   Assume (3.2), (3.3), (3.4), (3.5), (3.9). Let F and F_h be defined by (4.11) and (4.12) and assume (4.13). Then, for h sufficiently small, the following are true.

(i) For α near 0 there exists a unique smooth path $(\xi_h(\alpha), \alpha) \in \mathbb{R}^2$ satisfying

$$f_h(\alpha, \xi_h(\alpha)) = 0,$$

and hence a unique smooth path $(x_h(\alpha), \lambda_h(\alpha))$ satisfying

$$F_h(x_h(\alpha), \lambda_h(\alpha)) = 0.$$

(ii) Further,

$$|\lambda_h(\alpha) - \lambda(\alpha)| + \|x_h(\alpha) - x(\alpha)\| \leq C\|(T - T_h)G(x(\alpha), \lambda(\alpha))\|,$$

with similar bounds for the approximation of derivatives of $(x(\alpha), \lambda(\alpha))$ with respect to α.

(iii) If, in addition, (3.13) holds, that is, the fold is quadratic, then the approximate problem has a quadratic fold point (x_h^0, λ_h^0) satisfying

$$|\lambda_h^0 - \lambda_0| + \|x_h^0 - x_0\| \leq C \sum_{l=0}^{1} \left\| (T - T_h) \frac{d^l}{d\alpha^l} G(x(\alpha), \lambda(\alpha)) \Big|_{\alpha=0} \right\|,$$

with a (possibly sharper) bound available for $\lambda_h^0 - \lambda_0$, namely (cf. (4.11))

$$|\lambda_h^0 - \lambda_0| \leq \left\{ \left| \langle (T - T_h) G^0, \psi_0 \rangle \right| \right.$$
$$+ \|(T - T_h) G^0\| \cdot \left\| \left[(T - T_h) D_x G^0 \right]' \psi_0 \right\|$$
$$\left. + \sum_{l=0}^{1} \left\| (T - T_h) \frac{d^l}{d\alpha^l} G(x(\alpha), \lambda(\alpha)) \Big|_{\alpha=0} \right\|^2 \right\}.$$

As shown in Brezzi et al. (1981a), if the numerical method is a Galerkin method then the bound on $\lambda_h^0 - \lambda_0$ can be written as

$$|\lambda_h^0 - \lambda_0| \leq C \left\{ \left(\inf_{v_h \in V_h} \left\| \frac{dx(0)}{dt} - v_h \right\| \right)^2 + \left(\inf_{v_h \in V_h} \|x_0' - v_h\| \right)^2 \right.$$
$$\left. + \left(\inf_{v_h \in V_h} \|x_0 - v_h\| \right) \left(\inf_{\psi_h \in V_h} \|\eta - \psi_h\| \right) \right\} \qquad (4.18)$$

where $\eta = T' \psi_0$. This bound clearly shows the superconvergence result.

4.2. Numerical approximation of bifurcation from the trivial solution and Z_2-symmetry breaking

The analysis for the numerical approximation of these two cases is very similar and so we discuss both in a single subsection. The material here is based on Brezzi et al. (1981b) and relies on the numerical Lyapunov–Schmidt decomposition discussed in the previous subsection.

First, for the case of bifurcation from the trivial solution,

$$F(0, \lambda) = 0, \quad D_\lambda F(0, \lambda) = 0, \ldots, \forall \lambda, \qquad (4.19)$$

and we assume the same property for the discrete problem, namely

$$F_h(0, \lambda) = 0, \quad D_\lambda F_h(0, \lambda) = 0, \ldots, \forall \lambda. \qquad (4.20)$$

(Note: if $G(0, \lambda) = 0$ for all λ for G given in (4.10), then (4.19) and (4.20) clearly hold for F and F_h given by (4.11) and (4.12).)

The Lyapunov–Schmidt reduction for the continuous case provides

$$f(0, 0) = \frac{\partial f}{\partial \alpha}(0, 0) = \frac{\partial f}{\partial \xi}(0, 0) = 0, \quad \frac{\partial^2 f}{\partial \alpha \partial \xi}(0, 0) \neq 0$$

and so, as a function of ξ, $\frac{\partial f}{\partial \alpha}(0, \xi)$ changes sign at $\xi = 0$. For the discrete reduced problem $f_h(\alpha, \xi) = 0$, one readily deduces the existence of a unique $(0, \xi_h^0)$ such that

$$\frac{\partial f_h}{\partial \alpha}(0, \xi_h^0) = 0.$$

Since $f_h(0, \xi) = 0$, and $\frac{\partial f_h}{\partial \xi}(0, \xi) = 0$, $\forall \xi$ sufficiently small, we have the existence of a bifurcation point, $(0, \xi_h^0)$ in the discrete reduced problem with $|\lambda_h^0 - \lambda_0| \leq C|\frac{\partial f_h}{\partial \alpha}(0, 0)|$. As in the quadratic fold case there is a superconvergence result for the critical parameter if a Galerkin method is used as given by Brezzi et al. (1981b, Theorem 6):

$$\left| \lambda_h^0 - \lambda_0 \right| \leq C \left\{ \inf_{\phi \in V_h} \| \phi_0 - \phi_h \| \right. \tag{4.21}$$

$$\left. + \sum_{l=0}^{1} \inf_{w_h \in V_h} \left\| \frac{d^l u(0)}{dt^l} - w_h \right\| \right\} \left(\inf_{\psi_h \in V_h} \| \eta - \psi_h \| \right),$$

where $\eta = T' \psi_0$.

Let us now turn to the case of Z_2-symmetry breaking bifurcation. We assume

$$F_h(x, \lambda) \text{ satisfies the equivariance condition (3.30).} \tag{4.22}$$

This implies that the reduced equation inherits the equivariance condition

$$f_h(-\alpha, \xi) = -f_h(\alpha, \xi)$$

and hence

$$f_h(0, \xi) = 0 \text{ for all } \xi \text{ sufficiently small.}$$

Now the analysis is essentially the same as for the case of bifurcation from the trivial solution. Since $\frac{\partial f}{\partial \alpha}(0, 0) = 0$ and $\frac{\partial^2 f}{\partial \xi \partial \alpha}(0, 0) \neq 0$, we deduce the existence of ξ_h^0 such that $\frac{\partial f_h}{\partial \alpha}(0, \xi_h^0) = 0$. Also $f_h(0, \xi) = 0$, and $\frac{\partial f_h}{\partial \xi}(0, \xi) = 0$ for all ξ sufficiently small, and so bifurcation in $f(\alpha, \xi) = 0$ occurs at $(0, \xi_h^0)$, with

$$|\lambda_h^0 - \lambda_0| \leq C \left| \frac{\partial f_h}{\partial \alpha}(0, 0) \right| \tag{4.23}$$

$$\leq C \left| \left\langle D_x F_h^0 \left(\phi_0 + \frac{\partial v_h}{\partial \alpha}(0, 0) \right), \psi_0 \right\rangle \right|$$

$$= C \left| \left\langle (D_x F^0 - D_x F_h^0) \left(\phi_0 + \frac{\partial v_h}{\partial \alpha}(0, 0) \right), \psi_0 \right\rangle \right|,$$

and if F and F_h are given by (4.11) and (4.12) then

$$|\lambda_h^0 - \lambda_0| \leq C \left| \left\langle (T - T_h) G_x^0 \left(\phi_0 + \frac{\partial v_h}{\partial \alpha}(0, 0) \right), \psi_0 \right\rangle \right|.$$

Thus $(x_0 + v_h(0, \xi_h^0), \lambda_h^0)$ is a Z_2-symmetry breaking bifurcation point of $F_h(x_h, \lambda_h) = 0$. This result is in Theorem 4 of Brezzi et al. (1981b). That (4.23) provides superconvergence for $(\lambda_h^0 - \lambda_0)$ when a Galerkin method is used is not explicitly stated in Brezzi et al. (1981b), though the result is easily shown by repeating the steps on page 23 of that paper. In fact, though the superconvergence of λ_h^0 to λ_0 is well known, it appears difficult to find a formal statement in the literature for the symmetry breaking case, and so for completeness we state it below. Note that, in the theory, condition (3.5) and nondegeneracy condition $B_0 \neq 0$ in Theorem 3.4 provide the stability condition for this analysis in the same way that (3.5) and (3.9) implied stability in the discretized fold point case. If F and F_h are given by (4.11) and (4.12) then consistency comes from (4.13).

Theorem 4.3. Assume the conditions of Theorem 3.4, and so (x_0, λ_0) is a Z_2-symmetry breaking bifurcation point. Also assume $F_h(x_h, \lambda_h) = 0$ satisfies (4.22) and appropriate consistency conditions. Then there exists a Z_2-symmetry breaking bifurcation point (x_h^0, λ_h^0) of $F_h(x_h, \lambda_h) = 0$ with $x_h^0 = x_0 + v_h(0, \xi_h^0) \in X_s$ and with

$$|\lambda_h^0 - \lambda_0| \leq C \left| \frac{\partial f_h}{\partial \alpha}(0,0) \right| \leq C \left| \left\langle (T - T_h) D_x G^0 \left(\phi_0 + \frac{\partial v_h}{\partial \alpha}(0,0) \right), \psi_0 \right\rangle \right|.$$

If a Galerkin method is used to compute λ_h^0 then

$$|\lambda_h^0 - \lambda_0| \leq C \left\{ \inf_{\phi_h \in V_h} \|\phi_0 - \phi_h\| + \sum_{l=0}^{1} \inf_{w_h \in V_h} \|u^l(0) - w_h\| \right\} \left(\inf_{\psi_h \in V_h} \|\eta - \psi_h\| \right).$$

where $\eta = T'\psi_0$. Numerical results illustrating this superconvergence are shown in Section 6.

In problems with reflectional symmetry, the need to preserve symmetry in the discretization is well known (see, for example, Brezzi et al. (1981b), Descloux and Rappaz (1982), Moore et al. (1986)). In the simplest case of a Z_2-reflectional symmetry the codimension zero bifurcation is a symmetric pitckfork bifurcation. However, if a discretization is used that does not respect the Z_2-symmetry, then the symmetric pitchfork bifurcation is likely to be perturbed in the same way that a codimension two pitchfork bifurcation is perturbed in a setting with no symmetry (see, for example Golubitsky and Schaeffer (1985), p. 147). On the other hand, one must beware of imposing a symmetric structure on a discretized problem if one wishes to detect symmetry breaking bifurcations (see the discussion at the beginning of Section 7.3).

4.3. Numerical approximation near Hopf bifurcation

The theory of numerical approximation near a Hopf bifurcation is more difficult because of the interplay between the time and space discretizations, and there is surprisingly little numerical analysis on this very important topic. The work of Bernardi (1982) and Bernardi and Rappaz (1984) was the first to give a rigorous analysis of spatial discretization of nonlinear evolution equations near Hopf bifurcations. Questions on the creation of invariant curves after time discretization are considered by Brezzi et al. (1984) and these ideas are explored further in Hofbauer and Iooss (1984) and recently in Lubich and Ostermann (1998).

We outline the main ideas in Bernardi (1982). The theory of the numerical approximation is in the style of the analysis of the Brezzi, Rappaz, Raviart papers. An approximation $\mathcal{F}_h$ to $\mathcal{F}$ given by (3.45) is introduced as

$$\mathcal{F}_h(\lambda, w, v) := \mathcal{A}v + \mathcal{A}\mathcal{T}_h \left\{ (w - w_0)\frac{dv}{ds} + \tilde{G}(\lambda, v) \right\},$$

where $\mathcal{T}_h$ is an approximation of $\mathcal{T}$, the inverse of $\mathcal{A}$ given by (3.46). Consistency results follow under the assumption that $\|(\mathcal{T}_h - \mathcal{T})f\|_{\mathcal{X}} \to 0$ as $h \to 0$ for all $f \in \mathcal{X}$, where $\mathcal{X}$ is defined in Section 3.5. A Lyapunov–Schmidt analysis is then carried out on $\mathcal{F}_h$. The main result is as follows.

Theorem 4.4. (Bernardi 1982, Theorem VI.1) For sufficiently small h, $\mathcal{F}_h(\lambda, \omega, v) = 0$ has a unique branch of solutions $(\lambda_h(\epsilon), \omega_h(\epsilon), v_h(\epsilon))$ in a neighbourhood of the branch given in Theorem 3.5 which has the form

$$\begin{aligned}
\lambda_h(\epsilon) &= \lambda_0 + \sigma_h(\epsilon), \\
\omega_h(\epsilon) &= \omega_0 + \chi_h(\epsilon), \\
x_h(\epsilon) &= x_e(\lambda_h(\epsilon)) + v_h(\epsilon)(\omega_h(\epsilon)t),
\end{aligned}$$

where

$$v_h(\epsilon) = \epsilon(\zeta_0 + \bar{\zeta}_0) + \epsilon w_{1h}(\epsilon)$$

for smooth mappings σ_h, χ_h, w_{1h}. Moreover,

$$|\lambda_h(\epsilon) - \lambda(\epsilon)| + |\omega_h(\epsilon) - w(\epsilon)| + \|v_h(\epsilon) - v(\epsilon)\|_{\mathcal{X}} \le C \left\|(\mathcal{T} - \mathcal{T}_h)\tilde{E}(\epsilon)\right\|_{\mathcal{X}}$$

for some smooth $\tilde{E}(\epsilon)$ (see Bernardi (1982)) and where C is a constant independent of h. Bounds for the approximation of derivatives of $\lambda(\epsilon)$, $\omega(\epsilon)$ and $v(\epsilon)$ are also given.

An estimate which allows the possibility of superconvergence when using a Galerkin method is given by the following lemma.

Lemma 4.1. (Bernardi 1982, Lemma VI.3) For sufficiently small h,

$$|\lambda_h(0) - \lambda_0| + |\omega_h(0) - \omega_0| \le C \left| \left[\mathcal{A}(\mathcal{T} - \mathcal{T}_h) D_v G^0(\zeta_0 + \bar{\zeta}_0), \zeta_0^* \right] \right|,$$

where C is a constant independent of h, and $[\cdot, \cdot]$ denotes the duality pairing between $\mathcal{X}$ and $\mathcal{X}'$. The inner product term on the right-hand side indicates the possibility of superconvergence in the approximation of the critical values for λ and ω.

In Bernardi (1982) a numerical scheme based on a Galerkin approximation in V and Fourier approximation in time is analysed, and a superconvergence result for the quantity $|\lambda_h(0) - \lambda_0| + |\omega_h(0) - \omega_0|$ is explicitly stated (Proposition VII.1).

In nonlinear parabolic-type problems $\dot{x} + F(x, \lambda) = 0$ where the spatial derivatives in $F(x, \lambda)$ are discretized by the finite element method, we obtain a discretized problem of the form

$$M_h \dot{x}_h + F_h(x_h, \lambda) = 0,$$

where M_h denotes a mass matrix. If $x_h{}^e(\lambda)$ denotes the discretized equilibrium solution, then Hopf bifurcation points may be found by finding the value λ_h^0 such that $\det(\mu M_h + D_x F_h(x_h{}^e(\lambda_h^0), \lambda_h^0)) = 0$ has purely imaginary roots $\mu_\pm = \pm i\omega_h^0$, say. In the computations presented in Section 6 approximations to λ_h^0 and ω_h^0 were found using the Cayley transform method discussed in Section 7.4, and then the Hopf bifurcation was located using the system in Griewank and Reddien (1983) adapted to allow M_h rather than the identity matrix.

4.4. Numerical approximation by finite difference methods

Most of the convergence theory is carried out for projection-type methods, probably because of the important rôle played by the finite element method in continuum mechanics, the main application area for bifurcation theory. Nonetheless, finite difference methods are used in many applications and it is important to have a sound theory. Here we briefly discuss the approach in the papers by Moore and Spence (1981) and Moore et al. (1986) where the Keller approximation theory (Keller 1975) is extended to deal with fold points and Z_2-symmetry breaking bifurcation points.

Consider

$$F(x, \lambda) = 0, \quad F : V \times \mathbb{R} \to V \qquad (4.24)$$

and the approximating problem, for some $h > 0$,

$$F_h(x_h, \lambda_h) = 0, \quad F_h : V_h \times \mathbb{R} \to V_h \qquad (4.25)$$

where V_h is linked to V by the bounded restriction operator $r_h : V \to V_h$ such that $\|r_h x\| \to \|x\|$ as $h \to 0$ for appropriate norms. Natural extensions of the usual consistency conditions are assumed for $\{F_h\}$. The stability condition assumed is the spectral stability condition used by Chatelin (1973) for linear eigenvalue problems. Let L be a bounded linear operator on V, with approximating family $\{L_h\}$ which is consistent for all $x \in V$. Then $\{L_h\}$ is *spectrally stable* for a complex scalar z if, for h sufficiently small, $(L_h - zI)$ has a bounded inverse on V_h. If μ is an eigenvalue of L, and the circle of radius δ about μ in the complex plane is denoted by Γ, then we may define the spectral projection

$$\mathcal{P} := -\frac{1}{2\pi i} \int_\Gamma (L - zI)^{-1} \, dz.$$

$\mathcal{P}$ induces the decomposition $V = \mathcal{P}V \oplus (I - \mathcal{P})V$ with $\mathcal{P}V$ being the (generalized) eigenspace of L with respect to μ. Now if $\{L_h\}$ is consistent with L then $\mathcal{P}_h := -\frac{1}{2\pi i} \int_\Gamma (L_h - zI)^{-1} \, dz$ is known to be consistent (of same order) with $\mathcal{P}$.

The stability condition assumed in Theorem 4.5 below for (algebraically) simple fold points is

$$D_x F_h(r_h x_0, \lambda_0) \text{ is spectrally stable for } |z| = \delta \geq 0,$$
$$\delta \text{ sufficiently small, } and \text{ } \dim(\mathcal{P}_h X) = 1. \tag{4.26}$$

This implies (see Moore and Spence (1981), §3) that $D_x F_h(r_h x_0, \lambda_0)|_{(I-\mathcal{P}_h)V}$ is consistent with the $D_x F(x_0, \lambda_0)|_{(I-\mathcal{P})V}$, and has a uniformly bounded inverse for h sufficiently small. By comparison with condition (3.8), this is seen to be a very natural stability condition.

We now have the following convergence theorem for fold points.

Theorem 4.5. Assume (3.2), (3.3), (3.4), (3.5) and (3.9). Let H and H_h be defined by (3.10) and (4.8). If

(i) $\{F_h(\cdot, \lambda_h(t))\}$ is consistent with $F(\cdot, \lambda(t))$ at $x(t) \in V$ for t sufficiently small,

(ii) $\{D_x F_h(r_h x_0, \lambda_0)\}$ is consistent with $D_x F(x_0, \lambda_0)$, and $\{D_\lambda F_h(r_h x_0, \lambda_0)\}$ is consistent with $D_\lambda F(x_0, \lambda_0)$,

(iii) $D_x F_h$ and $D_\lambda F_h$ are uniformly Lipschitz in a ball centred on $(r_h x_0, \lambda_0)$,

(iv) $\{D_x F_h(r_h x_0, \lambda_0)\}$ is spectrally stable,

(v) $\dim(\mathcal{P}_h x_h) = 1$ for h sufficiently small,

then, for h sufficiently small,

(vi) $F_h(x_h, \lambda_h) = 0$ has a locally unique solution curve $(x_h(t), \lambda_h(t))$ with

$$\max \{|\lambda_h(t) - \lambda(t)|, \|x_h(t) - r_h x(t)\|\} \leq$$
$$C \max \{\|F_h(r_h x(t), \lambda(t)) - r_h F(x(t), \lambda(t))\|, \|(I - r_h)x\|\}.$$

Furthermore, if (3.13) holds and the second derivatives of F_h satisfy further consistency and smoothness conditions (see Moore and Spence (1981, Corollary 11)), then

(vii) $F_h(x_h, \lambda_h) = 0$ has a quadratic fold point (x_h^0, λ_h^0) and

$$\max\{|\lambda_h^0 - \lambda_0|, \|x_h^0 - r_h x_0\|\} \leq$$
$$C \max\{\|F_h(r_h x_0, \lambda_0) - r_h F(x_0, \lambda_0)\|, |\mu_h^0|\},$$

where μ_h^0 is the eigenvalue of smallest modulus of $D_x F_h(r_h x_0, \lambda_0)$.

Moore et al. (1986) treat the case of Z_2-symmetry breaking bifurcation. Since it is no longer assumed that $V_h \subset V$, a slightly different, but natural, invariance condition is required. With $F(x, \lambda) = 0$ satisfying (3.30) we assume the following equivariance condition. For each $h > 0$,

(a) there exists a uniformly bounded linear operator S_h on V_h (4.27)
    with $S_h^2 = I$, $S_h \neq I$ such that $F_h(S_h x_h, \lambda_h) = S_h F_h(x_h, \lambda_h)$,
    $\forall (x_h, \lambda_h) \in V_h \times \mathbb{R}$;

(b) symmetry and approximation commute, that is,

$$r_h S x = S_h r_h x, \quad \forall x \in V.$$

Since we are interested in symmetry breaking we assume $x_0 \in V_s$ and $\phi_0 \in V_a$, as in (3.34) in Section 3.4.

By restricting attention to V_s the Jacobian $D_x F(x_0, \lambda_0)|_{V_s}$ is singular, that is, there is a unique path of symmetric solutions $x^s(\lambda) \in V_s$ near $\lambda = \lambda_0$. Thus standard theory for paths of regular solutions applies, as is given in the following theorem.

Theorem 4.6. For the continuous problem $F(x, \lambda) = 0$, assume (3.2), (3.3), (3.4), (3.5), (3.30), (3.34) and

$$B_0 := \langle F_{x\lambda}\phi_0 + F_{xx}^0 \phi_0 w_0, \phi_0 \rangle \neq 0.$$

For the discrete problem $F_h(x_h, \lambda_h) = 0$ assume (4.27) and

(i) $\{F_h(x^s(\lambda), \lambda)\}$ is consistent with $\{F(x^s(\lambda), \lambda)\}$;
(ii) $D_x F_h(r_h x_0, \lambda_0)$ is spectrally stable and consistent with $D_x F(x_0, \lambda_0)$;
(iii) $D_x F_h$ is uniformly Lipschitz in a ball centred on $(r_h x_0, \lambda_0)$.

Then, for h small enough, $F_h(x_h, \lambda_h) = 0$ has a unique solution path $(x_h^s(\lambda), \lambda)$ near $\lambda = \lambda_0$ with

$$\|x_h^s(\lambda) - r_h x^s(\lambda)\| \leq C\|F_h(r_h x^s(\lambda), \lambda) - r_h F(x^s(\lambda), \lambda)\|.$$

In addition, under further smoothness and Lipschitz conditions on second derivatives of F_h, and if

(iv) $\dim \mathcal{P}_h x_h = 1$ for h small enough,

then there exists a Z_2-symmetry breaking bifurcation point (x_h^0, λ_h^0) of $F_h(x_h, \lambda_h) = 0$ with

$$|\lambda_h^0 - \lambda_0| \leq C|\mu_h(\lambda_0)| \tag{4.28}$$

where $\mu_h(\lambda)$ is the eigenvalue of smallest modulus of $D_x F_h(x_h{}^s(\lambda), \lambda)|_{V_a}$.

Remark. Though the analysis is aimed at finite difference methods it also applies to projection methods. Result (4.28) indicates the superconvergence result given earlier in Theorem 4.3 can also be proved using this analysis. We sketch the ideas. Let $\mu(\lambda)$ denote the eigenvalue of minimum modulus of $D_x F(x^s(\lambda), \lambda)$, so that $\mu(\lambda_0) = 0$, and introduce $\hat{\mu}(\lambda)$, the eigenvalue of minimum modulus of $D_x F(x_h{}^s(\lambda), \lambda)$. Thus

$$|\mu_h(\lambda_0)| \le |\mu_h(\lambda_0) - \hat{\mu}(\lambda_0)| + |\hat{\mu}(\lambda_0) - \mu(\lambda_0)|.$$

Now, if a Galerkin method is used for the numerical method, the first term on the right-hand side exhibits superconvergence using standard eigenvalue approximation theory (Mercier, Osborn, Rappaz and Raviart 1981). Using matrix perturbation ideas, the dominant term in $\hat{\mu}(\lambda_0) - \mu(\lambda_0)$ is $\langle (F_x(x_h{}^s(\lambda_0), \lambda_0) - F_x(x^s(\lambda_0), \lambda_0))\phi_0, \psi_0 \rangle$. This is a (nonlinear) functional of $(x_h{}^s - x^s)(\lambda_0)$ and standard Galerkin arguments provide superconvergence. Obviously technical details need to be provided in any given application.

4.5. Literature review

Here we briefly summarize or note some of the other literature on the convergence of numerical methods at bifurcation points.

One of the first rigorous accounts of the numerical analysis of the finite element method applied to fold (turning) point problems is given by Kikuchi (1977), where a numerical Lyapunov–Schmidt procedure is used, but expanding about the discrete solution (*cf.* Brezzi et al. (1981*a*) who expand about the exact solution). In Fujii and Yamaguti (1980) singularities arising in the shallow shell equations and their numerical approximation are analysed, with superconvergence results being obtained for symmetry breaking bifurcations. The papers by Beyn (1980) and Moore (1980) contain early accounts of convergence theory at transcritical bifurcation points with both showing the perturbation of the discretized branches near the bifurcation point. Closely related to the approach in Brezzi et al. (1981*a*) is the work of Li, Mei and Zhang (1986) who apply their results to the Navier–Stokes equations (see the next section). Other work on this topic is that by Paumier (1981).

The work of Fink and Rheinboldt (1983, 1984, 1985) provides a general framework for the numerical approximation of parameter-dependent nonlinear equations. A key theme throughout is the error estimation and in Fink and Rheinboldt (1985) the approach involves considering only a single discretized equation, rather than a family of approximations.

Another approach is provided in Griewank and Reddien (1984, 1989, 1996), where superconvergence results for general projection methods are obtained for classes of bifurcation points, called generalized turning points.

Finally, we do not address the question of spurious solutions in numerical approximations. The accepted wisdom is that after mesh refinement spurious solutions move away and physical solutions remain. There is little in the literature on this topic (but see Brezzi et al. (1984), Beyn and Doedel (1981) and Beyn and Lorenz (1982)).

5. Application to the Navier–Stokes equations

Here we apply the theory in Section 4.1 to the approximation of a fold point in the Navier–Stokes equations. In this section we consider the approximation of bifurcation points of the steady compressible Navier–Stokes equations discretized by the mixed finite element method (in the following section we present numerical results obtained using $Q_2 - P_1$ elements (Sani, Gresho, Lee and Griffiths 1981)). A general convergence theory at a fold point for a mixed finite element method is discussed by Li et al. (1986) based on the theory in Brezzi et al. (1981a), though a superconvergence result is not presented. Here we go through in some detail the convergence theory for a fold point and give a superconvergence result.

5.1. Function spaces and norms

We introduce the usual Sobolev spaces. Let $L^p(\Omega)$ be the space of Lebesgue-measurable, real-valued functions f defined on Ω such that

$$\|f\|_{L^p(\Omega)} = \int_\Omega |f|^p < \infty,$$

where $1 \leq p < \infty$ and

$$L_0^2(\Omega) = \left\{ v \in L^2(\Omega) : \int_\Omega v = 0 \right\}.$$

Let $a = (a_1, \ldots, a_N) \in \mathbb{N}^N$ and

$$|a| = \sum_{i=1}^N a_i.$$

For $v \in L^p(\Omega)$, let

$$\partial^a v = \frac{\partial^{|a|} v}{\partial^{a_1} x_1, \ldots, \partial^{a_N} x_N}$$

denote the weak derivative of v of order a.

For each integer $m \geq 0$, the Sobolev spaces $W^{m,p}(\Omega)$ are defined by

$$W^{m,p}(\Omega) = \left\{ v \in L^p(\Omega) : \partial^a v \in L^p(\Omega) \forall |a| \leq m \right\},$$

with the following norm:

$$\|v\|_{W^{m,p}(\Omega)} = \left(\sum_{|a| \le m} \int_\Omega |\partial^a v|^p \right)^{1/p}.$$

The space $W^{m,p}(\Omega)$ is also provided with the semi-norm

$$|v|_{W^{m,p}(\Omega)} = \left(\sum_{|a|=m} \int_\Omega |\partial^a v|^p \right)^{1/p}.$$

When $p = 2$, $W^{m,p}(\Omega)$ is denoted by $H^m(\Omega)$. $H^m(\Omega)$ is a Hilbert space with the scalar product

$$(u,v)_{H^m(\Omega)} = \sum_{|a| \le m} \int_\Omega \partial^a u\, \partial^a v.$$

As usual, $W_0^{m,p}(\Omega)$ denotes the closure of the space of smooth functions with compact support in Ω, with respect to the norm $\| \cdot \|_{W^{m,p}(\Omega)}$.

For $1 \le p < \infty$, the dual space of $W_0^{m,p}(\Omega)$ is denoted by $W^{-m,p'}(\Omega)$ with norm defined by

$$\|f\|_{W^{-m,p'}(\Omega)} = \sup_{v \in W_0^{m,p}(\Omega), v \ne 0} \frac{\langle f, v \rangle}{\|v\|_{W^{m,p}(\Omega)}},$$

where p' is given by

$$1/p + 1/p' = 1,$$

and $\langle \cdot, \cdot \rangle$ denotes the duality pairing between $W_0^{m,p}(\Omega)$ and $W^{-m,p'}(\Omega)$.

Let $m \ge 0$ be an integer and s and p be two real numbers such that $1 \le p < \infty$ and $s = m + \sigma$ with $0 < \sigma < 1$. Then $W^{s,p}(\Omega)$ denotes the space of functions, v, such that

$$v \in W^{m,p}(\Omega),$$

and

$$\int_\Omega \int_\Omega \frac{|\partial^a v(x) - \partial^a v(y)|^p}{|x - y|^{N+\sigma p}}\, dx\, dy < +\infty \quad \forall |a| = m.$$

$W^{s,p}(\Omega)$ is a Banach space with the norm

$$\|v\|_{W^{s,p}(\Omega)} = \left\{ \|v\|_{W^{m,p}(\Omega)}^p + \sum_{|a|=m} \int_\Omega \int_\Omega \frac{|\partial^a v(x) - \partial^a v(y)|^p}{|x - y|^{N+\sigma p}}\, dx\, dy \right\}^{1/p}.$$

5.2. Navier–Stokes equations

The Navier–Stokes equations govern the flow of viscous, incompressible fluids. The derivation of the equations may be found in a number of classical text books on fluid mechanics. See, for example, Landau and Lifshitz (1966), Batchelor (1970) or Chorin and Marsden (1979).

Consider a domain Ω that is an open, bounded subset of $\mathbb{R}^N$ with $N = 2$ or 3. The boundary of Ω is denoted by Γ. For the present purposes it is sufficient to assume that Γ is Lipschitz-continuous. For the precise definition of a Lipschitz-continuous boundary see Grisvard (1985). We say that the domain Ω is Lipschitz-continuous if it has a Lipschitz-continuous boundary. Lipschitz-continuous domains may have sharp corners, but domains with cuts or cusps are excluded. Many, but not all, of the domains encountered in modelling fluid flows are Lipschitz-continuous. Examples of flows in domains that are not Lipschitz-continuous include the flow over a thin plate, or the flow past a sphere or cylinder resting on a plane. All the flows considered in this review occur in Lipschitz-continuous domains.

The steady Navier–Stokes equations for flow in the domain Ω may be written in the form

$$R\mathbf{u} \cdot \nabla \mathbf{u} + \nabla p - \nabla^2 \mathbf{u} = \mathbf{f} \qquad \text{in } \Omega, \tag{5.1}$$

together with the continuity equation (or incompressibility constraint)

$$\nabla \cdot \mathbf{u} = 0 \qquad \text{in } \Omega, \tag{5.2}$$

and the boundary conditions

$$\mathbf{u} = \mathbf{g} \qquad \text{on } \Gamma, \tag{5.3}$$

where $\mathbf{u}$ denotes the velocity field, p the pressure, $\mathbf{f}$ the body force per unit mass acting on the fluid and R is the Reynolds number. All the quantities are assumed to have been nondimensionalized using suitable length and time-scales. The body force, $\mathbf{f}$, is assumed to be in $H^{-1}(\Omega)^N$ and the prescribed boundary velocity, $\mathbf{g}$, is assumed to be in $H^{1/2}(\Gamma)^N$ and to satisfy the compatibility condition

$$\int_\Gamma \mathbf{g} \cdot \mathbf{n} = 0, \tag{5.4}$$

where $\mathbf{n}$ is the outward pointing normal on Γ.

Consider the weak form of the Navier–Stokes equations, that is, find a pair $(\mathbf{u}, p) \in H^1(\Omega)^N \times L_0^2(\Omega)$ that satisfies

$$\int_\Omega R(\mathbf{u} \cdot \nabla \mathbf{u}) \cdot \mathbf{v} - p \nabla \cdot \mathbf{v} + \nabla \mathbf{u} : \nabla \mathbf{v} = \langle \mathbf{f}, \mathbf{v} \rangle, \quad \forall \mathbf{v} \in H_0^1(\Omega)^N, \tag{5.5}$$

$$\nabla \cdot \mathbf{u} = 0, \quad \text{in } \Omega, \tag{5.6}$$

$$\mathbf{u} = \mathbf{g}, \quad \text{on } \Gamma. \tag{5.7}$$

(Here, : denotes the standard double contraction of two matrices or two rank 2 tensors, so that $\nabla \mathbf{u} : \nabla \mathbf{v} = \sum_{i,j=1}^3 \frac{\partial u_i}{\partial x_j} \frac{\partial v_i}{\partial x_j}$.)

The following theorem states that there is always at least one solution to the steady-state Navier–Stokes equations, and, further, that if the Reynolds number is sufficiently small, the solution is unique.

Theorem 5.1. (See, *e.g.*, Girault and Raviart (1986).) Let $N \leq 3$ and let Ω be a bounded domain in $\mathbb{R}^N$ with a Lipschitz-continuous boundary Γ. Given $\mathbf{f} \in H^{-1}(\Omega)^N$ and $\mathbf{g} \in H^{\frac{1}{2}}(\Gamma)^N$, satisfying the condition (5.4), there exists at least one pair $(\mathbf{u}, p) \in H^1(\Omega)^N \times L_0^2(\Omega)$ that satisfies (5.5), (5.6) and (5.7).

Theorem 5.2. (See, *e.g.*, Girault and Raviart (1986).) Assume that the conditions of Theorem 5.1 hold. Then there is a positive number R^0 depending on Ω, $\mathbf{f}$ and $\mathbf{g}$, such that if $R < R^0$ the solution to (5.5), (5.6) and (5.7) is unique.

5.3. Stokes equations

The Stokes equations, which are obtained from the Navier–Stokes equations by removing the nonlinear term, are given by

$$\nabla p - \nabla^2 \mathbf{u} = \mathbf{f} \qquad \text{in } \Omega, \tag{5.8}$$

together with the continuity equation (or incompressibility constraint)

$$\nabla \cdot \mathbf{u} = r \qquad \text{in } \Omega, \tag{5.9}$$

and the boundary conditions

$$\mathbf{u} = \mathbf{g} \qquad \text{on } \Gamma,$$

where $\mathbf{u}$, p and $\mathbf{f}$ are as for the Navier–Stokes equations and $r \in L^2(\Omega)$ is the mass source term. For most practical flows there will be no mass source term. However, this term is need for the later results on the Navier–Stokes equations and so is retained here. The compatibility condition on the boundary velocity now becomes

$$\int_\Gamma \mathbf{g} \cdot \mathbf{n} = \int_\Omega r. \tag{5.10}$$

The compatibility condition (5.4) must also hold for the Stokes problem.

The Stokes problem has a unique solution in $H_0^1(\Omega)^N \times L_0^2(\Omega)$ (Girault and Raviart 1986, Theorem 5.1, p. 80). From now on, for the sake of simplicity, we concentrate on problems with homogeneous boundary conditions so that $\mathbf{g} = \mathbf{0}$.

We now consider approximate solutions to the Stokes problem. To do this we introduce finite element spaces $X_h \subset H_0^1(\Omega)$ and $M_h \subset L_0^2(\Omega)$, where M_h contains the constant functions. The approximate form of the Stokes problem is as follows.

Find a pair $(\mathbf{u}_h, p_h) \in X_h \times M_h$ such that

$$\int_\Omega \nabla \mathbf{u}_h : \nabla \mathbf{v}_h - p_h \nabla \cdot \mathbf{v}_h = \langle \mathbf{f}, \mathbf{v}_h \rangle, \quad \forall \mathbf{v}_h \in X_h, \qquad (5.11)$$

$$-\int_\Omega q_h \nabla \cdot \mathbf{u}_h = \int_\Omega r q_h, \quad \forall q_h \in M_h. \qquad (5.12)$$

It is assumed that the finite element spaces satisfy the following three hypotheses (Girault and Raviart 1986, p. 125).

Hypothesis 5.1. (Approximation property of X_h) There exists an operator $\Pi_{h,u} \in \mathcal{L}((H^2(\Omega) \cap H_0^1(\Omega))^N; X_h)$ and an integer l such that

$$\|\mathbf{v} - \Pi_{h,u}\mathbf{v}\|_{1,\Omega} \leq Ch^m \|\mathbf{v}\|_{m+1,\Omega} \quad \forall \mathbf{v} \in H^{m+1}(\Omega), \quad 0 \leq m \leq l. \qquad (5.13)$$

Here l depends on the finite element method being used and m on the smoothness of the solution to the problem.

Hypothesis 5.2. (Approximation property of M_h) There exists an operator $\Pi_{h,p} \in \mathcal{L}(L^2(\Omega); M_h)$ and an integer l such that

$$\|q - \Pi_{h,p}q\|_{0,\Omega} \leq Ch^m \|q\|_{m,\Omega} \quad \forall q \in H^m(\Omega), \quad 0 \leq m \leq l. \qquad (5.14)$$

Hypothesis 5.3. (Uniform inf-sup condition) For each $q_h \in M_h$, there exists a $\mathbf{v}_h \in X_h$ such that

$$\int_\Omega q_h \nabla \cdot \mathbf{v}_h = \|q_h\|_{0,\Omega}^2, \qquad (5.15)$$

$$|\mathbf{v}_h|_{1,\Omega} \leq C\|q_h\|_{0,\Omega}, \qquad (5.16)$$

with a constant $C > 0$ independent of h, q_h and $\mathbf{v}_h$.

The calculations presented in Section 6 were carried out using the $Q_2 - P_1$ element (Fortin 1993) which satisfies (5.16), and for which $l = 2$. Thus for sufficiently smooth v and q we have

$$\|\mathbf{v} - \Pi_{h,u}\mathbf{v}\|_{1,\Omega} \leq Ch^2 \|\mathbf{v}\|_{3,\Omega}, \quad \|q - \Pi_{h,p}q\|_{0,\Omega} \leq Ch^2 \|q\|_{2,\Omega}. \qquad (5.17)$$

The following theorem establishes the convergence of the finite element method for the Stokes problem.

Theorem 5.3. (Girault and Raviart 1986) Under Hypotheses 5.1, 5.2 and 5.3, equations (5.11) and (5.12) have a unique solution $(\mathbf{u}_h, p_h)$. In addition, $(\mathbf{u}_h, p_h)$ tends to the solution $(\mathbf{u}, p)$ of equations (5.8) and (5.9)

$$\lim_{h \to 0} \{|\mathbf{u}_h - \mathbf{u}|_{1,\Omega} + \|p_h - p\|_{0,\Omega}\} = 0. \qquad (5.18)$$

Furthermore, when $(\mathbf{u}, p)$ belongs to $H^{m+1}(\Omega) \times (H^m(\Omega) \cap L_0^2(\Omega))$ for some integer m with $1 \leq m \leq l$, the following error bound holds:

$$|\mathbf{u}_h - \mathbf{u}|_{1,\Omega} + \|p_h - p\|_{0,\Omega} \leq Ch^m \{\|\mathbf{u}\|_{m+1,\Omega} + \|p\|_{m,\Omega}\}. \qquad (5.19)$$

We now introduce the Stokes operator $\bar{T}$ which is defined as a map from $H^{-1}(\Omega)^N \times L_0^2(\Omega)$ to $H_0^1(\Omega)^N \times L_0^2(\Omega)$ by

$$\bar{T}(\mathbf{f}, r) = (\mathbf{u}, p), \qquad (5.20)$$

where

$$\int_\Omega \nabla \mathbf{u} : \nabla \mathbf{v} - p \nabla \cdot \mathbf{v} - q \nabla \cdot \mathbf{u}$$
$$= \int_\Omega \mathbf{f} \cdot \mathbf{v} + rq, \quad \forall (\mathbf{v}, q) \in H_0^1(\Omega)^N \times L_0^2(\Omega). \qquad (5.21)$$

The approximate Stokes operator $\bar{T}_h$ is defined as a map from $H^{-1}(\Omega) \times L_0^2(\Omega)$ to $X_h \times M_h$ by

$$\bar{T}_h(\mathbf{f}, r) = (\mathbf{u}_h, p_h), \qquad (5.22)$$

where

$$\int_\Omega \nabla \mathbf{u}_h : \nabla \mathbf{v}_h - p_h \nabla \cdot \mathbf{v}_h - q_h \nabla \cdot \mathbf{u}_h$$
$$= \int_\Omega \mathbf{f} \cdot \mathbf{v}_h + rq_h, \quad \forall (\mathbf{v}_h, q_h) \in X_h \times M_h. \qquad (5.23)$$

Theorem 5.3 implies that

$$\lim_{h \to 0} \|(\bar{T} - \bar{T}_h)(\mathbf{f}, r)\|_{H_0^1(\Omega)^N \times L_0^2(\Omega)} = 0, \quad \forall (\mathbf{f}, r) \in H^{-1}(\Omega)^N \times L_0^2(\Omega). \quad (5.24)$$

5.4. Convergence theory for the Navier–Stokes equations

We now consider the behaviour of the nonlinear terms in the Navier–Stokes equations. Our aim is to write these equations in a form to which we can apply the theory of Brezzi, Rappaz and Raviart. The basic idea is to write the Navier–Stokes equations so that the linearized equations take the form of a compact perturbation of the identity. The reason for doing this is so that the Fredholm alternative can be applied and the infinite-dimensional system behaves like a finite-dimensional system at a simple singular point. We will also see that the theory of Brezzi, Rappaz and Raviart allows us to analyse the behaviour of approximations to the Navier–Stokes equations at bifurcation points using the theory of the approximation of the Stokes problem. Since we are considering a mixed finite element approach to the discretization of the Navier–Stokes equations, the results for the Galerkin approximation described in Section 4 are not directly applicable. We define the map $G : H_0^1(\Omega)^N \times \mathbb{R} \mapsto L^{\frac{3}{2}}(\Omega)^N$ by

$$G(\mathbf{u}, R) = R\mathbf{u} \cdot \nabla \mathbf{u} - \mathbf{f}, \qquad (5.25)$$

where we now assume that $\mathbf{f} \in L^{\frac{3}{2}}(\Omega)^N$. The following lemmas hold.

Lemma 5.1. G is a bounded operator on all bounded subsets of $H_0^1(\Omega)^N$.

Lemma 5.2. The Navier–Stokes equations may be written in the form

$$(\mathbf{u}, p) + TG(\mathbf{u}, R) = 0. \tag{5.26}$$

Here T is the restriction of $\bar{T}$ to $L^{\frac{3}{2}}(\Omega)^N \times 0$. Furthermore, since $L^{\frac{3}{2}}(\Omega)^N$ is compactly embedded in $H^{-1}(\Omega)^N$ it follows from (5.24) that

$$\lim_{h \to 0} \|T - T_h\| = 0, \tag{5.27}$$

where the norm is taken to be the norm on $\mathcal{L}(L^{\frac{3}{2}}(\Omega)^N; H_0^1(\Omega)^N \times L_0^2(\Omega))$.

Suppose $(\mathbf{u}^0, R^0)$ is a solution of the Navier–Stokes equations at which there is a simple limit point. This means that the linearized Navier–Stokes equations have a simple zero eigenvalue, so that there exists $(\xi_0, \pi_0) \in H_0^1(\Omega)^N \times L_0^2(\Omega)$ such that

$$(\xi_0, \pi_0) + T \cdot DG^0 \cdot \xi_0 = 0, \quad \|\xi_0\|_{1,\Omega} + \|\pi_0\|_{0,\Omega} = 1. \tag{5.28}$$

By compactness and the Fredholm alternative, there exists $(\eta_0', \rho_0') \in H^{-1}(\Omega)^N \times L_0^2(\Omega)$ such that

$$[I + T \cdot DG^0]'(\eta_0', \rho_0') = 0, \tag{5.29}$$

where $[I + T \cdot DG^0]'$ denotes the adjoint operator of $I + T \cdot DG^0$, and

$$\int_\Omega \xi_0 \cdot \eta_0' + \pi_0 \rho_0' = 1. \tag{5.30}$$

Setting

$$(\eta_0, \rho_0) = \bar{T}(\eta_0', \rho_0'), \tag{5.31}$$

it is easy to check that $(\eta_0, \rho_0) \in H_0^1(\Omega)^N \times L_0^2(\Omega)$ is the eigenvector for the adjoint variationally posed eigenproblem and satisfies

$$\int_\Omega \nabla \eta_0 : \nabla \mathbf{v} - \rho_0 \nabla \cdot \mathbf{v} - q \nabla \cdot \eta_0$$
$$+ \int_\Omega (\mathbf{u}^0 \cdot \nabla \mathbf{v} + \mathbf{v} \cdot \nabla \mathbf{u}^0) \cdot \eta_0 = 0, \quad \forall (\mathbf{v}, q) \in H_0^1(\Omega)^N \times L_0^2(\Omega). \tag{5.32}$$

Now we can apply the results from Section 4 to the Navier–Stokes equations.

Theorem 5.4. Assume that the Navier–Stokes equations have a simple fold point at $(\mathbf{u}^0, p^0, R^0)$ and that a mixed finite element method satisfying Hypotheses 5.1, 5.2 and 5.3 is used to discretize the equations. Assume further that the solution branch in the neighbourhood of $(\mathbf{u}^0, p^0, R^0)$ lies in $H^{m+1}(\Omega)^N \times (H^m(\Omega) \cap L_0^2(\Omega)) \times \mathbb{R}$, and is parametrized by α. Then, for h sufficiently small, we obtain the following properties.

(i) The solution branch is of class C^∞ with respect to α and the following error estimate holds:

$$\|\mathbf{u}_h(\alpha) - \mathbf{u}(\alpha)\|_{1,\Omega} + \|p_h - p\|_{0,\Omega} + |R_h(\alpha) - R(\alpha)|$$
$$\leq Ch^m\{\|\mathbf{u}(\alpha)\|_{m+1,\Omega} + \|p(\alpha)\|_{m,\Omega}\} \quad (5.33)$$

with similar estimates for the approximation of the derivatives.

(ii) If, in addition, the fold point is quadratic, then the discrete problem has a quadratic fold point, $(\mathbf{u}_h{}^0, p_h{}^0, R_h{}^0)$ say, and the following error estimate holds:

$$\|\mathbf{u}_h{}^0 - \mathbf{u}^0\|_{1,\Omega} + \|p_h{}^0 - p^0\|_{0,\Omega} + |R_h{}^0 - R^0|$$
$$\leq Ch^m \sum_{k=0}^{1}\{\|\mathbf{u}^{0(k)}\|_{m+1,\Omega} + \|p^{0(k)}\|_{m,\Omega}\}. \quad (5.34)$$

(iii) Furthermore,

$$|R_h{}^0 - R^0| \leq C_1 h^{2m}\sum_{k=0}^{1}\{\|\mathbf{u}^{0(k)}\|_{m+1,\Omega} + \|p^{0(k)}\|_{m,\Omega}\}$$
$$+ C_2 h^m\{\|\mathbf{u}^0\|_{m+1,\Omega} + \|p^0\|_{m,\Omega}\} \times$$
$$\left\{\inf_{\eta_h \in X_h}\|\eta_0 - \eta_h\|_{1,\Omega} + \inf_{\rho_h \in M_h}\|\rho_0 - \rho_h\|_{0,\Omega}\right\}, \quad (5.35)$$

where (η_0, ρ_0) is the eigenvector for the variationally posed adjoint eigenproblem (5.32).

Proof. **(i)** Define $(\hat{\mathbf{u}}_h^{(k)}(\alpha), \hat{p}_h^{(k)}(\alpha))$ by

$$(\hat{\mathbf{u}}_h^{(k)}(\alpha), \hat{p}_h^{(k)}(\alpha)) = -T_h\frac{d^k}{d\alpha^k}G(\mathbf{u}(\alpha), R(\alpha)). \quad (5.36)$$

Clearly, since T does not depend on α,

$$(\mathbf{u}^{(k)}(\alpha), p^{(k)}(\alpha)) = -T\frac{d^k}{d\alpha^k}G(\mathbf{u}(\alpha), R(\alpha)). \quad (5.37)$$

This gives

$$\|(T - T_h)\frac{d^k}{d\alpha^k}G(\mathbf{u}(\alpha), R(\alpha))\|$$
$$= \|\hat{\mathbf{u}}_h^{(k)}(\alpha) - \mathbf{u}^{(k)}(\alpha)\|_{1,\Omega} + \|\hat{p}_h^{(k)}(\alpha) - p^{(k)}(\alpha)\|_{0,\Omega}. \quad (5.38)$$

The result now follows directly from Theorems 4.2 and 5.3.

(ii) Essentially the same as for part (i).

(iii) This result follows from Theorem 4.2(iii). We need to estimate

$$\left|\int_\Omega (T - T_h)G^0 \cdot (\eta_0', \rho_0')\right| \quad (5.39)$$

and

$$\|[(T - T_h) \cdot DG^0]' \cdot (\eta_0', \rho_0')\|_{H^{-1}(\Omega)^N \times L_0^2(\Omega)}. \tag{5.40}$$

Let

$$(\mathbf{u}, p) = TG^0, \tag{5.41}$$

$$(\hat{\mathbf{u}}_h, \hat{p}_h) = T_h G^0. \tag{5.42}$$

Then

$$\left| \int_\Omega (T - T_h) G^0 \cdot (\eta_0', \rho_0') \right|$$

$$= \left| \int_\Omega \nabla(\mathbf{u} - \hat{\mathbf{u}}_h) : \nabla \eta_0 - (p - \hat{p}_h)\nabla \cdot \eta_0 - \rho_0 \nabla \cdot (\mathbf{u} - \hat{\mathbf{u}}_h) \right| \tag{5.43}$$

$$= \left| \int_\Omega \nabla(\mathbf{u} - \hat{\mathbf{u}}_h) : \nabla(\eta_0 - \eta_h) - (p - \hat{p}_h)\nabla \cdot (\eta_0 - \eta_h) \right.$$

$$\left. - (\rho_0 - \rho_h)\nabla \cdot (\mathbf{u} - \hat{\mathbf{u}}_h) \right| \tag{5.44}$$

$$\leq \{|\mathbf{u} - \hat{\mathbf{u}}_h|_{1,\Omega} + \|p - \hat{p}_h\|_{0,\Omega}\} \{|\eta_0 - \eta_h|_{1,\Omega} + \|\rho_0 - \rho_h\|_{0,\Omega}\}. \tag{5.45}$$

By definition,

$$\|[(T - T_h) \cdot DG^0]' \cdot (\eta_0', \rho_0')\|_{H^{-1}(\Omega)^N \times L_0^2(\Omega)} =$$

$$\sup_{\substack{\mathbf{v} \in H_0^1(\Omega)^N \\ \|\mathbf{v}\|_{H_0^1(\Omega)^N} = 1}} \left| \int_\Omega (T - T_h) DG^0 \mathbf{v} \cdot (\eta_0', \rho_0') \right|, \tag{5.46}$$

which can be estimated using a technique similar to that used in the first part of the proof. □

Numerical results illustrating the superconvergence at a quadratic fold point are given in the next section. We have gone into considerable detail in the fold point case, but this theory can also be applied to symmetry breaking bifurcation points and, provided the discrete equations possess a Z_2-symmetry, they will have a symmetry breaking bifurcation point close to the bifurcation point in the continuous problem (see Brezzi et al. (1981b) and Theorem 4.3). Both branches in the neighbourhood of the symmetry breaking bifurcation point can be approximated to the same order of accuracy as for the fold point, and the parameter value at the bifurcation point exhibits the same superconvergent behaviour as in the case of the fold point. Numerical results illustrating this are given in the next section.

6. Demonstration of superconvergence results

Three nontrivial flows are used to illustrate the superconvergence results developed in the previous section, at a Z_2-symmetry breaking bifurcation point, at a quadratic fold point and at a Hopf bifurcation point. In all three cases, the primitive variable formulation of the incompressible 2D Navier–Stokes equations was discretized and solved using the Galerkin finite element method. The isoperimetric quadrilateral $Q_2 - P_1$ elements introduced in Sani et al. (1981) were used in all computations. This element is widely recognized as being the most accurate element to use for two-dimensional calculation (see Section 3.13.2 of Gresho and Sani (1998)). In this case, for sufficiently smooth functions $\mathbf{u}$ and p one expects

$$\|\mathbf{u} - \Pi_{h,u}\mathbf{u}\| \leq Ch^2\|\mathbf{u}\|_{3,\Omega} \text{ and } \|p - \Pi_{h,p}p\|_{0,\Omega} \leq Ch^2\|p\|_{2,\Omega}$$

(see Hypotheses 5.1 and 5.2 in the previous section) and $O(h^4)$ superconvergence for eigenvalues and bifurcation parameters.

6.1. Flow in a symmetric smoothly expanding channel

The flow of a Newtonian fluid in a two-dimensional channel with a sudden symmetric expansion is a conceptually simple example in which symmetry breaking plays a key rôle in the hydrodynamic stability problem at moderate Reynolds number. As such it has attracted considerable attention, and examples of recent numerical work include Drikakis (1997), Alleborn, Nandakumar, Raszillier and Durst (1997) and Battaglia, Tavener, Kulkarni and Merkle (1997). For small flow rates the flow is symmetric about the midchannel, and two equally sized recirculating eddies exist downstream of the expansion. In the laboratory, above a critical Reynolds number the flow remains steady, but one of the eddies (the left-hand, say) is seen to be clearly larger than the other, and the flow is therefore asymmetric about the midplane. A second steady flow, in which the right-hand eddy is the larger of the two, can also be observed but cannot be obtained by a gradual increase in flow rate. Instead, it must be established by suddenly starting the flow in the apparatus. Using the extended system technique described in Werner and Spence (1984), Fearn, Mullin and Cliffe (1990) located the Z_2-symmetry breaking bifurcation point responsible for this phenomenon in a channel with a 1 to 3 symmetric expansion, convincingly supporting their laboratory experiments.

We examine the flow in a 2D channel with a smooth 1 to 3 expansion in order to avoid the complications associated with the two concave corners in the more usual flow domain. Our expansion is a cubic polynomial chosen so that the sides of the channel have continuous first derivatives. Figure 6 is a plot of the streamlines of the symmetric flow at a Reynolds number of 30, below the critical value at which the symmetry breaking bifurcation point exists. The Reynolds number is based on the inlet width and the mean flow rate.

Beyond the critical Reynolds number the symmetric solution is unstable to antisymmetric disturbances. Figure 7 shows a plot of the streamlines

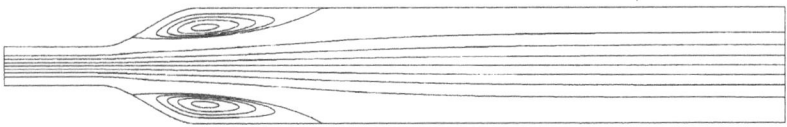

Fig. 6. Stable symmetric flow at $Re = 30$

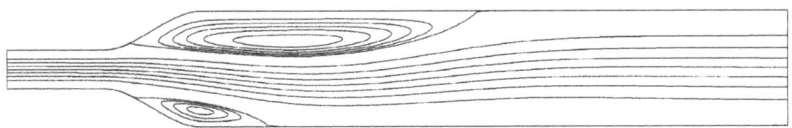

Fig. 7. Stable asymmetric flow at $Re = 40$

of the flow along one of the asymmetric branches at a Reynolds number of 40. The streamlines along the other asymmetric branch can be obtained by reflection about the midchannel.

A bifurcation diagram using the cross-stream velocity at a point on the centreline near the beginning of the expansion as a measure of the asymmetry of the flow, is given in Figure 8. It shows a pitchfork bifurcation with a critical Reynolds number near 33. Stable and unstable branches are indicated by 's' and 'u' respectively. The streamlines in Figures 6 and 7 correspond to points A and B in Figure 8, respectively.

Our best estimate of the critical Reynolds number $Re^\star_{sbbp}$ is 33.557505814927, for reasons that will be discussed later. We computed the symmetric flow at $Re^\star_{sbbp}$ on five geometrically similar meshes and used inverse iteration to determine the eigenvalue of the Jacobian matrix restricted to the antisymmetric subspace that lies closest to the origin. The first column in Table 1 indicates the number of quadrilateral elements on half of the flow domain on which the computations were performed. The eigenvalue nearest the origin converges towards zero as the mesh is refined, making it easy to determine the convergence rate by evaluating the ratio of the smallest eigenvalues on meshes with N and $4N$ elements. This ratio is clearly approaching 16 and the convergence rate is therefore approaching h^4. This is consistent with the last result in Theorem 4.5, namely that $|\lambda_h^0 - \lambda_0|$ is

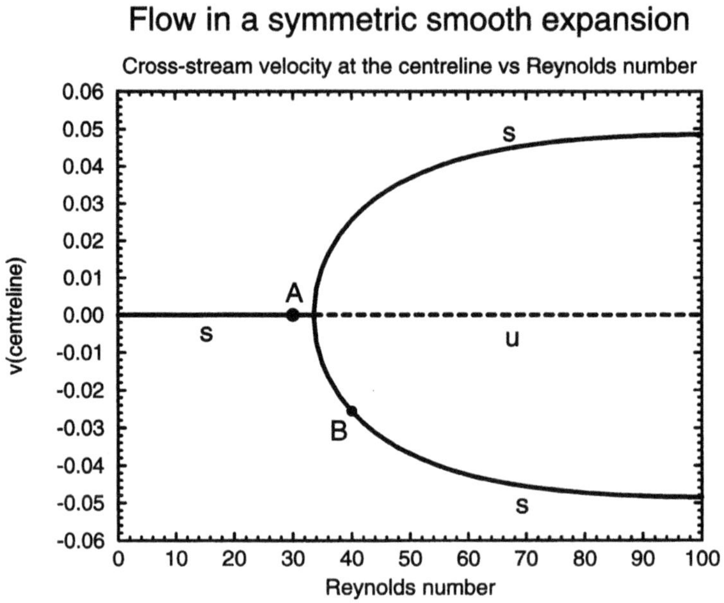

Flow in a symmetric smooth expansion

Cross-stream velocity at the centreline vs Reynolds number

Fig. 8. Bifurcation diagram for the symmetric smooth expansion

proportional to the eigenvalue of smallest modulus, which, as indicated in the remark after Theorem 4.5, should tend to zero with rate $O(h^4)$.

The critical Reynolds numbers at the symmetry breaking bifurcation points computed on six geometrically similar meshes are listed in Table 2. The first column again indicates the number of quadrilateral elements on half of the flow domain on which the computations were performed. The third and fourth columns give the change in critical Reynolds number from one mesh to the next finest and the ratio of these increments respectively.

Table 1. *Convergence of the eigenvalue closest to the origin at the symmetry breaking bifurcation point with mesh refinement*

No. of elements	Smallest eigenvalue	Ratio
32	−2.3426E-07	
128	−2.9898E-04	anomalous
512	−3.1113E-05	9.6
2048	−2.2912E-06	13.6
8192	−1.5600E-07	14.7
32768	−1.0177E-08	15.3

The convergence rate of the critical Reynolds number based on the final two mesh halvings is $h^{4.00}$, where h is a measure of the mesh size, which is the superconvergent rate expected from the result of Theorem 4.3. Extrapolating the values computed on the finest two meshes assuming an h^4 convergence rate gives $Re^\star_{sbbp} = 33.557505814927$ as our best estimate of the converged value of the critical Reynolds number.

6.2. Flow in a non-symmetric smoothly expanding channel

A closely related asymmetric problem was constructed by stretching the domain below the 'centreline' of the symmetric channel in the previous example by 10 per cent. The pitchfork bifurcation in the symmetric problem is thereby disconnected. One of the asymmetric branches, the 'primary' branch, is continuously connected to the small Reynolds number flow, while the other asymmetric flow occurs as a disconnected 'secondary' flow. The lower limit of stability of the secondary flow occurs at a quadratic limit point.

The cross-stream velocity at the same point along the centreline that was used to construct Figure 8 was used to produce the corresponding bifurcation diagram for the asymmetric smooth expansion which is given in Figure 9. The pitchfork has been disconnected and there is a fold at a Reynolds number near 39.

The critical Reynolds numbers at the limit points computed on six geometrically similar meshes are listed in Table 3. The first column indicates the number of quadrilateral elements on the entire flow domain. This computation cannot be performed on half of the domain since the flow at the limit point is asymmetric. The third and fourth columns give the change in critical Reynolds number from one mesh to the next finest and the ratio of these increments respectively.

Table 2. *Convergence of the symmetry breaking bifurcation points with mesh refinement*

N	$Re_{sbbp}(N)$	$Re_{sbbp}(N) - Re_{sbbp}(4N)$	$\frac{Re_{sbbp}(N/4) - Re_{sbbp}(N)}{Re_{sbbp}(N) - Re_{sbbp}(4N)}$
32	33.55741732	0.06344998	
128	33.49396734	−0.05874498	−1.08
512	33.55271231	−0.00449556	13.07
2048	33.55720788	−0.00027932	16.09
8192	33.55748719	−0.00001746	16.00
32768	33.55750465		

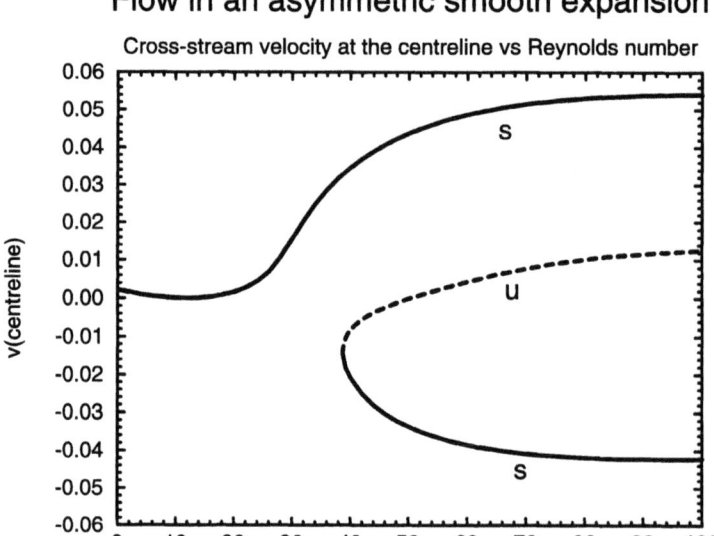

Fig. 9. Bifurcation diagram for the asymmetric smooth expansion

The convergence rate of the critical Reynolds number based on the final two mesh halvings is $h^{4.09}$, where h is a measure of the mesh size. This value is very close to the super-convergent rate predicted in Theorems 4.1, 4.2 and 5.4. Extrapolating the values computed on the finest two meshes assuming an h^4 convergence rate gives $Re^{\star}_{\text{fold}} = 38.740407816531$ as our best estimate of the converged value of the critical Reynolds number.

Table 3. *Convergence of the quadratic fold points with mesh refinement*

N	$Re_{\text{fold}}(N)$	$Re_{\text{fold}}(N) - Re_{\text{fold}}(4N)$	$\frac{Re_{\text{fold}}(N/4) - Re_{\text{fold}}(N)}{Re_{\text{fold}}(N) - Re_{\text{fold}}(4N)}$
64	39.38265205	0.69629546	
256	38.68635659	−0.04692872	−14.84
1024	38.73328531	−0.00663299	7.07
4096	38.73991830	−0.00046072	14.40
16384	38.74037902	−0.00002699	17.07
65536	38.74040674		

6.3. Flow past a cylinder in a channel

The flow past a bluff body and the appearance of a Karman vortex street above a large enough Reynolds number is a classical problem in hydro-dynamic stability and has been studied using a large number of different experimental, analytical and numerical approaches. As just one example, Jackson (1987) considered the 2D flow past a cylinder in an unbounded do-main and computed the location of the Hopf bifurcation point where the low Reynolds number steady symmetric flow loses stability to a time-dependent flow. We choose to locate the cylinder along the centreline of a 2D channel so that the lateral boundary conditions are well determined. The symmet-ric flow that exists for low flow rates loses stability at a Hopf bifurcation point whose critical Reynolds number and angular frequency depend upon the blockage ratio, that is, the ratio of the cylinder diameter to channel width. (See Chen, Pritchard and Tavener (1995) for a full description of the problem.) For a blockage ratio of 0.3, the critical Reynolds number and angular frequency at the Hopf bifurcation computed on three geometrically similar meshes are listed in Table 4. Due to the complicated nature of both the neutrally stable, steady symmetric flow and the null eigenvector, a large number of elements are required before the asymptotic convergence rate is observed.

We assume that, for each of the three computations reported in Table 4.

$$(Re_{\text{Hopf}})_i - (Re_{\text{Hopf}})^\star = Ah_i^p \quad i = 1, 2, 3,$$
$$(\omega_{\text{Hopf}})_i - (\omega_{\text{Hopf}})^\star = Bh_i^q, \quad i = 1, 2, 3,$$

where h_i is a measure of the discretization. Solving for the unknowns $(Re_{\text{Hopf}})^\star$, $(\omega_{\text{Hopf}})^\star$, A, B, p and q, we find $p = 3.89$ and $q = 4.53$, indic-ating the expected superconvergence in both these quantities, as predicted by Lemma 4.1

Table 4. *Convergence of the Hopf bifurcation points with mesh refinement*

No. of elements	Re_{Hopf}	ω_{Hopf}
10368	10.47482559098	3.443201635738
15552	10.47913658919	3.445954684625
20736	10.47989122928	3.446334713924

7. Numerical implementation

In this section we discuss numerical techniques for the computation of paths of regular solutions, singular points and the paths of singular points. We shall describe the techniques used to produce the numerical solutions of the Taylor–Couette problem described in Section 8. We shall also describe briefly in Section 7.5 the alternative approach of *minimal extended systems* using bordered matrices.

Consider the nonlinear system

$$\boldsymbol{F}(\boldsymbol{x}, \lambda) = \boldsymbol{0} \qquad \boldsymbol{x} \in \mathbb{R}^N, \ \lambda \in \mathbb{R}, \tag{7.1}$$

where we assume $\boldsymbol{F}$ has been obtained by the discretization of a nonlinear PDE. (To avoid cumbersome notation we change from the form $F_h(x_h, \lambda) = 0$ used in Section 4, which is more suited to convergence analysis.)

7.1. Computation of solution paths

Here we give a brief account of the main ideas in a simple continuation algorithm. This is based on Keller's pseudo-arclength method (Keller 1977, 1987). There are many continuation methods and several packages are available free. We refer the reader to the review article on continuation and path following by Allgower and Georg (1993) which contains an extensive listing of the software available in 1992.

As in Section 3 we denote the solution set of (7.1) by

$$S := \{(\boldsymbol{x}, \lambda) \in \mathbb{R}^{N+1} : \boldsymbol{F}(\boldsymbol{x}, \lambda) = \boldsymbol{0}\}. \tag{7.2}$$

Often in applications one is interested in computing the whole set S or a continuous portion of it, but in practice S is computed by finding a discrete set of points on S and then using some graphics package to interpolate. So the basic numerical question to consider is: Given a point $(\boldsymbol{x}_0, \lambda_0) \in S$ how would we compute a nearby point on S? Throughout we use the notation $\boldsymbol{F}^0 = \boldsymbol{F}(\boldsymbol{x}_0, \lambda_0)$, $\boldsymbol{F}_{\boldsymbol{x}}^0 = \boldsymbol{F}_{\boldsymbol{x}}(\boldsymbol{x}_0, \lambda_0)$, etc.

If $\boldsymbol{F}_{\boldsymbol{x}}^0$ is nonsingular, then a simple strategy for computing a point of S near $(\boldsymbol{x}_0, \lambda_0)$ is to choose a steplength $\Delta\lambda$, set $\lambda_1 = \lambda_0 + \Delta\lambda$ and solve $\boldsymbol{F}(\boldsymbol{x}, \lambda_1) = \boldsymbol{0}$ by Newton's method with starting guess $\boldsymbol{x}_0$. We know from the IFT that this will work if $\Delta\lambda$ is sufficiently small. However, this method will fail as a fold point is approached, and for this reason the pseudo-arclength method was introduced in Keller (1977).

In this section we shall assume that there is an arc of S such that at all points in the arc

$$\mathrm{Rank}[\boldsymbol{F}_{\boldsymbol{x}}|\boldsymbol{F}_\lambda] = N, \tag{7.3}$$

and so any point in the arc is either a regular point or fold point of S. The IFT implies that the arc is a smooth curve in $\mathbb{R}^{N+1}$, and so there is a unique tangent direction at each point of the arc. Let t denote any parameter used to describe the arc, that is, $(\boldsymbol{x}(t), \lambda(t)) \in S$, with $(\boldsymbol{x}_0, \lambda_0) = (\boldsymbol{x}(t_0), \lambda(t_0)) \in S$, and denote the unit tangent at $(\boldsymbol{x}_0, \lambda_0)$ by $\boldsymbol{\tau}_0 = \left(\frac{d\boldsymbol{x}}{dt}(0), \frac{d\lambda}{dt}(0)\right)$.

Since $\boldsymbol{F}(\boldsymbol{x}(t), \lambda(t)) = \boldsymbol{0}$, differentiating with respect to t gives

$$\boldsymbol{F}_{\boldsymbol{x}}(\boldsymbol{x}(t), \lambda(t))\frac{d\boldsymbol{x}}{dt}(t) + \boldsymbol{F}_{\lambda}(\boldsymbol{x}(t), \lambda(t))\frac{d\lambda}{dt}(t) = \boldsymbol{0},$$

$$\boldsymbol{\tau}_0 \in \mathrm{Ker}[\boldsymbol{F}_{\boldsymbol{x}}^0 | \boldsymbol{F}_{\lambda}^0]. \tag{7.4}$$

Suppose now that $\boldsymbol{\tau}_0 = [\boldsymbol{c}^T, d]^T$. We use this vector to devise an extended system which can be solved by Newton's method without fail for a point $(\boldsymbol{x}_1, \lambda_1)$ on S near $(\boldsymbol{x}_0, \lambda_0)$ for $\lambda_1 - \lambda_0$ small enough. The appropriate extended system is

$$\boldsymbol{H}(\boldsymbol{y}, t) = \boldsymbol{0} \tag{7.5}$$

where $\boldsymbol{y} = (\boldsymbol{x}, \lambda) \in \mathbb{R}^{N+1}$ and $\boldsymbol{H} : \mathbb{R}^{N+2} \to \mathbb{R}^{N+1}$ is given by

$$\boldsymbol{H}(\boldsymbol{y}, t) = \left[\begin{array}{c} \boldsymbol{F}(\boldsymbol{x}, \lambda) \\ \boldsymbol{c}^T(\boldsymbol{x} - \boldsymbol{x}_0) + d(\lambda - \lambda_0) - (t - t_0) \end{array}\right]. \tag{7.6}$$

The last equation in system (7.6) is the equation of the plane perpendicular to $\boldsymbol{\tau}_0$ a distance $\Delta t = (t - t_0)$ from t_0 (see Figure 10). So in (7.6) we in fact implement a specific parametrization local to $(\boldsymbol{x}_0, \lambda_0)$, namely parametrization by the length of the projection of $(\boldsymbol{x}, \lambda)$ onto the tangent direction at $(\boldsymbol{x}_0, \lambda_0)$.

With $\boldsymbol{y}_0 = (\boldsymbol{x}_0, \lambda_0)$, we have $\boldsymbol{H}(\boldsymbol{y}_0, t_0) = \boldsymbol{0}$ and

$$\boldsymbol{H}_{\boldsymbol{y}}(\boldsymbol{y}_0, t_0) = \left[\begin{array}{cc} \boldsymbol{F}_{\boldsymbol{x}}^0 & \boldsymbol{F}_{\lambda}^0 \\ \boldsymbol{c}^T & d \end{array}\right].$$

Since $(\boldsymbol{c}^T, d)^T$ is orthogonal to each of the rows of $[\boldsymbol{F}_{\boldsymbol{x}}^0, \boldsymbol{F}_{\lambda}^0]$, the matrix $\boldsymbol{H}_{\boldsymbol{y}}(\boldsymbol{y}_0, t_0)$ is nonsingular and so by the IFT there exist solutions of (7.6) satisfying $\boldsymbol{y} = (\boldsymbol{x}^T, \lambda)^T = (\boldsymbol{x}(t)^T, \lambda(t))^T$ for t near t_0. For $t_1 = t_0 + \Delta t$ and Δt sufficiently small we know that $\boldsymbol{F}(\boldsymbol{y}, t_1) = \boldsymbol{0}$ has a unique solution $\boldsymbol{y} = \boldsymbol{y}(t_1) = (\boldsymbol{x}_1, \lambda_1)$ and $\boldsymbol{H}_{\boldsymbol{y}}(\boldsymbol{y}(t_1), t_1)$ is nonsingular. Thus Newton's method will converge for sufficiently small Δt. If we take as starting guess $\boldsymbol{y}_0 = \boldsymbol{y}_0 = (\boldsymbol{x}_0^T, \lambda_0)^T$, it is a straightforward exercise to show that (i) the first Newton iterate is $(\boldsymbol{x}_0, \lambda_0) + \Delta t(\boldsymbol{c}, d)$, that is, the first iterate 'steps out' along the tangent, as one might expect, and (ii) all the Newton iterates lie in the plane shown in Figure 10. Since length along the tangent at $(\boldsymbol{x}_0, \lambda_0)$ is used as parameter this technique is called *pseudo-arclength continuation* (Keller 1977). In fact the arclength normalization given by the second equation in (7.6) is usually altered to

$$\theta\boldsymbol{c}^T(\boldsymbol{x} - \boldsymbol{x}_0) + (1 - \theta)d(\lambda - \lambda_0) - (t - t_0) = 0 \tag{7.7}$$

where θ is a weighting parameter, chosen to give more importance to λ

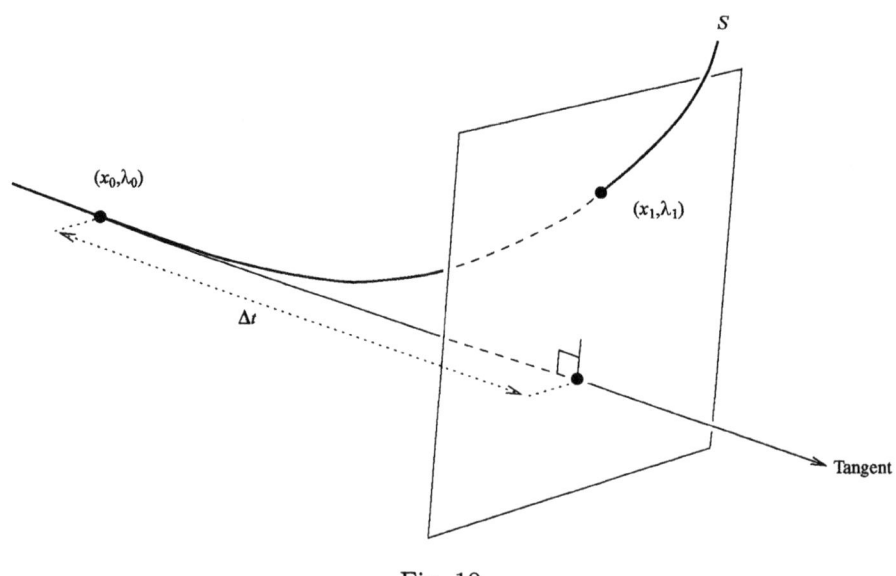

Fig. 10.

when N is large. As a consequence the resulting continuation algorithm moves round fold points more quickly. A different modification, but with similar aims is discussed in Mittelmann (1986) where the scalar $r = \|x\|$ is used instead of x in (7.7).

For a discussion of the details of an efficient continuation algorithm, including the solution of the bordered matrices that arise and step control we refer the reader to Allgower and Georg (1993) and the references cited therein. An introductory account is given in Spence and Graham (1999). For our purposes it is sufficient to note that an efficient solution procedure is needed for $(N+1) \times (N+1)$ systems with 'bordered' coefficient matrices of the form

$$H_y(y, t) := \begin{bmatrix} F_x & F_\lambda \\ c^T & d \end{bmatrix}. \tag{7.8}$$

A common approach is to use a block Gaussian elimination algorithm (see Keller (1977)). This works well if F_x is well conditioned, but near a fold point it may fail to produce reliable results, as was discussed by Moore (1987) and Govaerts (1991). An iterative refinement approach (see Govaerts (1991)) works well in such cases.

7.2. Fold points: detection and computation

The theory for fold points is given in Section 3.1. If the fold is quadratic then Theorem 3.1 has several important numerical implications. Assume S is defined by (7.2) and that (7.3) holds. At an algebraically simple quadratic fold point, an eigenvalue of the Jacobian F_x passes through zero, and so

$\det(\boldsymbol{F_x})$ changes sign. Also, in the notation of Theorem 3.1, λ_t changes sign. So a quadratic fold point can be detected by monitoring $\det(\boldsymbol{F_x})$ or λ_t along the solution path $(x(t), \lambda(t))$. In fact, other tests are possible and this aspect is discussed in Seydel (1994), Section 5.2.

Once located roughly a quadratic fold point may be computed in a number of ways. We shall briefly describe its direct calculation using the extended system

$$\boldsymbol{T}(\boldsymbol{y}) := \begin{pmatrix} \boldsymbol{F}(\boldsymbol{x}, \lambda) \\ \boldsymbol{F_x}(\boldsymbol{x}, \lambda)\boldsymbol{\phi} \\ \boldsymbol{l}^T\boldsymbol{\phi} - 1 \end{pmatrix}, \quad \boldsymbol{y} = \begin{pmatrix} \boldsymbol{x} \\ \boldsymbol{\phi} \\ \lambda \end{pmatrix} \in \mathbb{R}^{2N+1} \tag{7.9}$$

with $\boldsymbol{l} \in \mathbb{R}^N \setminus \{\boldsymbol{0}\}$, since Theorem 3.1 shows that $(\boldsymbol{x}_0, \boldsymbol{\phi}_0, \lambda_0)$ is a regular point of (7.9) provided $\boldsymbol{l}^T\boldsymbol{\phi}_0 \neq 0$.

Let $\tilde{\boldsymbol{\phi}}$ be the approximation to $\boldsymbol{\phi}$ with $\boldsymbol{l}^T\tilde{\boldsymbol{\phi}} = 1$. Reordering the unknowns, the Newton system for the correction $(\Delta\boldsymbol{x}, \Delta\lambda, \Delta\boldsymbol{\phi}) \in \mathbb{R}^{N+1}$ is

$$\begin{pmatrix} \boldsymbol{F_x} & \boldsymbol{F_\lambda} & \boldsymbol{O} \\ \boldsymbol{F_{xx}}\tilde{\boldsymbol{\phi}} & \boldsymbol{F_{x\lambda}}\tilde{\boldsymbol{\phi}} & \boldsymbol{F_x} \\ \boldsymbol{0}^T & 0 & \boldsymbol{l}^T \end{pmatrix} \begin{pmatrix} \Delta\boldsymbol{x} \\ \Delta\lambda \\ \Delta\boldsymbol{\phi} \end{pmatrix} = \begin{pmatrix} \boldsymbol{r}_1 \\ \boldsymbol{r}_2 \\ 0 \end{pmatrix}, \tag{7.10}$$

where $\boldsymbol{O}$ is the $N \times N$ zero matrix, and $\boldsymbol{r}_1, \boldsymbol{r}_2 \in \mathbb{R}^N$. It is important to realize that one need not solve this system directly. An efficient approach based on solving four N-dimensional systems with the same coefficient matrix was given in Moore and Spence (1980). Here we outline a slightly different algorithm.

Introduce the new unknowns $\Delta\hat{\boldsymbol{x}} \in \mathbb{R}^N$, $\xi \in \mathbb{R}$ defined by

$$\Delta\hat{\boldsymbol{x}} := \Delta\boldsymbol{x} - \xi\tilde{\boldsymbol{\phi}}, \quad \xi := \boldsymbol{l}^T\Delta\boldsymbol{x} \tag{7.11}$$

and so the unknown $\Delta\hat{\boldsymbol{x}}$ satisfies

$$\boldsymbol{l}^T\Delta\hat{\boldsymbol{x}} = 0 \tag{7.12}$$

since $\boldsymbol{l}^T\tilde{\boldsymbol{\phi}} = 1$. Thus (7.10) becomes

$$\begin{pmatrix} \boldsymbol{F_x} & \boldsymbol{F_\lambda} & \boldsymbol{O} & \boldsymbol{F_x}\tilde{\boldsymbol{\phi}} \\ \boldsymbol{l}^T & 0 & \boldsymbol{0}^T & 0 \\ \boldsymbol{F_{xx}}\tilde{\boldsymbol{\phi}} & \boldsymbol{F_{x\lambda}}\tilde{\boldsymbol{\phi}} & \boldsymbol{F_x} & \boldsymbol{F_{xx}}\tilde{\boldsymbol{\phi}}\tilde{\boldsymbol{\phi}} \\ \boldsymbol{0}^T & 0 & \boldsymbol{l}^T & 0 \end{pmatrix} \begin{pmatrix} \Delta\hat{\boldsymbol{x}} \\ \Delta\lambda \\ \Delta\boldsymbol{\phi} \\ \xi \end{pmatrix} = \begin{pmatrix} \boldsymbol{r}_1 \\ 0 \\ \boldsymbol{r}_2 \\ 0 \end{pmatrix}. \tag{7.13}$$

Note that, as $\tilde{\boldsymbol{\phi}}$ tends to the true null vector, the (1,4)-element tends to the zero vector, and so the matrix becomes block lower triangular. One could base a quasi-Newton method for solving (7.10) on this observation. However, we take a different approach here. The matrix in (7.13) has the (nonstandard) block LU factorization

$$\left(\begin{array}{cc} L_{11} & 0 \\ L_{21} & I \end{array}\right)\left(\begin{array}{cc} I & U_{12} \\ 0 & U_{22} \end{array}\right), \tag{7.14}$$

where L_{11} and U_{22} are $(N+1) \times (N+1)$ matrices given by

$$L_{11} = \left(\begin{array}{cc} \boldsymbol{F_x} & \boldsymbol{F_\lambda} \\ \boldsymbol{l}^T & 0 \end{array}\right), \qquad U_{22} = \left(\begin{array}{cc} \boldsymbol{F_x} & \boldsymbol{F_{xx}}\tilde{\phi}\tilde{\phi} + \boldsymbol{\delta} \\ \boldsymbol{l}^T & 0 \end{array}\right)$$

with $\boldsymbol{\delta} \in \mathbb{R}^N$ satisfying $\|\boldsymbol{\delta}\| \leq C\|\boldsymbol{F_x}\phi\|$, for some constant C. Hence $\boldsymbol{\delta}$ tends to zero as the fold point is approached. Note that $\boldsymbol{l}^T\phi \neq 0$, and at a quadratic fold point $\boldsymbol{F}_\lambda^0$, $\boldsymbol{F}_{xx}^0\phi^0\phi^0 \notin \mathrm{Range}(\boldsymbol{F}_x^0)$. Thus Lemma 3.1 shows that L_{11} and U_{22} are nonsingular at a quadratic fold point. Forward and back substitution shows that the solution of (7.13) requires the solution of 2 systems with coefficient matrix L_{11}, and 2 systems with coefficient matrix U_{22}. However, U_{22} differs from L_{11} only in the last column, and simple manipulation converts a system with coefficient matrix U_{22} into a system with coefficient matrix L_{11}. One is left with a 2×2 system to solve for $\Delta\lambda$ and ξ, from which $\Delta\boldsymbol{x}$ and $\Delta\phi$ are recovered. Overall the main work is the solution of 4 systems with coefficient matrix L_{11}. This is comparable to the main work in the 'minimal system' approach in Section 7.5.

In a two-parameter problem $\boldsymbol{F}(\boldsymbol{x}, \lambda, \alpha) = \boldsymbol{0}$ we may wish to compute a path of quadratic fold points. To do this using the pseudo-arclength method in Section 7.1 we require the solution of systems with the $(2N+2) \times (2N+2)$ coefficient matrix

$$\left(\begin{array}{cccc} \boldsymbol{F_x} & \boldsymbol{F_\lambda} & \boldsymbol{O} & \boldsymbol{F_\alpha} \\ \boldsymbol{F_{xx}}\phi & \boldsymbol{F_{x\lambda}}\phi & \boldsymbol{F_x} & \boldsymbol{F_{x\alpha}}\phi \\ \boldsymbol{0}^T & 0 & \boldsymbol{l}^T & 0 \\ \boldsymbol{c}_1^T & c_2 & \boldsymbol{c}_3^T & d \end{array}\right).$$

To do this, a block LU factorization is employed similar to that for the quadratic fold point and an efficient algorithm requiring the solution of 6 linear systems all with coefficient matrix L_{11} can be derived. (The details are left to the reader.)

The concept that the solution of the $(2N+1) \times (2N+1)$ linear Jacobian system is accomplished using solutions of $(N+1) \times (N+1)$ systems has been extended to many other situations: see, for example, the Hopf bifurcation algorithms of Griewank and Reddien (1983) and Jepson (1981).

7.3. Z_2-symmetry breaking bifurcations: detection and computation

Consider the Z_2-symmetry breaking case. In the finite-dimensional setting S is an $N \times N$ matrix with

$$S\boldsymbol{F}(\boldsymbol{x}, \lambda) = \boldsymbol{F}(S\boldsymbol{x}, \lambda), \quad S^2 = I, \ \boldsymbol{x} \in \mathbb{R}^N,$$

and we may introduce symmetric and antisymmetric subspaces

$$\mathbb{R}^N = \mathbb{R}_s^N \oplus \mathbb{R}_a^N.$$

The theory is given in Section 3.4, where we see that, provided the non-degeneracy condition $B_0 \neq 0$ holds (see Theorem 3.4), then a simple eigenvalue of the antisymmetric Jacobian $\boldsymbol{F_x}(\boldsymbol{x}^s(\lambda), \lambda)|_{\mathbb{R}_a^N}$ changes sign. Hence $\det(\boldsymbol{F_x}(\boldsymbol{x}^s(\lambda), \lambda)|_{\mathbb{R}_a^N})$ changes sign, but $\boldsymbol{F_x}(\boldsymbol{x}^s(\lambda), \lambda)|_{\mathbb{R}_s^N}$ has no singularity. In applications it is often the case that the Z_2-symmetry is used to reduce computational cost, so that to compute the path $(\boldsymbol{x}^s(\lambda), \lambda) \in \mathbb{R}_s^N \times \mathbb{R}$ one forms only the symmetric Jacobian $\boldsymbol{F_x}(\boldsymbol{x}^s(\lambda), \lambda)|_{\mathbb{R}_s^N}$. It is important to note that in order to detect the symmetry breaking bifurcation one also needs to form $\boldsymbol{F_x}(\boldsymbol{x}^s(\lambda), \lambda)|_{\mathbb{R}_a^N}$, and then monitor the determinant or smallest eigenvalue.

The computation of the Z_2-symmetry breaking bifurcation using the extended system (3.39) is considerably easier than for a quadratic fold. If $(\boldsymbol{x}_0, \boldsymbol{\phi}_0, \lambda_0) \in \mathbb{R}_s^N \times \mathbb{R}_a^N \times \mathbb{R}$ is an estimate of the bifurcation point, then one step of Newton's method applied to (3.39) produces a coefficient matrix

$$\begin{pmatrix} \boldsymbol{F}_x^s & \boldsymbol{O} & \boldsymbol{F}_\lambda^s \\ \boldsymbol{F}_{xx}^a \boldsymbol{\phi}_0 & \boldsymbol{F}_x^a & \boldsymbol{F}_{x\lambda}^a \boldsymbol{\phi}_0 \\ \boldsymbol{0}^T & \boldsymbol{l}^T & 0 \end{pmatrix}$$

where $\boldsymbol{F}_x^s$ denotes $\boldsymbol{F_x}(\boldsymbol{x}_0, \lambda_0)|_{\mathbb{R}_s^N}$, etc., and $\boldsymbol{z}_x^s \in \mathbb{R}_s^N$, $\boldsymbol{z}_\phi^a \in \mathbb{R}_a^N$. Now, since $\boldsymbol{F}_x^s$ is nonsingular, this matrix has the block LU factorization

$$\begin{pmatrix} \boldsymbol{F}_x^s & \boldsymbol{O} & \boldsymbol{0} \\ \boldsymbol{F}_{xx}^a \boldsymbol{\phi}_0 & \boldsymbol{I} & \boldsymbol{0} \\ \boldsymbol{0}^T & \boldsymbol{0}^T & 1 \end{pmatrix} \begin{pmatrix} \boldsymbol{I} & \boldsymbol{O} & -\boldsymbol{w}_s \\ \boldsymbol{O} & \boldsymbol{F}_x^a & \boldsymbol{F}_{xx}^a \boldsymbol{\phi}_0 \boldsymbol{w}_s + \boldsymbol{F}_{x\lambda}^0 \boldsymbol{\phi}_0 \\ \boldsymbol{0}^T & \boldsymbol{l}^T & 0 \end{pmatrix}$$

where $\boldsymbol{F}_x^s \boldsymbol{w}_s + \boldsymbol{F}_\lambda^s = \boldsymbol{0}$. At the bifurcation point the matrix

$$\begin{bmatrix} \boldsymbol{F}_x^a & \boldsymbol{F}_{xx}^a \boldsymbol{\phi}_0 \boldsymbol{w}^s + \boldsymbol{F}_{x\lambda}^a \boldsymbol{\phi}_0 \\ \boldsymbol{l}^T & 0 \end{bmatrix} \tag{7.15}$$

is nonsingular. (This is proved using Lemma 3.1, noting that $B_0 \neq 0$ (see Theorem 3.4) and $\boldsymbol{l}^T \boldsymbol{\phi}_0 \neq 0$.) Standard forward and back substitution produces the solution of the Newton system. The main work involves the solution of two systems with coefficient matrix $\boldsymbol{F}_x^s$ and two systems with coefficient matrix (7.15).

The computation of *paths* of Z_2-symmetry breaking bifurcation points in a two-parameter problem $\boldsymbol{F}(\boldsymbol{x}, \lambda, \alpha) = \boldsymbol{0}$ is accomplished in a similar manner. The main work requires the solution of *three* systems with coefficient matrix $\boldsymbol{F}_x^s$ and *two* systems with coefficient matrix (7.15).

7.4. Hopf bifurcation: detection and computation

For the finite-dimensional problem (obtained perhaps after discretization of an ODE/PDE) $\dot{x} + F(x, \lambda) = 0$, a Hopf bifurcation point may arise on a branch of steady solutions when the Jacobian matrix has a pair of pure imaginary eigenvalues (Theorem 3.5). This fact was used by Jepson (1981) and Griewank and Reddien (1983) to compute Hopf bifurcations using extended systems of the form

$$H(y) = 0, \qquad (7.16)$$

where

$$H(y) := \begin{pmatrix} F(x, \lambda) \\ F_x(x, \lambda)\phi - \omega\psi \\ c^T\phi - 1 \\ F_x(x, \lambda)\psi + \omega\phi \\ c^T\psi \end{pmatrix}, \quad y := \begin{pmatrix} x \\ \phi \\ \lambda \\ \psi \\ \beta \end{pmatrix} \in \mathbb{R}^{3N+2} \qquad (7.17)$$

with $H : \mathbb{R}^{3N+2} \to \mathbb{R}^{3N+2}$. This is the real form of the extended system (3.47) with $V = \mathbb{R}^N$. Note that there are two conditions on the eigenvector $\phi + i\psi$ since a complex vector requires two real normalizations. Theorem 3.6 shows that under certain assumptions $y_0 = (x_0{}^T, \phi_0{}^T, \lambda_0, \psi_0{}^T, \omega_0) \in \mathbb{R}^{3N+2}$ is a regular solution of (7.16).

Note that fold points also satisfy (7.16) since if (x_0, λ_0) is a fold point and $\phi_0 \in \mathrm{Ker}(F_x(x_0, \lambda_0))$ then $y_0 = (x_0, \phi_0, \lambda_0, 0, 0)$ satisfies $H(y_0) = 0$. In fact y_0 is a regular solution if the conditions of Theorem 3.1 hold.

Jepson (1981) and Griewank and Reddien (1983) showed that the linearization of (7.16) could be reduced to solving systems with a bordered form of $F_x^2(x, \lambda) + \beta^2 I$. This is natural since an alternative system for a Hopf bifurcation can be derived by using the fact that the second and fourth equations of (7.16) can be written as $(F_x(x, \lambda) + \beta^2 I)v = 0$ with $v = \phi$ or ψ.

To eliminate the possibility of computing a fold point rather than a Hopf bifurcation point, Werner and Janovsky (1991) used the system

$$R(y) = 0, \qquad (7.18)$$

where

$$R(y) = \begin{pmatrix} F(x, \lambda) \\ (F_x^2(x, \lambda) + \nu I)\phi \\ c^T\phi \\ c^T F_x(x, \lambda)\phi - 1 \end{pmatrix}, \quad y = \begin{pmatrix} x \\ \phi \\ \lambda \\ \nu \end{pmatrix} \in \mathbb{R}^{2N+2} \qquad (7.19)$$

where $R : \mathbb{R}^{2N+2} \to \mathbb{R}^{2N+2}$, and where c is a constant vector. The last equation in (7.18) ensures that the solution cannot be a fold point. The system $R(y) = 0$ is closely related to a system derived by Roose and Hlavacek (1985), but (7.18) has several advantages when computing paths of Hopf bifurcations if a second parameter is varying (see Werner and Janovsky (1991)).

The extended systems in (7.16) and (7.18) can only be used when we know we are near a Hopf point, and obtaining good starting values is difficult in large systems. In fact the reliable and efficient detection of Hopf bifurcations in large systems arising from discretized PDEs remains an important and challenging problem.

When computing a path of steady solutions of $\dot{x} + F(x, \lambda) = 0$, using a numerical continuation method it is easy to pass over a Hopf bifurcation point without 'noticing' it, since when a complex pair of eigenvalues crosses the imaginary axis there is no easy detection test based on the linear algebra of the continuation method. In particular, the sign of the determinant of F_x does not change. If N is small then the simplest test is merely to compute all the eigenvalues of F_x during the continuation. For large N, say when F arises from a discretized PDE, such an approach will usually be out of the question. The review article Garratt, Moore and Spence (1991) discusses in detail both classical techniques from complex analysis and linear algebra-based methods. It is natural to try to use classical ideas from complex analysis for this problem since one then seeks an *integer*, namely the number of eigenvalues in the unstable half-plane, and counting algorithms are applicable. This is explored for large systems in Govaerts and Spence (1996) but there is still work to be done in this area. A certain bialternate product (see Guckenheimer, Myers and Sturmfels (1997) and Govaerts (2000)) of $F_x(x, \lambda)$ is singular at a Hopf bifurcation. This is a very nice theoretical property, but the bialternate product is of dimension $N(N - 1)/2$, and this is likely to limit its usefulness significantly when N is large.

The *leftmost* eigenvalues of $F_x(x, \lambda)$ determine the (linearized) stability of the steady solutions of $\dot{x} + F(x, \lambda) = 0$ and one strategy for the detection of Hopf bifurcation points is to monitor a few of the leftmost eigenvalues as the path of steady state solutions is computed. (Note that the leftmost eigenvalue is not a continuous function of λ: see Neubert (1993).) Standard iterative methods, for instance Arnoldi's method and simultaneous iteration, compute extremal or dominant eigenvalues, and there is no guarantee that the leftmost eigenvalue will be computed by direct application of these methods to F_x. The approach in Christodoulou and Scriven (1988), Garratt et al. (1991) and Cliffe, Garratt and Spence (1993) is first to transform the eigenvalue problem using the *generalized Cayley transform*

$$C(A) = (A - \alpha_1 I)^{-1}(A - \alpha_2 I), \quad \alpha_1, \alpha_2 \in \mathbb{R},$$

which has the key property that if $\mu \neq \alpha_1$ is an eigenvalue of A then $\theta :=$

$(\mu - \alpha_1)^{-1}(\mu - \alpha_2)$ is an eigenvalue of $C(A)$. Also, $\text{Re}(\mu) \leq (\geq)(\alpha_1 + \alpha_2)/2$ if and only if $|\theta| \leq (\geq)1$. Thus eigenvalues to the right of the line $\text{Re}(\mu) = (\alpha_1 + \alpha_2)/2$ are mapped outside the unit circle and eigenvalues to the left of the line mapped inside the unit circle. In Garratt et al. (1991) and Cliffe et al. (1993), algorithms based on computing dominant eigenvalues of $C(\boldsymbol{F_x})$ using Arnoldi or simultaneous iteration are presented, with consequent calculation of rightmost eigenvalues of $\boldsymbol{F_x}$. These algorithms were tested on a variety of problems, including systems arising from mixed finite element discretizations of the Navier–Stokes equations. Quite large problems can in fact be tackled. Indeed, in Gresho, Gartling, Torczynski, Cliffe, Winters, Garratt, Spence and Goodrich (1993) the problem of the stability of flow over a backward facing step is discussed in detail and the rightmost eigenvalues of a system with over 3×10^5 degrees of freedom are found using the generalized Cayley transform allied with simultaneous iteration.

However, it was later noted (see Meerbergen, Spence and Roose (1994)) that

$$C(A) = I + (\alpha_1 - \alpha_2)(A - \alpha_1 I)^{-1}$$

and so Arnoldi's method applied to $C(A)$ builds the same Krylov subspace as Arnoldi's method applied to the shift-invert transformation $(A - \alpha_1 I)^{-1}$. Thus, if Arnoldi's method is the eigenvalue solver, it would appear that there is no advantage in using the Cayley transform, which needs two parameters, over the standard shift-invert transformation (see Meerbergen et al. (1994)). This is indeed the case if a direct method is used to solve the systems with coefficient matrix $(A - \alpha_1 I)$. However, it turns out that if an iterative method is used then the Cayley transform is superior to the shift-invert transformation because the spectral condition number of the Cayley transform can be more tightly bounded (see Meerbergen and Roose (1997), Lehoucq and Meerbergen (1999) and Lehoucq and Salinger (1999)).

One can think of the approach in Cliffe et al. (1993) as the computation of the subspace containing the eigenvectors corresponding to the rightmost eigenvalues of $\boldsymbol{F_x}$. A similar theme, derived using a completely different approach, is described by Schroff and Keller (1993) and refined by Davidson (1997). In these papers the subspace corresponding to a set of (say rightmost) eigenvalues is computed using a hybrid iterative process based on a splitting technique. Roughly speaking, a small subspace is computed using a Newton-type method and the solution in the larger complementary space is found using a Picard (contraction mapping) approach. One advantage is that the Jacobian matrix $\boldsymbol{F_x}$ need never be evaluated.

When using mixed finite element methods to solve the incompressible Navier–Stokes equations a special block structure arises in the matrices due to the discretization of the incompressibility condition. After linearization about a steady solution one obtains a generalized eigenvalue problem of the form $A\phi = \mu B\phi$ where A and B have the block structure (see Cliffe et al. (1993)),

$$A = \left(\begin{array}{cc} K & C \\ C^T & 0 \end{array} \right), \qquad \left(\begin{array}{cc} M & 0 \\ 0 & 0 \end{array} \right),$$

with K nonsymmetric, M symmetric positive definite, and C of full rank. The shift invert transformation has the form $(A - \alpha B)^{-1}B$ and the Cayley transform, $(A-\alpha_1 B)^{-1}(A-\alpha_2 B)$. Though most of the linear algebra theory for the transformations is unaltered, the fact that B is singular means that there is a multiple eigenvalue at zero for the shift-invert transformation and at one for the Cayley transform. Care is needed in the implementation of numerical algorithms to ensure that these multiple eigenvalues do not give spurious results (see Meerbergen and Spence (1997)).

Finally we note that Chapter 5 of Seydel (1994) contains an overview of Hopf detection techniques.

7.5. Minimally extended systems

Griewank and Reddien (1984, 1989) (and with improvements Govaerts (1995)) suggested an alternative way of calculating fold points (and other higher-order singularities). This involves setting up a 'minimal' defining system

$$\mathbf{T}(\mathbf{y}) = \left[\begin{array}{c} \mathbf{F}(\mathbf{x}, \lambda) \\ g(\mathbf{x}, \lambda) \end{array} \right] = \mathbf{0}, \qquad \mathbf{y} \in \mathbb{R}^{n+1}, \tag{7.20}$$

where $g(\mathbf{x}, \lambda) : \mathbb{R}^n \times \mathbb{R} \to \mathbb{R}$ is implicitly defined through the equations

$$M(\mathbf{x}, \lambda) \left[\begin{array}{c} v(\mathbf{x}, \lambda) \\ g(\mathbf{x}, \lambda) \end{array} \right] = \left[\begin{array}{c} \mathbf{0} \\ 1 \end{array} \right], \tag{7.21}$$

and

$$(\mathbf{w}^T(\mathbf{x}, \lambda), g(\mathbf{x}, \lambda))M(\mathbf{x}, \lambda) = (\mathbf{0}^T, 1), \tag{7.22}$$

where

$$M(\mathbf{x}, \lambda) = \left[\begin{array}{cc} \mathbf{F}_{\mathbf{x}}(\mathbf{x}, \lambda) & \mathbf{b} \\ \mathbf{c}^T & d \end{array} \right], \tag{7.23}$$

for some $\mathbf{b}, \mathbf{c} \in \mathbb{R}^n$, $d \in \mathbb{R}$. (The fact that $g(\mathbf{x}, \lambda)$ is defined uniquely by *both* (7.21) and (7.22) may be seen since both equations imply that $g(\mathbf{x}, \lambda) = [M^{-1}(\mathbf{x}, \lambda)]_{n+1,n+1}$.) Note that $M(\mathbf{x}, \lambda)$ is a bordering of $\mathbf{F}_{\mathbf{x}}$, as arises in the numerical continuation method (Section 4). Assuming $\mathbf{b}, \mathbf{c}$, and d are chosen so that $M(\mathbf{x}, \lambda)$ is nonsingular (see the ABCD Lemma 3.1) then $g(\mathbf{x}, \lambda)$ and $v(\mathbf{x}, \lambda)$ in (7.21) are uniquely defined. (Note: if S is parametrized by t near $(\mathbf{x}_0, \lambda_0)$, that is, $(\mathbf{x}(t), \lambda(t))$ near $t = t_0$, then $v = v(\mathbf{x}(t), \lambda(t))$ and $g = g(\mathbf{x}(t), \lambda(t))$, and these functions may be differentiated with respect to t.) Also, if we apply Cramer's rule in (7.21) we have (with $M(x, \lambda)$ nonsingular)

$$g(x, \lambda) = \det(F_x(x, \lambda)) / \det(M(x, \lambda)), \qquad (7.24)$$

and so

$$g(x, \lambda) = 0 \iff F_x(x, \lambda) \text{ is singular.}$$

It is easily shown that quadratic fold points are regular solutions of (7.20). To apply Newton's method to (7.20), derivatives of $g(x, \lambda)$ are required and these can be found by differentiation of (7.21). When the details of an efficient implementation of Newton's method applied to (7.20) are worked out then the main cost is two linear solves with M and one with M^T. This should be compared with the costs of computing a fold point using the method described in Section 7.2. A nice summary of this approach is given in Beyn (1991). A numerically convenient Lyapunov–Schmidt reduction procedure can be accomplished using bordered systems (see, for example, Janovsky (1987), Janovský and Plecháč (1992), Govaerts (1997)), and a complete account is in the recent book by Govaerts (2000).

Finally, a difficulty that arises when implementing all types of extended systems is that they require the evaluation of derivatives of the discretized equations with respect to both the state variables and the parameters. The higher the codimension of the singularity, the higher the order of the derivative required. Evaluating these derivatives is both tedious and error-prone. An efficient method for computing the necessary derivatives for Galerkin finite element discretizations which makes use of the symbolic manipulation package REDUCE (Hearn 1987) is presented in Cliffe and Tavener (2000).

8. Taylor–Couette flow

The flow of a viscous incompressible fluid in the annular gap between two concentric cylinders, commonly known as Taylor–Couette flow, has been intensively studied for many decades and has served as an important vehicle for developing ideas in hydrodynamic stability and bifurcation, and in transition to turbulence. The flow is most commonly driven by the rotation of the inner cylinder with the top and bottom surfaces and outer cylinder held stationary, and we will concentrate on this case exclusively. A number of variants do exist in which the outer cylinder co-rotates or counter-rotates with the inner cylinder (see, *e.g.*, Andereck, Liu and Swinney (1986), Nagata (1986), Golubitsky and Stewart (1986), Iooss (1986)), or in which one or both of the two ends rotate with the inner cylinder (see, *e.g.*, Cliffe and Mullin (1986) and Tavener, Mullin and Cliffe (1991), respectively), but we will not consider them here.

In long Taylor–Couette devices, the flow at small rotation rates has no obvious structure, as is illustrated by the photograph of a flow visualization experiment shown in Figure 11. Over the great majority of the length of the cylinders the particle paths are essentially circles about the central axis, although some small-scale, more complicated features always exist near the ends of the apparatus. As the rotation rate of the inner cylinder is gradually increased, a number of toroidally shaped 'Taylor' cells develop, which can be readily observed using flow visualization techniques as shown in Figure 12.

When the cylinders are long, the onset of cellular flow apparently occurs rapidly over a short range of Reynolds number, suggesting that they arise as the result of a bifurcation. However, using laser Doppler measurements of the radial velocities near the centre of the cylinders and also near one end, Mullin and Kobine (1996) demonstrate that the rate at which cellular motion develops depends strongly on how its presence or absence is determined. For the boundary conditions discussed here, Taylor cells are steady and axisymmetric when they first appear. The hydrodynamic problem for a Newtonian fluid is given below:

$$R\left(u_r\frac{\partial u_r}{\partial r} + u_z\frac{\partial u_r}{\partial z} - \frac{u_\theta^2}{(r+\beta)}\right)$$
$$+ \frac{\partial p}{\partial r}$$
$$- \left(\frac{1}{(r+\beta)}\frac{\partial}{\partial r}\left[(r+\beta)\frac{\partial u_r}{\partial r}\right] + \frac{1}{\Gamma^2}\frac{\partial^2 u_r}{\partial z^2} - \frac{u_r}{(r+\beta)^2}\right) = 0, \quad (8.1)$$

$$R\left(u_r\frac{\partial u_\theta}{\partial r} + u_z\frac{\partial u_\theta}{\partial z} + \frac{u_r u_\theta}{(r+\beta)}\right)$$
$$- \left(\frac{1}{(r+\beta)}\frac{\partial}{\partial r}\left[(r+\beta)\frac{\partial u_\theta}{\partial r}\right] + \frac{1}{\Gamma^2}\frac{\partial^2 u_\theta}{\partial z^2} - \frac{u_\theta}{(r+\beta)^2}\right) = 0, \quad (8.2)$$

$$R\left(u_r\frac{\partial u_z}{\partial r} + u_z\frac{\partial u_z}{\partial z}\right)$$
$$+ \frac{1}{\Gamma^2}\frac{\partial p}{\partial z}$$
$$- \left(\frac{1}{(r+\beta)}\frac{\partial}{\partial r}\left[(r+\beta)\frac{\partial u_z}{\partial r})\right] + \frac{1}{\Gamma^2}\frac{\partial^2 u_z}{\partial z^2}\right) = 0, \quad (8.3)$$

$$\frac{1}{(r+\beta)}\frac{\partial}{\partial r}\left[(r+\beta)u_r\right] + \frac{\partial u_z}{\partial z} = 0. \quad (8.4)$$

The steady, axisymmetric Navier–Stokes equations have been nondimensionalized using

$$r = \frac{r^\star}{d} - \beta, \quad z = \frac{z^\star}{h}, \quad \boldsymbol{u} = \frac{1}{r_1\Omega}\left(u_r^\star, u_\theta^\star, \frac{u_z^\star}{\Gamma}\right), \quad p = \frac{dp^\star}{\mu r_1\Omega},$$

Fig. 11. Visualization of the flow in the
Taylor apparatus at a slow rotation rate.
(Thanks to T. Mullin)

Fig. 12. Visualization of the flow in the
Taylor apparatus at a larger rotation rate
for which cellular flows are observed.
(Thanks to T. Mullin)

where $d = r_2 - r_1$, $\beta = r_1/d = \eta/(1 - \eta)$ and $\star$ denotes dimensional quantities. Here r_1 and r_2 are the radii of the inner and outer cylinders, respectively, h is the height of the cylinders and Ω is the rotation rate of the inner cylinder. The three nondimensional parameters are the Reynolds number, $R = \rho \Omega r_1 d/\mu$, aspect ratio, $\Gamma = h/d$ and radius ratio, $\eta = r_2/r_1$, where ρ is the density and μ is the molecular viscosity. Equations (8.1) to (8.4) pertain in a region $D = \{(r, z), 0 \leq r \leq 1, -1/2 \leq z \leq 1/2\}$.

8.1. The infinite cylinder model

The earliest recorded experimental work is by Couette (1890) in which he held the inner cylinder fixed and rotated the outer. Mallock (1896) repeated Couette's findings and also considered the case with a rotating inner cylinder and stationary outer cylinder. Rayleigh (1916) developed a stability criterion for inviscid fluids and co-rotating cylinders, namely $\omega_2 r_2^2 > \omega_1 r_1^2$ to ensure stability (*i.e.*, the angular momentum must increase radially), and so explained the gross differences observed between Couette's and Mallock's experiments.

The hugely influential work by Taylor (1923) compared laboratory experiments with the results of a linear stability analysis. Taylor's analysis was based on the assumption that both cylinders were infinitely long. The first advantage of this assumption is that it supports a simple exact solution for all values of the Reynolds number. In this solution the axial and radial velocities are zero and the azimuthal velocity is a function of the radius, r only. Specifically $v = Ar + B/r$ where $A = (\Omega_2 r_2^2 - \Omega_1 r_1^2)/(r_2^2 - r_1^2)$ and $B = (\Omega_1 - \Omega_2)r_1^2 r_2^2/(r_2^2 - r_1^2)$. Further, nearby solutions can be sought in which the perturbation is periodic in the axial direction z, with the period treated as a free parameter. The axially periodic disturbance that becomes unstable at the lowest critical Reynolds number is assumed to be the one that will occur in practice. Using this approach, Taylor was able for the first time to obtain excellent agreement with experiment, both with respect to the critical Reynolds number for the onset of Taylor cells, and with regard to the axial wavelength of the cellular flow. Roberts (1965) has subsequently tabulated the critical Reynolds numbers for a range of radius ratios. Synge (1933) has extended Rayleigh's result to viscous flows.

Taylor also recognized that the steady axisymmetric cellular flows that are observed first on steadily increasing the rotation rate of the inner cylinder, themselves lose stability to an increasingly complicated series of time-dependent flows as the rotation rate of the inner cylinder is further increased. Much of the subsequent work on Taylor–Couette flow has concentrated on ideas of transition to turbulence in this experimentally simple and (apparently) theoretically accessible device. Reviews of much of the early work are given by DiPrima and Swinney (1981) and Stuart (1986). For a more recent discussion of the complex time-dependent phenomena observed in the Taylor–Couette system see Mullin (1995) and the references therein.

8.2. Finite cylinder models

Benjamin (1978), invoking a number of abstract results regarding the properties of viscous incompressible flows in arbitrary bounded domains D with boundary ∂D (with $\mathbf{v} \cdot \mathbf{n} = 0$ on ∂D), departed from the established infinite cylinder assumption. He sought to address the following three observations that cannot be explained by the infinite cylinder model.

1. Cellular motion does not occur at a specific Reynolds number, but cells develop near the top and bottom surfaces and spread inwards. While this process may occur over a small range of rotation rates for long cylinders, there is never an unambiguous critical Reynolds number. However, a simple 'softening' or disconnection of the bifurcation in the infinite cylinder model is insufficient to explain the following observations.

2. For a given length of cylinder, a unique flow exists upon a slow (quasi-static) increase in the Reynolds number from zero. The number of cells in this 'primary' flow is a function of the length of the cylinders. If $2N$ cells develop for a certain length then $(2N + 2)$ cells develop for some greater length.

3. 'Secondary' cellular flows also exist. These flows differ from the primary flow and cannot be obtained by a gradual increase in the Reynolds number, but are stable above a critical finite Reynolds number. Golubitsky and Schaeffer (1983) show that when a pitchfork of revolution arising in $O(2)$-symmetric problems is disconnected the secondary branches are unstable.

Using results from the general existence theory for the Navier–Stokes equations, Leray–Schauder degree theory, and bifurcation theory, Benjamin proposed the sequence of bifurcation diagrams shown in Figure 13 to explain the process whereby the primary flow changes from a $2N$-cell flow to a $(2N + 2)$-cell flow as the aspect ratio is increased. The primary branch is seen to develop hysteresis as the aspect ratio is varied. At a critical aspect ratio, the folded primary and secondary branches connect at a transcritical bifurcation point. As the aspect ratio is varied further, the transcritical bifurcation point disconnects in the opposite manner and the formerly secondary branch is now continuously connected to the unique solution at small Reynolds number.

Benjamin then examined the exchange of stability between primary two-cell and four-cell flows in an annulus whose aspect ratio could be varied

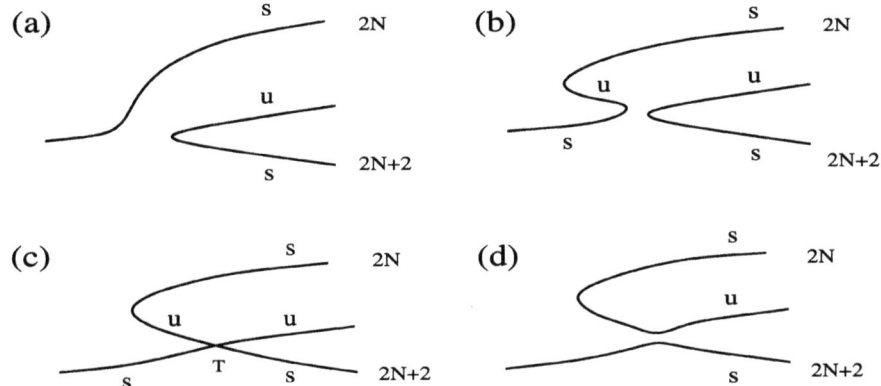

Fig. 13. Schematic sequence of bifurcation diagrams for the exchange between a $2N$-cell primary flow and a $(2N + 2)$-cell flow with a continuous increase in the aspect ratio. Stability and instability are indicated by 's' and 'u' respectively

between 3.4 and 4.0, and observed the predicted folding of the primary two-cell flow and the attendant hysteresis and (downward-facing) cusp-shaped locus of fold points. Benjamin also observed 'anomalous' flows containing an odd or an even number of cells, in which one or both of the end cells rotate outwardly along one (or both) of the end walls. Such flows contradict the usual argument based on Ekman boundary layers, which suggests that the flow should always be inward along the stationary top and bottom surfaces. Flows with outwardly spiralling flow along the end walls had never been reported previously.

Since his arguments were based on properties in the abstract, Benjamin contended that his ideas were relevant for cylinders of any length. Considerable subsequent experimental and numerical work supports his conjecture, for example the experimental study of Lorenzon and Mullin (1985) which compares stability properties of anomalous flows at aspect ratios 10 and 40.

The origin of even-celled anomalous modes was addressed by Schaeffer (1980) who considered the exchange between $2N$-cell and $(2N+2)$-cell flows as the aspect ratio is varied. He constructed a model using a homotopy parameter τ, to continuously connect flows with non-flux stress-free boundary conditions ($\tau = 0$) to those with realistic non-slip boundary conditions ($\tau = 1$). When $\tau = 0$, an axially independent flow exists and pairs of cellular flows can bifurcate from this 'trivial' flow. The cellular flows along each of the two bifurcating branches differ by a translation of one-half wavelength. These supercritical pitchfork bifurcations become disconnected for nonzero τ. The 'primary' flow that is continuously connected to the flow at low Reynolds number has normal (inward) flow along the top and bottom walls. The flow with anomalous (outward) flow along the top and bottom walls remains as a disconnected 'secondary' flow. Schaeffer's model is strictly

valid for $N \geq 2$ only and, by employing perturbation techniques and applying Schaeffer's boundary conditions at the top and bottom surfaces, Hall
(1980) examined the two-cell/four-cell interaction studied experimentally by
Benjamin (1978). Later, Hall (1982) determined explicit values for the constants in Schaeffer's model for the four-cell/six-cell exchange and correctly
predicted the cusp to face upwards in this case. Benjamin and Mullin (1981)
extended Schaeffer's argument to consider flows with an odd number of cells.

Numerical bifurcation techniques have made a significant contribution to
this radical re-evaluation of a classic problem in hydrodynamic stability in
a number of ways. We consider first the four-cell/six-cell exchange and then
briefly discuss anomalous modes.

Four-cell/six-cell exchange mechanism

At a fixed radius ratio, two adjustable parameters remain: the Reynolds
number and the aspect ratio. The cusp-shaped locus of the limits of stability of the 'normal' four-cell and six-cell flows at a radius ratio of 0.6 was
determined experimentally by Mullin (1985), and his data are reproduced in
Figure 14. It is worthwhile mentioning the experimental technique used by
Mullin to investigate the four-cell/six-cell exchange mechanism (and also the
6/8, 8/10 and 10/12-cell exchange mechanisms). At aspect ratios above and
below the cusp-shaped exchange region, the secondary four-cell and six-cell
flows are only stable above a critical Reynolds number and were established
via sudden starts within certain narrow speed ranges. The Reynolds number
was then decreased in small steps allowing ample time for re-equilibration
between speed changes. The secondary flows eventually collapsed at a critical value of the Reynolds number along AT for four-cell flows and along
HE for six-cell flows. Within the cusp-shaped region a gradual increase in
speed from a small value resulted in a four-cell flow which suddenly jumped
to become a six-cell flow when HT was crossed. This six-cell flow remained
stable upon decreasing the Reynolds number until HE was reached, at which
point the six-cell flow collapsed back to a four-cell flow. In this manner a
definite hysteresis was observed, although smaller and smaller speed changes
and longer and longer settling times were required to obtain repeatable observations as the non-degenerate hysteresis point H was neared. Point T is
a transcritical bifurcation point (see Figure 13(c)) where the upper (subcritical) fold point on the folded primary branch and the (supercritical) fold
point on the secondary branch merge.

Complementary finite element computations by Cliffe (1988) were performed as follows. At a fixed aspect ratio of 4, the four-cell primary branch
was computed up to a Reynolds number of 300 using arclength continuation. The aspect ratio was then increased to 6 at which aspect ratio the
four-cell flow is no longer the primary flow, but a secondary flow that exists only above a finite Reynolds number. This lower limit was found by

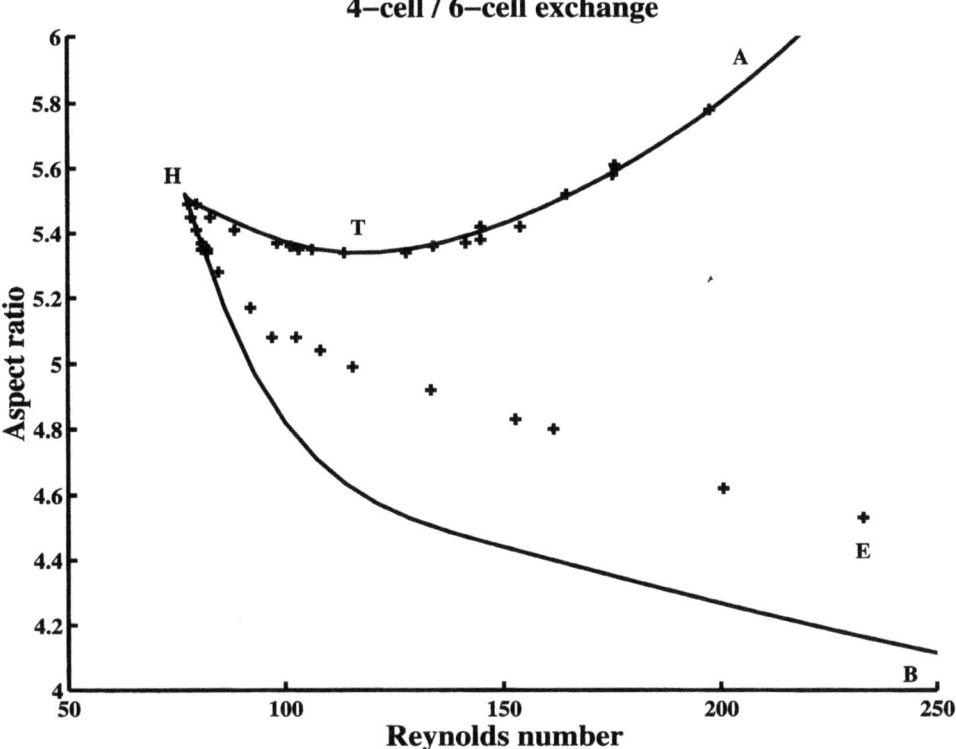

Fig. 14. Numerical and experimental comparison of the four-cell/six-cell exchange. The solid line is the computed locus of fold points on the 'normal' four-cell and six-cell branches. Experimental measurements of the collapse of the four-cell and six-cell flows are indicated by '+'. T is a transcritical bifurcation point and H is a hysteresis point

decreasing the Reynolds number (at fixed aspect ratio 6) until a fold point was encountered. The locus of fold points was then computed in the Reynolds number/aspect ratio plane using the extended system described in Section 7.2 and appears in Figure 14. The cusp-shaped exchange region is delimited by non-degenerate hysteresis point H and transcritical bifurcation point T, which were subsequently located more precisely using extended system techniques similar to those discussed in Section 7 and described in detail in Spence and Werner (1982) and Jepson and Spence (1985a), respectively. It can be seen from Figure 14 that the computed lower limit of stability for the four-cell flow agrees well with the experimental data for aspect ratios exceeding that at the transcritical bifurcation point, T. There is also excellent agreement in the hysteretic region, but for aspect ratios less than approximately 5.35, the experimentally determined lower limit of stability

4–cell / 6–cell exchange

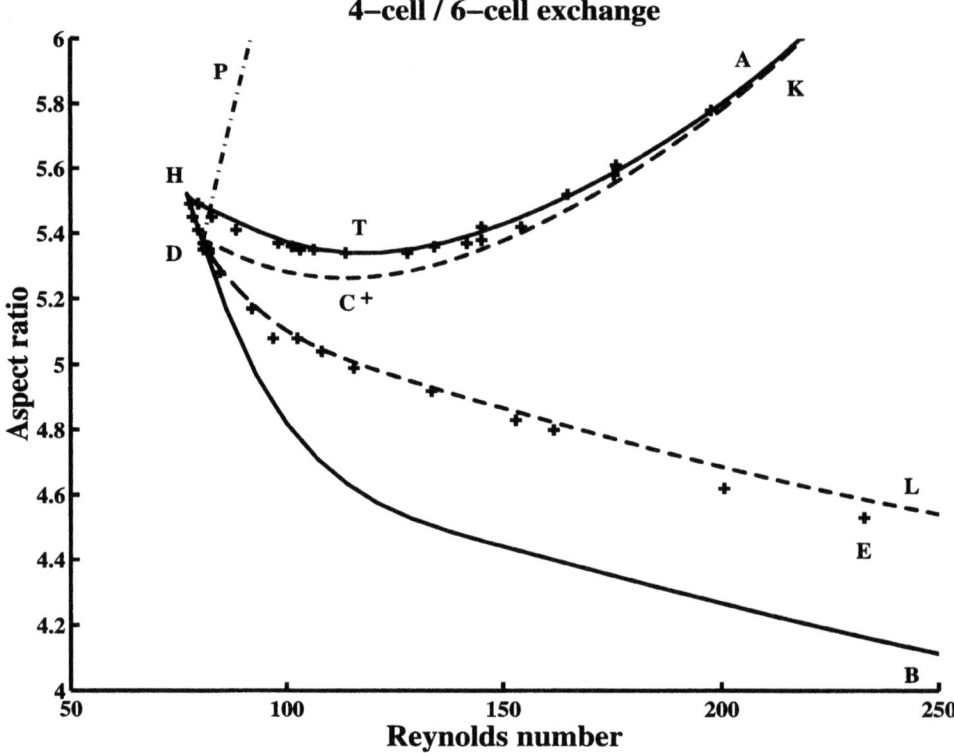

Fig. 15. Numerical and experimental comparison of the four-cell/six-cell exchange. The solid line is the locus of fold points on the 'normal' four-cell and six-cell branches. The dashed line is a path of symmetry breaking bifurcation points and the chained line is the a path of turning points on the asymmetric branches. Experimental measurements of the collapse of the four-cell and six-cell flows are indicated by '+'. T is a transcritical bifurcation point and H is a hysteresis point. C^+ is a coalescence point and D is a double singular point

for six-cell flows lies considerably above the computed path of fold points on the six-cell secondary branches.

As shown by Cliffe (1983) and Cliffe and Spence (1984), the primary six-cell flow is invariant with respect to a Z_2-symmetry operator which is essentially a reflection about the midplane of the cylinders. A symmetry breaking bifurcation point was found on the six-cell secondary branch at an aspect ratio of 5. The locus of symmetry breaking bifurcation points in the Reynolds number/aspect ratio plane was computed using the Werner–Spence extended system, and is shown in Figure 15. It agrees convincingly with the experimentally determined points at which the six-cell flow collapses with decreasing Reynolds number.

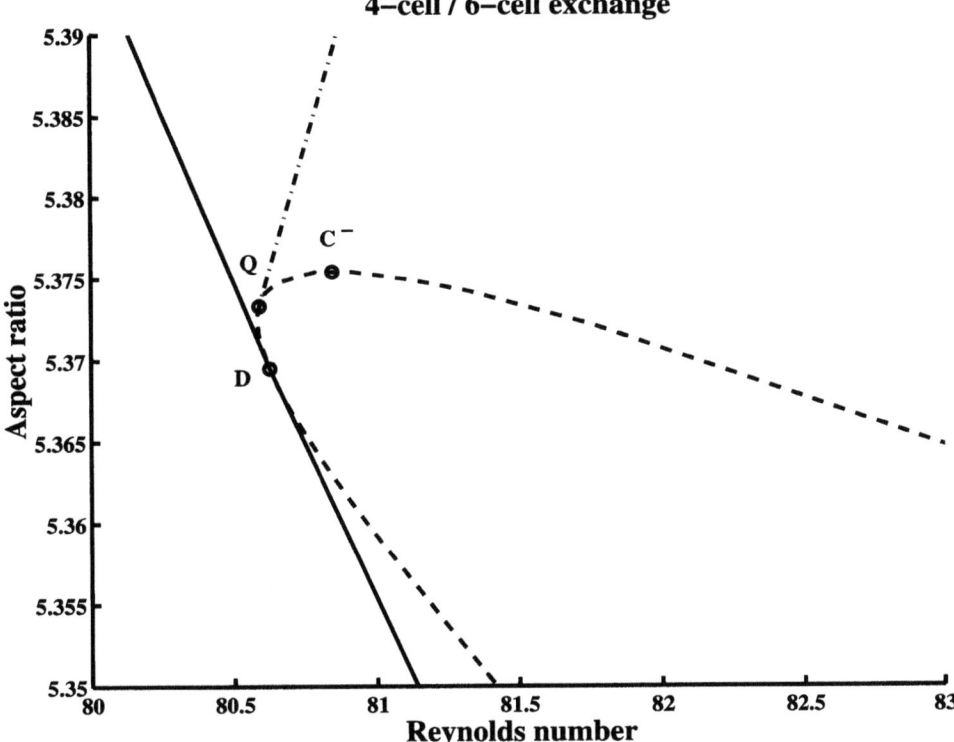

4–cell / 6–cell exchange

Fig. 16. Detail near the double singular point. The solid line is a path of fold points, the dashed line a path of symmetry breaking bifurcation points and the chained line a path of fold points on the asymmetric branches. D is a double singular point, Q is a quartic bifurcation point and C^- is a coalescence point

Details near the double-singular points D appear in Figure 16. The secondary bifurcation point crosses the lower (supercritical) fold point on the 'S'-shaped six-cell branch at a double singular point D. At the quartic bifurcation point Q the symmetry breaking bifurcation changes from being supercritical to subcritical with increasing aspect ratio. A pair of asymmetric solution branches intersect on the symmetric solution branch at the coalescence point C^-, then merge and disconnect from the symmetric branch with increasing aspect ratio. An isola (see Golubitsky and Schaeffer (1985), p. 133) of asymmetric solutions develops with increasing aspect ratio from the coalescence point C^+, in Figure 15. All three singularities are described on page 268 of Golubitsky and Schaeffer (1985). Full details of the exchange mechanism appear in Cliffe (1988). For our purposes it suffices to observe that the experimentally determined loss of stability of six-cell flows is clearly associated with the breaking of their midplane symmetry.

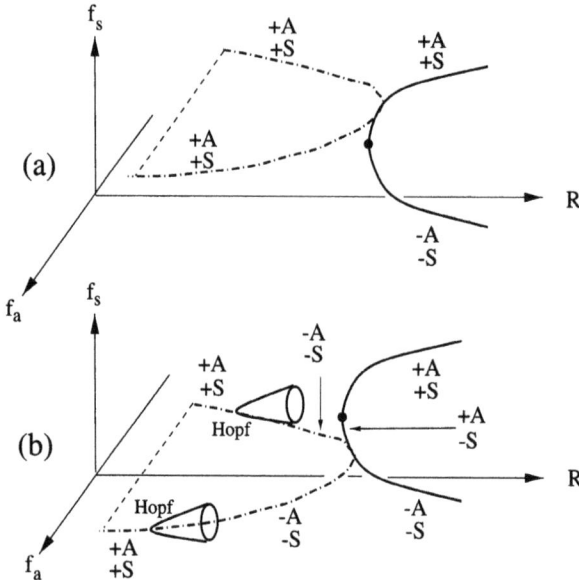

Fig. 17. Schematic bifurcation diagrams illustrating the emergence of a path of Hopf bifurcation points from a double singular point. f_s is a measure of the symmetric component of the solution and f_a is a measure of the antisymmetric component of the solution. The solid line is a branch of symmetric solutions. The chained line is a branch of asymmetric solutions. The symbols +S +A, (–S, –A) indicate stability (instability) with respect to symmetric and antisymmetric disturbances

A second double singular point arises far from the cusp region. It lies just outside the region of parameter space shown in Figures 14 and 15, where the locus of symmetry breaking bifurcation points is tangential to the locus of fold points and the symmetry breaking bifurcation point crosses from the unstable to the stable symmetric solution surface. Langford (1979), Guckenheimer (1981) and Golubitsky et al. (1988, Chapter 19) all discuss a more general class of problems, but here we simply repeat the argument of Mullin and Cliffe (1986) to explain why a path of axisymmetric Hopf bifurcations may be expected to emanate from such a double singular point.

Consider the situation in which a Z_2-symmetry breaking bifurcation point crosses from the stable to the unstable side of the symmetric solution surface as a parameter (e.g., aspect ratio) is varied as shown in Figure 17. The symbols +S, –S, +A and –A along the symmetric solution branches indicate the signs of the smallest (most unstable) symmetric and antisymmetric eigenvalues respectively. While the eigenvectors on the asymmetric surface need not be strictly symmetric or antisymmetric, we retain this notation on the asymmetric solution branches for clarity. By continuity arguments, there must be two unstable eigenvalues along asymmetric branches near the symmetry breaking bifurcation point in Figure 17(b). However, sufficiently far from the bifurcation point, these branches retain the stability they enjoyed when the symmetry breaking bifurcation point lay on the stable symmetric solution surface in Figure 17(a). The simplest way to resolve this conflict is via Hopf bifurcation on the asymmetric branches. A path of Hopf bifurcation points emanates from the double singular point and the frequency of the periodic orbit approaches zero at the double singular point.

The path of Hopf bifurcation points arising from the double singular point remote from the hysteresis point in the four-cell/six-cell exchange, was computed for a (different) radius ratio of 0.5 using the extended system technique outlined in Section 7.4 and described in detail in Griewank and Reddien (1983). Both the stable asymmetric flows and the axisymmetric singly periodic flows were subsequently observed in experiments. Figures 18 and 19, reproduced from Mullin, Cliffe and Pfister (1987), demonstrate the excellent agreement between finite element computations and experiments both with respect to the critical Reynolds number and in regard to the frequency of the periodic flow near the Hopf bifurcation point.

This example argues strongly for complementary theoretical, experimental and numerical studies. Benjamin's theoretical arguments suggested that conducting experiments in short cylinders would be profitable in order to limit the multiplicity of the solution set. The disparity between experimental and numerical work then showed that a consideration of the symmetric solution set alone was insufficient to explain the details of the exchange mechanism and that the hitherto ignored possibility of symmetry breaking must be addressed. Once confidence in the numerical approach was established by quantitative comparison with experiment, numerical predictions of stable periodic flows were subsequently confirmed by careful experiment.

The special case of one-cell/two-cell interaction was first examined by Benjamin and Mullin (1981) and was re-examined in greater detail by Cliffe (1983) and Pfister, Schmidt, Cliffe and Mullin (1988). This study again provides an excellent example of the mutual reinforcement of experimental and computational approaches. Numerical work led to the experimental discovery of a new two-cell flow and new axisymmetric periodic flows, and experimental work suggested the presence of a Takens–Bogdanov point which was subsequently located numerically.

Anomalous modes

Benjamin and Mullin (1981) examined the N-cell/$(N+1)$-cell exchange process and experimentally determined the lower limits of stability for anomalous flows with 2 to 7 cells. Cliffe and Mullin (1985) presented experimental and numerical comparisons of the ranges of stability of anomalous four-cell flows and streamline comparisons for anomalous three, four and five-cell flows and the normal four-cell flow. Cliffe, Kobine and Mullin (1992) demon-

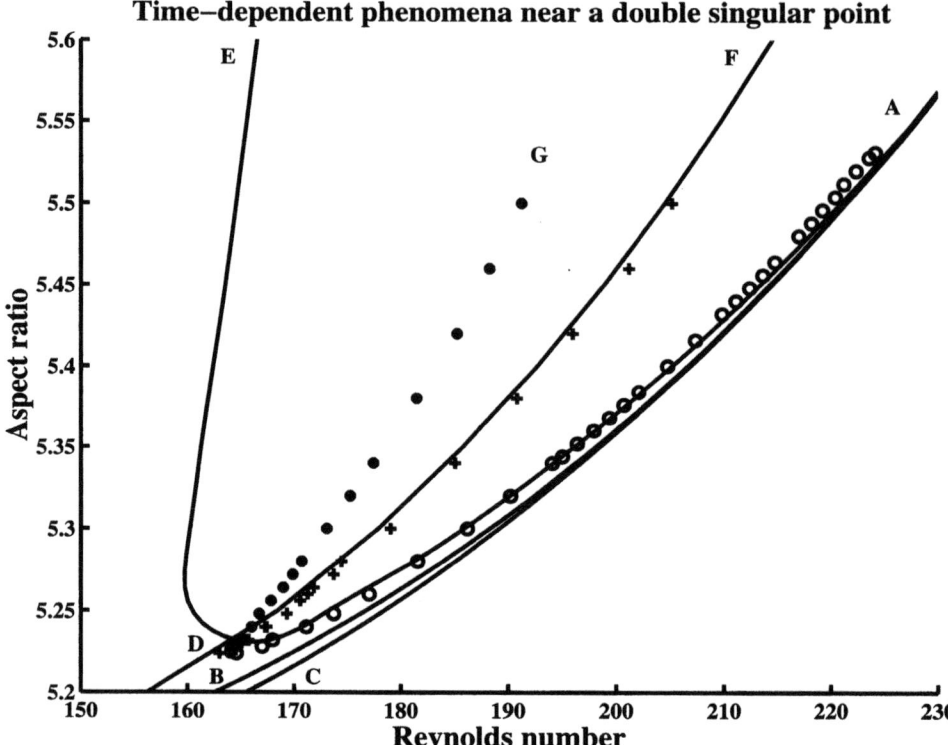

Fig. 18. Numerical and experimental comparison of the critical Reynolds numbers for the onset of time-dependent flows. AB is a path of supercritical fold points, AC is a path of subcritical symmetry breaking bifurcation points, ADE is the locus of axisymmetric Hopf bifurcation points, and DF is a path of subcritical Hopf bifurcation points for a non-axisymmetric time-dependent flow with azimuthal wavenumber $m = 1$. 'o' denotes the experimentally observed onset of axisymmetric periodic flows, '+' the onset of wavy periodic flows with azimuthal wave number 1, and '•' the axisymmetric modulation of the $m = 1$ periodic mode

strated that the lower limit of stability of the N-cell anomalous modes in cylinders with aspect ratios N are essentially the same. This suggests that their stability is governed by the stability of the two anomalous cells near the top and bottom boundary. Benjamin and Mullin (1982) showed that for $N > 10$ the interior parts of the normal and anomalous flows are very similar. Cliffe et al. (1992) also demonstrated how the range of stability of anomalous modes decreases dramatically with radius ratio, and thereby suggested why they had not been observed in experiments that attempted to approximate the infinite cylinder model.

To summarize, numerical computations have made significant contributions to our understanding in the following areas.

Frequency of axisymmetric periodic flows

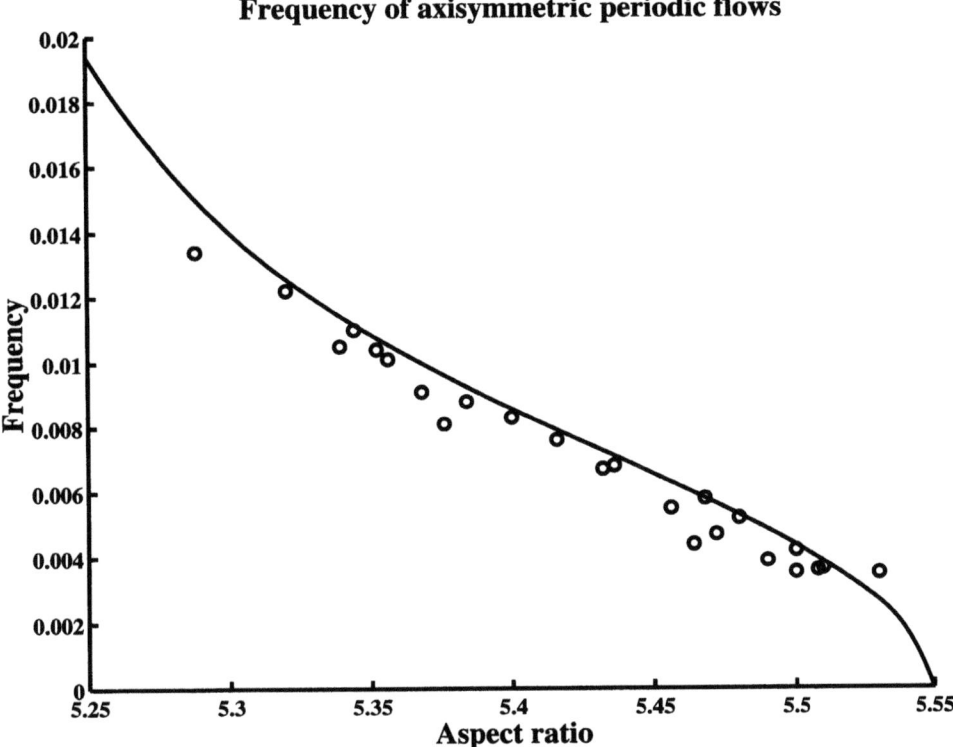

Fig. 19. Numerical and experimental comparison of the nondimensionalized frequency of the axisymmetric singly periodic flow. (The nondimensionalization was performed with respect to the frequency of the inner cylinder.)

1. The mechanisms by which the primary flow changes as the aspect ratio is varied when non-slip boundary conditions are imposed at the top and bottom surfaces.
2. The existence and stabilities of anomalous modes.
3. The connection between time-dependent phenomena and bifurcations of the steady solution set.

The Taylor–Couette system has also been studied extensively using numerical bifurcation techniques by H. B. Keller and his collaborators. Bolstad and Keller (1987) used a streamfunction-vorticity formulation of the axisymmetric Navier–Stokes equations, a finite difference discretization and a multigrid solver to compute the stability of anomalous modes having two to six cells. Anomalous modes were obtained by performing continuation in the Schaffer homotopy parameter as described above. Good quantitative agreement with the results of Cliffe and Mullin (1985) was obtained. Meyer-Spasche and Keller (1980), Frank and Meyer-Spasche (1981), Meyer-Spasche and Keller (1985) and Specht, Wagner and Meyer-Spasche (1989) all used continuation techniques to compute flows in the Taylor–Couette apparatus applying axially periodic boundary conditions. They examined the onset of cellular flow, and the exchange between flows with different even numbers of cells as the axial period was varied. The later problem was also investigated numerically by Tavener and Cliffe (1991). Schröder and Keller (1990) considered the Navier–Stokes equations in a rotating reference frame and used continuation methods to compute the onset and nonlinear development of wavy-Taylor vortex flows, again assuming axially periodic boundary conditions. They compared their results very favourably with the long cylinder experiments of Coles (1965).

A parallel research effort has concentrated on spherical Couette flow in which the fluid is confined between concentric spheres and driven by the rotation of the inner sphere. Schrauf (1983, 1986), Marcus and Tuckerman (1987) and Mamun and Tuckerman (1995) have all performed computations of the flow using at first continuation and then more sophisticated numerical bifurcation techniques. All studies have shown the breaking of the midplane symmetry to play a crucial rôle in the stability of steady solutions. Yang (1996) uses a time-dependent code to solve the flow between concentric spheres with additional, non-physical symmetry constraints until convergence is almost achieved. He then removes the extra constraints and allows the time-dependent code to converge. In this manner, by essentially choosing the initial condition sufficiently close to the desired solution, he is able to compute multiple solutions at a single point in parameter space, if, of course, the symmetries of the various flows are known *a priori*.

9. Other applications

Recognizing that the literature on hydrodynamic stability and bifurcation phenomena is vast, we attempt simply to give a flavour of some of the other applications of numerical bifurcation techniques to problems in fluid mechanics and apologize to those authors whose work we do not mention.

9.1. Rayleigh–Bénard–Marangoni convection

The study of the motion arising in a thin layer of fluid heated from below has a long history originating with the work of Bénard (1901). This is a fundamental problem in heat transfer with widespread applications for both large- and small-scale industrial processes. Block (1956) and Pearson (1958) showed that the hexagonal rolls observed by Bénard (1901) were due in fact to temperature-dependent surface tension forces, rather than to buoyancy forces as assumed by Rayleigh (1916). The origin of the forces driving convection depends upon the depth of the fluid, with surface tension forces dominating in sufficiently thin layers. Motion due to buoyancy forces is now commonly known as Rayleigh–Bénard convection and motion due to surface tension forces as Marangoni–Bénard convection.

Cliffe and Winters (1986) used the techniques described in Section 7 and Spence and Werner (1982) and Jepson and Spence (1985a) to compute the onset of (buoyancy-driven) Rayleigh–Bénard convection in finite two-dimensional rectangular domains. Their study focuses on the rôle played by the symmetries of the problem and on the exchange of stability between flows with different number of convection rolls as the aspect ratio is changed. Winters, Plesser and Cliffe (1988) computed the onset of cellular flows due to the combined effects of buoyancy- and temperature-dependent surface tension in two-dimensional rectangular domains. They consider both limiting cases: when the surface tension is independent of temperature and flow is driven by buoyancy alone, and when the density is independent of temperature and the flow is driven by surface tension forces alone. They examined the exchange of primary flows with aspect ratio for the latter case. Dijkstra (1992) computed the complex exchange processes between competing cellular states in two-dimensional rectangular domains as their aspect ratio varied. He considered the effect of surface tension forces in the absence of gravity. Relaxing the assumption that the free surface be horizontal and rigid, Cliffe and Tavener (1998) examined the effect of surface deformations on two-dimensional Marangoni–Bénard convection. Locally varying chemical species concentrations can also give rise to surface tensions gradients, and Bergeon, Henry, BenHadid and Tuckerman (1998) examined the combined effects of temperature and species concentration on two-dimensional free-surface flows.

9.2. Double diffusive convection

Rayleigh–Bénard convection arises due to the inverse relationship between temperature and the local density of the fluid. Chemical species concentrations can also affect the local density and complicated instability phenomena can arise when there is a competition between the effects of temperature and species concentrations. Both heat and chemical species are convected with the flow and both diffuse, but usually at very different rates. The fluid motion that arises due to a combination of these two effects is commonly known as double-diffusive convection. It is a nonlinear, multiparameter problem in which multiple solutions exist and is particularly suitable for study by numerical bifurcation techniques. Xin, LeQuere and Tuckerman (1998) and Dijkstra and Kranenborg (1996) considered two different realizations of doubly diffusive systems in rectangular containers. Motivated by material processing concerns, Xin et al. (1998) considered the case of horizontal temperature and concentration gradients. Dijkstra and Kranenborg (1996) considered a stable vertical chemical species (salt) concentration and a horizontal temperature gradient, with a view to understanding vertically layered temperature and salinity structures found in the ocean. Dijkstra and Molemaker (1997) later examined thermohaline-driven flows in a more complicated model problem arising from considerations of ocean circulations. Implementation details of the methods used by Dijkstra and his colleagues are discussed in Dijkstra, Molemaker, Vanderploeg and Botta (1995).

9.3. The Dean problem

A fundamental understanding of the fluid flow in curved tubes has considerable biological as well as industrial interest. The nature and stability of flows in curved tubes or ducts is commonly known as the Dean problem, following the pioneering work of Dean (1928). Centrifugal forces produce a component of flow perpendicular to the tube axis and spiralling motions result. Flows with both two and four counter-rotating cells have been observed. Winters (1987) performed a computational study of flows in curved ducts with rectangular cross-section using the numerical bifurcation techniques outlined in Section 7. He demonstrated the connection between the two-cell and four-cell flows and predicted that, amongst the multiple solution set, only two-cell flows should be stable with respect to both symmetric and antisymmetric disturbances. An experimental study by Bara, Nandakumar and Masliyah (1992) investigated Winter's predictions using a duct with square cross-section and a fixed radius of curvature. Direct comparisons proved to be difficult as Winter's computations assumed a fully developed flow and it was apparent that the flow was still developing when it reached the end of the curved experimental section which extended some 240 degrees. It is also hard to design an experiment to test satisfactorily whether a flow is stable with respect to symmetric disturbances but unstable with respect to antisymmetric ones.

Nandakumar and co-workers have applied extended system techniques of the type discussed in Section 7 to examine a number of related hydrodynamic stability problems. Implementation details (of at least one of their approaches) appear in Weinitschke (1985). Nandakumar, Raszillier and Durst (1991) computed the fully developed pressure-driven flows arising in a square duct that is rotating about an axis perpendicular to the axis of the duct. Spiralling flows similar to those in curved ducts are obtained. An understanding of such flows has application to the design of turbomachinery. With heat exchangers in mind, Nandakumar and Weinitschke (1991) computed the fully developed pressure-driven flows in a horizontal rectangular duct along which a constant temperature gradient is imposed. Flow perpendicular to the axis of the duct is driven in this instance not by centrifugal forces but by buoyancy. The combined effects of duct curvature and duct rotation was studied in Selmi, Nandakumar and Finlay (1994).

10. Concluding remarks

It is inevitable that in a review article of this type many aspects of the numerical solution of bifurcation problems will be omitted or at best are covered only briefly. We discuss some of these omissions now and also suggest areas for further research.

We do not provide any details of numerical continuation for multiparameter problems, say following paths of codimension 1 singularities (*e.g.*, hysteresis points) or calculation of 'organizing centres' (see Golubitsky and Schaeffer (1985), Golubitsky et al. (1988)) – these techniques can be inferred for the discussion in Section 7 and the quoted references. Also, we do not discuss mode interactions, like Takens–Bogdanov points (see, for example, Werner and Janovsky (1991), Spence, Cliffe and Jepson (1989)) and double singular points (Aston, Spence and Wu 1997), even though they are important in understanding complex phenomena in applications. Complicated symmetries have also been avoided (see Vanderbauwhede (1982), Golubitsky et al. (1988)) but an understanding of the relationship between symmetry and the presence of multiple solutions is vital especially in problems in fluid mechanics and nonlinear elasticity.

Also, our discussion of the important topic of minimally extended systems (in Section 7.5) is very brief. Our defence is that, to the best of our knowledge, these methods have not yet been used in large-scale fluid mechanics applications.

In the applications discussed in this paper stability assignments have been determined by eigenvalue techniques, that is, by checking when a certain linearization has real or complex eigenvalues which cross the imaginary axis. As seen in Section 8 on the Taylor–Couette problem these techniques can give excellent agreement with laboratory experiments. However, as is discussed in Trefethen, Trefethen, Reddy and Driscoll (1993) there are several classical fluids problems when eigenvalues do not predict the correct stability results. For very non-normal problems one should consider whether the pseudo-spectra (Trefethen 1997) of the linearization provide a more accurate tool to analyse stability. Certainly, in cases where numerical methods are to be used as a design tool, without complementary experimental results, it would only be prudent to consider estimating the pseudo-spectra of the linearization to help provide estimates of the reliability of any stability prediction.

Tuckerman and co-workers, for example Mamun and Tuckerman (1995) and Bergeon et al. (1998), use software developed for time-dependent problems in a novel and effective way to implement Newton's method for the calculation of stable and unstable steady states and periodic orbits in the discretized Navier–Stokes equations. The key idea involves the interplay between an implicit/explicit discretization of the time-dependent problem and the discretized steady state equation. A short paper (Hawkins and Spence 2000) discusses the problems of the detection of Hopf bifurcations using these techniques. We refer to the above references for details, though we mention that their approach is closely related to that in Davidson (1997), where a preconditioned version of the recursive projection method in Schroff and Keller (1993) is analysed. The power of the recursive projection method applied to fluids problems has been illustrated in von Sosen (1994) and Love (1999).

The reliable detection of Hopf bifurcations in large-scale problems is an important subject only briefly touched on in Section 7.4. Efficient and reliable methods for the detection of the loss of stability due to a complex pair of eigenvalues crossing the imaginary axis still need to be developed for problems arising from discretized partial differential equations. Though preconditioned iterative techniques (e.g., domain decomposition) are the norm for general 3D problems, these techniques are rarely used for eigenvalue calculation (but see the recent review by Knyazev (1998) where symmetric eigenvalue problems are discussed).

The calculation of periodic orbits in large-scale problems, say, arising from discretized partial differential equations, remains a major challenge. One possible approach is described by Lust, Roose, Spence and Champneys (1998). For small-scale problems, numerical methods are well developed for 'long time' dynamical phenomena, for example homoclinic orbits and Lyapunov exponents, and there are several software packages available which compute such phenomena (see, for example, Khibnik, Yu, Levitin and Nikoleav (1993), Kuznetsov and Levitin (1996), Doedel et al. (1997)). A major area for future work is to extend these techniques to large-scale problems, perhaps by first projecting onto a small-dimensional subspace of 'active' variables and then applying the standard small-scale techniques.

A review of this type inevitably reflects the interests, expertise and bias of the authors. There are certainly omissions and some important topics and results receive only a brief mention. However, we hope that this review will inform the reader of the mathematical tools needed and the challenges awaiting those who attempt to provide a rigorous numerical analysis of bifurcation problems. We also aim to stimulate further interest in this interesting and hugely important area. Finally, we hope to show by our detailed discussion of the Taylor–Couette problem that reliable numerical methods provide an essential tool when attempting to solve challenging problems from applications.

REFERENCES

J. P. Abbot (1978), 'An efficient algorithm for determination of certain bifurcation points', *J. Comput. Appl. Math.* **4**, 19–27.

N. Alleborn, K. Nandakumar, H. Raszillier and F. Durst (1997), 'Further contributions on the two-dimensional flow in a sudden expansion', *J. Fluid Mech.* **330**, 169–188.

E. L. Allgower and K. Georg (1993), Continuation and path following, in *Acta Numerica*, Vol. 2, Cambridge University Press, pp. 1–64.

C. D. Andereck, S. S. Liu and H. L. Swinney (1986), 'Flow regimes in a circular Couette system with independently rotating cylinders', *J. Fluid Mech.* **164**, 155–183.

P. M. Anselone and R. H. Moore (1966), 'An extension of the Newton–Kantorovich method for solving nonlinear equations with an application to elasticity', *J. Math. Anal. Appl.* **13**, 467–501.

P. J. Aston, A. Spence and W. Wu (1997), 'Hopf bifurcation near a double singular point', *J. Comput. Appl. Math.* **80**, 277–297.

B. Bara, K. Nandakumar and J. H. Masliyah (1992), 'An experimental and numerical study of the Dean problem: flow development towards two-dimensional multiple solutions', *J. Fluid Mech.* **244**, 339–376.

G. K. Batchelor (1970), *An Introduction to Fluid Mechanics*, Cambridge University Press, Cambridge.

F. Battaglia, S. J. Tavener, A. K. Kulkarni and C. L. Merkle (1997), 'Bifurcation of low Reynolds number flows in symmetric channels', *AIAA J.* **35**, 99–105.

L. Bauer, H. B. Keller and E. L. Reiss (1975), 'Multiple eigenvalues lead to secondary bifurcation', *SIAM J. Appl. Math.* **17**, 101–122.

H. Bénard (1901), 'Les tourbillons cellulaires dans une nappe de liquide transportant de la chaleur par convection en régime permanent', *Ann. Chem. Phys.* **23**, 62–144.

T. B. Benjamin (1978), 'Bifurcation phenomena in steady flows of a viscous liquid', *Proc. R. Soc. Lond. A* **359**, 1–26.

T. B. Benjamin and T. Mullin (1981), 'Anomalous modes in the Taylor experiment', *Proc. R. Soc. Lond. A* **377**, 221–249.

T. B. Benjamin and T. Mullin (1982), 'Notes on the multiplicity of flows in the Taylor experiment', *J. Fluid Mech.* **121**, 219–230.

A. Bergeon, D. Henry, H. BenHadid and L. S. Tuckerman (1998), 'Marangoni convection in binary mixtures with Soret effect', *J. Fluid Mech.* **375**, 143–174.

C. Bernardi (1982), 'Approximation of Hopf bifurcation', *Numer. Math.* **39**, 15–37.

C. Bernardi and J. Rappaz (1984), Approximation of Hopf bifurcation for semilinear parabolic equations, in *Numerical Methods for Bifurcation Problems* (T. Küpper, H. D. Mittelmann and H. Weber, eds), Birkhäuser, Boston, pp. 29–41.

W.-J. Beyn (1980), On discretization of bifurcation problems, in *Bifurcation Problems and their Numerical Solution* (H. D. Mittelmann and H. Weber, eds), Birkhäuser, Basel, pp. 46–75.

W.-J. Beyn (1984), Defining equations for singular solutions and numerical applications, in *Numerical Methods for Bifurcation Problems* (T. Küpper, H. D. Mittelmann and H. Weber, eds), Birkhäuser, Boston, pp. 42–56.

W. J. Beyn (1991), Numerical methods for dynamical systems, in *Advances in Numerical Analysis* (W. Light, ed.), Clarendon Press, Oxford, pp. 175–227.

W.-J. Beyn and E. Doedel (1981), 'Stability and multiplicity of solutions to discretizations of nonlinear ordinary differential equations', *SIAM J. Sci. Statist. Comput.* **2**, 107–120.

W.-J. Beyn and J. Lorenz (1982), 'Spurious solutions for discrete superlinear boundary value problems', *Computing* **28**, 43–51.

M. J. Block (1956), 'Surface tension as the cause of Bénard cells and surface deformation in a liquid film', *Nature* **178**, 650–651.

J. H. Bolstad and H. B. Keller (1987), 'Computation of anomalous modes in the Taylor experiment', *J. Comput. Phys.* **69**, 230–251.

A. Bossavit (1986), 'Symmetry, groups and boundary value problems: A progressive introduction to noncommutative harmonic analysis of partial differential equations in domains with geometrical symmetry', *Comput. Meth. Appl. Mech. Engrg* **56**, 167–215.

F. Brezzi and H. Fujii (1982), Numerical imperfections and perturbations in the approximation of nonlinear problems, in *Mathematics of Finite Elements and Applications IV* (J. R. Whiteman, ed.), Academic Press.

F. Brezzi, J. Rappaz and P. A. Raviart (1980), 'Finite dimensional approximation of nonlinear problems, Part I: Branches of nonsingular solutions', *Numer. Math.* **36**, 1–25.

F. Brezzi, J. Rappaz and P. A. Raviart (1981a), 'Finite dimensional approximation of nonlinear problems, Part II: Limit points', *Numer. Math.* **37**, 1–28.

F. Brezzi, J. Rappaz and P. A. Raviart (1981b), 'Finite dimensional approximation of nonlinear problems, Part III: Simple bifurcation points', *Numer. Math.* **38**, 1–30.

F. Brezzi, S. Ushiki and H. Fujii (1984), Real and ghost bifurcation dynamics in difference schemes for ordinary differential equations, in *Numerical Methods for Bifurcation Problems* (T. Küpper, H. D. Mittelmann and H. Weber, eds), Birkhäuser, Boston, pp. 79–104.

F. Chatelin (1973), 'Convergence of approximation methods to compute eigenvalues of linear operators', *SIAM J. Numer. Anal.* **10**, 939–948.

J.-H. Chen, W. G. Pritchard and S. J. Tavener (1995), 'Bifurcation for flow past a cylinder between parallel plates', *J. Fluid Mech.* **284**, 23–41.

A. J. Chorin and J. E. Marsden (1979), *A Mathematical Introduction to Fluid Mechanics*, Springer.

S.-N. Chow and J. K. Hale (1982), *Methods of Bifurcation Theory*, Springer, New York.

K. N. Christodoulou and L. E. Scriven (1988), 'Finding leading modes of a viscous free surface flow: an asymmetric generalised eigenvalue problem', *J. Sci. Comput.* **3**, 355–406.

K. A. Cliffe (1983), 'Numerical calculations of two-cell and single-cell Taylor flows', *J. Fluid Mech.* **135**, 219–233.

K. A. Cliffe (1988), 'Numerical calculations of the primary-flow exchange process in the Taylor problem', *J. Fluid Mech.* **197**, 57–79.

K. A. Cliffe and T. Mullin (1985), 'A numerical and experimental study of anomalous modes in the Taylor experiment', *J. Fluid Mech.* **153**, 243–258.

K. A. Cliffe and T. Mullin (1986), A numerical and experimental study of the Taylor problem with asymmetric end conditions, Technical Report TP1179, AERE.

K. A. Cliffe and A. Spence (1984), The calculation of high order singularities in the finite Taylor problem, in *Numerical Methods for Bifurcation Problems* (T. Küpper, H. D. Mittelmann and H. Weber, eds), Birkhäuser, Boston, pp. 129–144.

K. A. Cliffe and S. J. Tavener (1998), 'Marangoni–Bénard convection with a deformable free surface', *J. Comput. Phys.* **145**, 193–227.

K. A. Cliffe and S. J. Tavener (2000), 'Implementation of extended systems using symbolic algebra', *Notes in Numerical Fluid Mechanics.* To appear.

K. A. Cliffe and K. H. Winters (1986), 'The use of symmetry in bifurcation calculations and its application to the Bénard problem', *J. Comput. Phys.* **67**, 310–326.

K. A. Cliffe, T. J. Garratt and A. Spence (1993), 'Eigenvalues of the discretized Navier–Stokes equations with application to the detection of Hopf bifurcations', *Adv. Comput. Math.* **1**, 337–356.

K. A. Cliffe, J. J. Kobine and T. Mullin (1992), 'The role of anomalous modes in Taylor–Couette flow', *Proc. R. Soc. Lond. A* **439**, 341–357.

K. A. Cliffe, A. Spence and S. J. Tavener (2000), '$\mathcal{O}(2)$-symmetry breaking bifurcation: with application to flow past a sphere in a pipe', *Internat. J. Numer. Methods in Fluids.* To appear.

D. Coles (1965), 'Transition in circular Couette flow', *J. Fluid Mech.* **21**, 385–425.

M. Couette (1890), 'Études sur le frottement des liquides', *Ann. Chem. Phys.* **21**, 433–510.

M. G. Crandall and P. H. Rabinowitz (1971), 'Bifurcation from a simple eigenvalue', *J. Funct. Anal.* **8**, 321–340.

M. G. Crandall and P. H. Rabinowitz (1973), 'Bifurcation, perturbation of simple eigenvalues and linearised stability', *Arch. Rat. Mech. Anal.* **52**, 161–180.

M. G. Crandall and P. H. Rabinowitz (1977), 'The Hopf bifurcation thoerem in infinite dimensions', *Arch. Rat. Mech. Anal.* **67**, 53–72.

B. D. Davidson (1997), 'Large-scale continuation and numerical bifurcation for partial differential equations', *SIAM J. Numer. Anal.* **34**, 2008–2027.

W. R. Dean (1928), 'The stream-line motion of fluid in a curved pipe', *Phil. Mag.* **5**, 673–695.

J. Descloux and J. Rappaz (1982), 'Approximation of solution branches of nonlinear equations', *RAIRO* **16**, 319–349.

H. A. Dijkstra (1992), 'On the structure of cellular solutions in Rayleigh–Bénard–Marangoni flows in small-aspect-ratio containers', *J. Fluid Mech.* **243**, 73–102.

H. A. Dijkstra and E. J. Kranenborg (1996), 'A bifurcation study of double diffusive flows in a laterally heated stably stratified liquid layer', *Int. J. Heat Mass Trans.* **39**, 2699–2710.

H. A. Dijkstra and M. J. Molemaker (1997), 'Symmetry breaking and overturning oscillation in thermohaline driven flows', *J. Fluid Mech.* **331**, 169–198.

H. A. Dijkstra, M. J. Molemaker, A. Vanderploeg and E. F. F. Botta (1995), 'An efficient code to compute nonparallel steady flows and their linear-stability', *Computers and Fluids* **24**, 415–434.

R. C. DiPrima and H. L. Swinney (1981), Instabilities and transition in flow between concentric roataing cylinders, in *Hydrodynamic Instability and the Transition to Turbulence* (H. L. Swinney and J. P. Gollub, eds), Springer, pp. 139–180.

E. J. Doedel and J. P. Kernevez (1986), AUTO: Software for continuation and bifurcation problems in ordinary differential equations, Technical report, Caltech, Pasadena.

E. J. Doedel, A. R. Champneys, T. F. Fairgrieve, Y. A. Kuznetsov, B. Sandstede and X. J. Wang (1997), AUTO: *Continuation and Bifurcation Software with Ordinary Differential Equations (with HomCont), User's Guide*, Concordia University, Montreal, Canada.

D. Drikakis (1997), 'Bifurcation phenomena in incompressible sudden expansion flows', *Phys. Fluids* **9**, 76–87.

R. M. Fearn, T. Mullin and K. A. Cliffe (1990), 'Nonlinear flow phenomena in a symmetric sudden expansion', *J. Fluid Mech.* **211**, 595–608.

J. P. Fink and W. C. Rheinboldt (1983), 'On discretization error of parametrized nonlinear equations', *SIAM J. Numer. Anal.* **20**, 732–746.

J. P. Fink and W. C. Rheinboldt (1984), 'Solution manifolds and submanifolds of parametrized equations and their discretization errors', *Numer. Math.* **45**, 323–343.

J. P. Fink and W. C. Rheinboldt (1985), 'Local error estimates for parametrized nonlinear equations', *SIAM J. Numer. Anal.* **22**, 729–735.

M. Fortin (1993), Finite element solutions of the Navier–Stokes equations, in *Acta Numerica*, Vol. 2, Cambridge University Press, pp. 239–284.

G. Frank and R. Meyer-Spasche (1981), 'Computation of transitions in Taylor vortex flows', *ZAMP* **32**, 710–720.

H. Fujii and M. Yamaguti (1980), 'Structure of singularities and its numerical realisation in nonlinear elasticity', *J. Math. Kyoto Univ.* **20**, 489–590.

T. J. Garratt, G. Moore and A. Spence (1991), Two methods for the numerical detection of Hopf bifurcations, in *Bifurcation and Chaos: Analysis, Algorithms, Applications* (R. Seydel, T. Küpper, F. W. Schneider and H. Troger, eds), Birkhäuser, Basel, pp. 129–133.

V. Girault and P. A. Raviart (1986), *Finite Element Methods for Navier–Stokes Equations*, Springer, New York.

M. Golubitsky and D. Schaeffer (1979a), 'Imperfect bifurcation theory in the presence of symmetry', *Commun. Math. Phys.* **67**, 205–232.

M. Golubitsky and D. Schaeffer (1979b), 'A theory for imperfect bifurcation via singularity theory', *Commun. Pure. Appl. Math.* **32**, 21–98.

M. Golubitsky and D. G. Schaeffer (1983), 'A discussion of symmetry and symmetry breaking', *Proc. Symp. Pure Math.* **40**, 499–515.

M. Golubitsky and D. G. Schaeffer (1985), *Singularities and Groups in Bifurcation Theory*, Vol. I, Springer, New York.

M. Golubitsky and I. Stewart (1986), 'Symmetry and stability in Taylor–Couette flow', *SIAM J. Math. Anal.* **17**, 249–288.

M. Golubitsky, I. Stewart and D. G. Schaeffer (1988), *Singularities and Groups in Bifurcation Theory*, Vol. II, Springer, New York.

W. Govaerts (1991), 'Stable solvers and block elimination for bordered systems', *SIAM J. Matrix Anal. Appl.* **12**, 469–483.

W. Govaerts (1995), 'Bordered matrices and singularities of large nonlinear systems', *Int. J. Bifurcation and Chaos* **5**, 243–250.

W. Govaerts (1997), 'Computation of singularities in large nonlinear systems', *SIAM J. Numer. Anal.* **34**, 867–880.

W. Govaerts (2000), *Numerical Methods for Bifurcations of Dynamic Equilibria*, SIAM. To appear.

W. Govaerts and A. Spence (1996), 'Detection of Hopf points by counting sectors in the complex plane', *Numer. Math.* **75**, 43–58.

P. M. Gresho and R. L. Sani (1998), *Incompressible Flow and the Finite Element Method: Advection-Diffusion and Isothermal Laminar Flow*, Wiley, Chichester, UK.

P. M. Gresho, D. K. Gartling, J. R. Torczynski, K. A. Cliffe, K. H. Winters, T. J. Garratt, A. Spence and J. W. Goodrich (1993), 'Is the steady viscous incompressible 2D flow over a backward facing step at Re=800 stable?', *Int. J. Numer. Meth. Fluids* **17**, 501–541.

A. Griewank and G. Reddien (1983), 'The calculation of Hopf points by a direct method', *IMA J. Numer. Anal.* **3**, 295–303.

A. Griewank and G. Reddien (1984), 'Characterization and computation of generalized turning points', *SIAM J. Numer. Anal.* **21**, 176–185.

A. Griewank and G. Reddien (1989), 'Computation of cusp singularities for operator equations and their discretizations', *J. Comput. Appl Math.* **16**, 133–153.

A. Griewank and G. W. Reddien (1996), 'The approximate solution of defining equations for generalised turning points', *SIAM J. Numer. Anal.* **33**, 1912–1920.

P. Grisvard (1985), *Elliptic Problems in Nonsmooth Domains*, Pitman, Boston.

J. Guckenheimer (1981), On a codimension two bifurcation, in *Dynamical Systems and Turbulence* (D. A. Rand and L.-S. Young, eds), Vol. 898 of *Lecture Notes in Mathematics*, Springer, pp. 99–142.

J. Guckenheimer, M. Myers and B. Sturmfels (1997), 'Computing Hopf bifurcations I', *SIAM J. Numer. Anal.* **34**, 1–21.

P. Hall (1980), 'Centrifugal instabilities in finite container: a periodic model', *J. Fluid Mech.* **99**, 575–596.

P. Hall (1982), 'Centrifugal instabilities of circumferential flows in finite cylinders: the wide gap problem', *Proc. R. Soc. Lond.* A **384**, 359–379.

B. D. Hassard, N. D. Kazarinoff and Y.-H. Wan (1981), *Theory and Applications of Hopf Bifurcation*, Vol. 41 of *London Math. Soc. Lecture Note Series*.

S. C. Hawkins and A. Spence (2000), 'The detection of Hopf bifurcations in discretizations of certain nonlinear partial differential equations', *Notes in Numerical Fluid Mechanics*. To appear.

A. C. Hearn (1987), REDUCE *Users' Manual*, CP78, Rand Publications.

J. Hofbauer and G. Iooss (1984), 'A Hopf bifurcation theorem for difference equations approximating a differential equation', *Mh. Math.* **98**, 99–113.

G. Iooss (1986), 'Secondary bifurcations of Taylor vortices into wavy inflow or outflow boundaries', *J. Fluid. Mech.* **173**, 273–288.

C. P. Jackson (1987), 'A finite-element study of the onset of vortex shedding in flow past variously shaped bodies', *J. Fluid Mech.* **182**, 23–45.

V. Janovsky (1987), 'Minimally extended defining conditions for singularities of codim less-than-or-equal-to 2', *Numer. Funct. Anal. Opt.* **9**, 1309–1349.

V. Janovský and P. Plecháč (1992), 'Computer aided analysis of imperfect bifurcation diagrams, I: Simple bifurcation point and isola formation centre', *SIAM J. Numer. Anal.* **21**, 498–512.

A. Jepson and A. Spence (1985a), 'Folds in solutions of two parameter systems and their calculation, Part I', *SIAM J. Numer. Anal.* **22**, 347–368.

A. D. Jepson (1981), Numerical Hopf bifurcation, PhD thesis, Caltech, Pasadena.

A. D. Jepson and A. Spence (1984), Singular points and their computation, in *Numerical Methods for Bifurcation Problems* (T. Küpper, H. D. Mittelmann and H. Weber, eds), Birkhäuser, Boston, pp. 195–209.

A. D. Jepson and A. Spence (1985b), 'The numerical solution of nonlinear equations having several parameters, I: scalar equations', *SIAM J. Numer. Anal.* **22**, 736–759.

H. B. Keller (1975), 'Approximate methods for nonlinear problems with application to two-point boundary value problems', *Math. Comput.* **29**, 464–474.

H. B. Keller (1977), Numerical solution of bifurcation and nonlinear eigenvalue problems, in *Applications of Bifurcation Theory* (P. Rabinowitz, ed.), Academic Press, New York, pp. 359–384.

H. B. Keller (1987), *Numerical Methods in Bifurcation Problems*, Springer.

J. B. Keller and S. Antman, eds (1969), *Bifurcation Theory and Nonlinear Eigenvalue Problems*, Benjamin, New York.

A. Khibnik, K. Yu, V. V. Levitin and E. Nikoleav (1993), LOCBIF *Version 2: Interactive LOCal BIFurcation Analyser*, CAN Expertise Centre, Amsterdam.

F. Kikuchi (1977), 'Finite element approximations to bifurcation problems of turning point type', *Theoretical and Applied Mechanics* **27**, 99–114.

A. V. Knyazev (1998), 'Preconditioned eigensolvers: an oxymoron', *ETNA* pp. 104–123.

M. A. Krasnosel'skii, G. M. Vainikko, P. P. Zabreiko, Y. B. Rutitskii and V. Y. Stetsenko (1972), *Approximate Solution of Operator Equations*, Wolters-Noordhoff, Groningen.

M. Kubíček and M. Marek (1983), *Computational Methods in Bifurcation Theory and Dissipative Structures*, Springer.

T. Küpper, H. D. Mittelmann and H. Weber, eds (1984), *Numerical Methods for Bifurcation Problems*, Vol. 70 of *International Series in Numerical Mathematics*, Birkhäuser, Basel.

T. Küpper, R. Seydel and H. Troger, eds (1987), *Bifurcation: Analysis, Algorithms and Applications*, Vol. 79 of *International Series in Numerical Mathematics*, Birkhäuser, Basel.

P. H. Rabinowitz (1977), *Applications of Bifurcation Theory*, Academic Press, New York.

Lord Rayleigh (1916), On the dynamics of revolving fluids, Vol. 6 of *Scientific Papers*, Cambridge University Press, pp. 447–453.

E. L. Reiss (1969), Column buckling: an elementary example of bifurcation, in *Bifurcation Theory and Nonlinear Eigenvalue Problems* (J. B. Keller and S. Antman, eds), Benjamin, New York, pp. 1–16.

W. C. Rheinboldt (1978), 'Numerical methods for a class of finite dimensional bifurcation problems', *SIAM J. Numer. Anal.* **15**, 1–11.

W. C. Rheinboldt (1986), *Numerical Analysis of Parametrized Nonlinear Equations*, Wiley-Interscience, New York.

E. Ricks (1972), 'The application of Newton's method to the problem of elastic stability', *J. Appl. Mech.* pp. 1060–1065.

P. H. Roberts (1965), 'Appendix in "Experiments on the stability of viscous flow between rotating cylinders"', *Proc. R. Soc. Lond. A* **283**, 550–555.

D. Roose and V. Hlavacek (1985), 'A direct method for the computation of Hopf bifurcation points', *SIAM J. Appl. Math.* **45**, 879–894.

D. Roose, B. D. Dier and A. Spence (1990), *Continuation and Bifurcations: Numerical Techniques and Applications*, Vol. 313 of *NATO ASI Series C: Mathematical and Physical Sciences*, Kluwer, Dordrecht.

R. L. Sani, P. M. Gresho, R. L. Lee and D. F. Griffiths (1981), 'On the cause and cure(?) of the spurious pressures generated by certain FEM solutions of the incompressible Navier–Stokes equations, Parts 1 and 2', *Int. J. Numer. Meth. Fluids.* Part 1: pp. 17–43; Part 2: pp. 171–204.

D. H. Sattinger (1973), *Topics in Stability and Bifurcation Theory*, Vol. 309 of *Lecture Notes in Mathematics*, Springer, Berlin.

D. G. Schaeffer (1980), 'Qualitative analysis of a model for boundary effects in the Taylor problem', *Math. Proc. Camb. Phil. Soc.* **87**, 307–337.

G. Schrauf (1983), 'Numerical investigation of Taylor-vortex flows in a spherical gap', *ZAMM* **63**, T282–T286.

G. Schrauf (1986), 'The 1st instability in spherical Taylor–Couette flow', *J. Fluid Mech.* **166**, 287–303.

W. Schröder and H. B. Keller (1990), 'Wavy Taylor-vortex flows via multigrid-continuation methods', *J. Comput. Phys.* **91**, 197–227.

G. Schroff and H. B. Keller (1993), 'Stabilization of unstable procedures: The recursive projection method', *SIAM J. Numer. Anal.* **30**, 1099–1120.

M. Selmi, K. Nandakumar and W. H. Finlay (1994), 'A bifurcation study of viscous-flow through a rotating curved duct', *J. Fluid Mech.* **262**, 353–375.

R. Seydel (1979a), 'Numerical computation of branch points in nonlinear equations', *Numer. Math.* **32**, 339–352.

R. Seydel (1979b), 'Numerical computation of branch points in ordinary differential equations', *Numer. Math.* **32**, 51–68.

R. Seydel (1994), *Practical Bifurcation and Stability Analysis: From Equilibrium to Chaos*, Springer, New York.

R. Seydel, T. Küpper, F. W. Schneider and H. Troger, eds (1991), *Bifurcation and Chaos: Analysis, Algorithms, Applications*, Vol. 97 of *International Series in Numerical Mathematics*, Birkhäuser, Basel.

H. von Sosen (1994), Part I: Folds and bifurcations in solutions of semi-explicit differential-algebraic equations; Part II: The Recursive Projection Method applied to differential-algebraic systems and incompressible fluid mechanics, PhD thesis, California Institute of Technology.

H. Specht, M. Wagner and R. Meyer-Spasche (1989), 'Interactions of secondary branches of Taylor vortex solutions', *ZAMM* **69**, 339–352.

A. Spence and I. G. Graham (1999), Numerical methods for bifurcation problems, in *The Graduate Student's Guide to Numerical Analysis '98* (M. Ainsworth, J. Levesley and M. Marletta, eds), Vol. 26 of *Springer Series in Computational Mathematics*, Springer, Berlin, pp. 177–216.

A. Spence and B. Werner (1982), 'Non-simple turning points and cusps', *IMA J. Numer. Anal.* **2**, 413–427.

A. Spence, K. A. Cliffe and A. D. Jepson (1989), 'A note on the calculation of Hopf bifurcations', *J. Comput. Appl. Math.* **26**, 125–131.

I. Stackgold (1971), 'Branching of solutions of nonlinear equations', *SIAM Rev.* **13**, 289–332.

J. T. Stuart (1986), 'Taylor-vortex flow: A dynamical system', *SIAM Review* **28**, 315–342.

J. L. Synge (1933), 'The stability of heterogeneous liquids', *Trans. R. Soc. Canada* **27**, 1–18.

S. J. Tavener and K. A. Cliffe (1991), 'Primary flow exchange mechanisms in the Taylor apparatus applying impermeable stress-free boundary conditions', *IMA J. Appl. Math.* **46**, 165–199.

S. J. Tavener, T. Mullin and K. A. Cliffe (1991), 'Novel bifurcation phenomena in a rotating annulus', *J. Fluid Mech.* **229**, 483–497.

G. I. Taylor (1923), 'Stability of a viscous liquid between two rotating cylinders', *Proc. R. Soc. Lond. A* **223**, 289–343.

L. N. Trefethen (1997), 'Pseudospectra of linear operators', *SIAM Review* **39**, 383–406.

L. N. Trefethen, A. E. Trefethen, S. C. Reddy and T. A. Driscoll (1993), 'Hydrodynamical stability without eigenvalues', *Science* **3**, 578–584.

A. Vanderbauwhede (1982), *Local bifurcation and symmetry*, Vol. 75 of *Research Notes in Mathematics*, Pitman, London.

H. J. Weinitschke (1985), 'On the calculation of limit and bifurcation points in stability problems of elastic shells', *Int. J. Solids Structures* **21**, 79–95.

R. Weiss (1974), 'On the approximation of fixed points of nonlinear compact operators', *SIAM J. Numer. Anal.* **11**, 550–553.

B. Werner and V. Janovsky (1991), Computation of Hopf branches bifurcating from Takens–Bogdanov points for problems with symmetry, in *Bifurcation and Chaos: Analysis, Algorithms, Applications* (R. Seydel, T. Küpper, F. W. Schneider and H. Troger, eds), Birkhäuser, Basel, pp. 377–388.

B. Werner and A. Spence (1984), 'The computation of symmetry breaking bifurcation points', *SIAM J. Numer. Anal.* **21**, 388–399.

S. Wiggins (1990), *Introduction to Applied Nonlinear Dynamical Systems and Chaos*, Springer, New York.

K. H. Winters (1987), 'A bifurcation study of laminar-flow in a curved tube of rectangular cross-section', *J. Fluid Mech.* **180**, 343–369.

K. H. Winters, T. Plesser and K. A. Cliffe (1988), 'The onset of convection in a finite container due to surface tension', *Physica D* **29**, 387–401.

J. Wloka (1987), *Partial Differential Equations*, Cambridge University Press. Translated by C. B. and M. L. Thomas.

S. H. Xin, P. LeQuere and L. S. Tuckerman (1998), 'Bifurcation analysis of double-diffusive convection with opposing horizontal thermal and solutal gradients', *Phys. Fluids* **10**, 850–858.

R. J. Yang (1996), 'A numerical procedure for predicting multiple solutions of a spherical Taylor–Couette flow', *Int. J. Numer. Meth. Fluids* **22**, 1135–1147.

Acta Numerica (2000), pp. 133–213

Triangulations and meshes in computational geometry

Herbert Edelsbrunner*

Department of Computer Science,
Duke University, Durham, NC 27708
and
Raindrop Geomagic, Research Triangle Park,
North Carolina, NC 27709, USA

The Delaunay triangulation of a finite point set is a central theme in computational geometry. It finds its major application in the generation of meshes used in the simulation of physical processes. This paper connects the predominantly combinatorial work in classical computational geometry with the numerical interest in mesh generation. It focuses on the two- and three-dimensional case and covers results obtained during the twentieth century.

* Research is partially supported by the Army Research Office under grant DAAG55-98-1-0177 and by the National Science Foundation under grants CCR-96-19542 and CCR-97-12088.

CONTENTS

1. Introduction

This is a paper about computational geometry and its connection to science and engineering. We argue that computational geometry draws its motivation from applications to various areas including mesh generation and that it can maintain its livelihood only if it fulfils the promise of advancing these applications in a significant manner.

History

The beginning of computational geometry as an independent intellectual discipline is usually dated around 1975, when Michael Shamos and Dan Hoey proposed algorithmic solutions for a host of basic geometric tasks (Shamos 1975, Shamos and Hoey 1975, 1976). They defined computational geometry as the study of the computational complexity of geometric problems. It is important to notice the implicit but significant shift from a continuous to a discrete conception of geometry. Application areas use geometry to model a presumably continuous reality, while computational complexity relates the finite amount of time it takes to solve a problem with the finite size in which the problem presents itself. Within a few years after its inception, computational geometry developed a strong affinity to discrete geometry as practised by combinatorialists (Erdős 1979, Pach and Agarwal 1995). This affinity was natural and helped the field to mature to a point where it is ready for a reorientation back to its continuous roots.

The intellectual development in computational geometry can be traced fairly well through the series of proceedings documenting the annual Symposium on Computational Geometry, first held in 1985. The breadth of the field is evident from the textbooks, which all take different views and explore different aspects of the field (de Berg, van Kreveld, Overmars and Schwarzkopf 1997, Edelsbrunner 1987, Mulmuley 1994, Klein 1997, Okabe, Boots and Sugihara 1992, O'Rourke 1987, 1994, Preparata and Shamos 1985). We also refer to a recent handbook, which organizes the combined field of discrete and computational geometry in 52 chapters (Goodman and O'Rourke 1997).

Outline

We illustrate the claimed function of computational geometry as a bridge between continuous and discrete methods with a focus on geometric triangulations and in particular Delaunay triangulations. Half the paper studies combinatorial properties of and algorithms for Delaunay triangulations. The other half explores questions that arise in the use of Delaunay triangulations as a representation of pieces of continuous space. To emphasize the shift in focus, we then refer to the triangulation as a mesh, which is the traditional engineering term for space decompositions used in numerical analysis (Bern and Eppstein 1992).

There is an orthogonal way of structuring this paper in two halves. Sections 2 to 8 deal with triangulations in the Euclidean plane, and Sections 9 to 16 study tetrahedrizations in three-dimensional Euclidean space.

In the predominantly discrete block consisting of Sections 2 to 5, we see a progression from geometric/structural to algorithmic considerations, and we see the same in the block consisting of Sections 9 to 12. The move towards a continuous and numerical viewpoint is pursued in the block consisting of Sections 6 to 8 and in the block consisting of Sections 13 to 16.

Style

The style of this paper is representative of the dominant style in computational geometry. Understanding is sought through formulating general claims and proving them. Similarly, algorithms are described in detail and the running time is analysed under worst-case and average assumptions. We make a conscious effort to concentrate on the two- and three-dimensional cases and with a few exceptions avoid discussions of the general d-dimensional case. While identifying properties that hold independent of the particular dimension is generally commendable, it seems counterproductive in the study of meshes whose properties vary significantly with changing dimension.

Each section is designed as a lecture in a graduate course. Whenever there is a choice, we prefer topics that have a general appeal over more specialized ones, and topics that are easy to explain over more complicated ones. Each section ends with bibliographic notes collecting references to the literature and comments on important related developments.

2. Voronoi and Delaunay

This section introduces Delaunay triangulations as duals of Voronoi diagrams. It discusses the role of general position in the definition and explains some of the basic properties of Delaunay triangulations.

Voronoi diagrams

Given a finite set of points in the plane, the idea is to assign to each point a region of influence in such a way that the regions decompose the plane. To describe a specific way to do that, let $S \subseteq \mathbb{R}^2$ be a set of n points and define the *Voronoi region* of $p \in S$ as the set of points $x \in \mathbb{R}^2$ that are at least as close to p as to any other point in S, that is,

$$V_p = \{x \in \mathbb{R}^2 : \|x - p\| \leq \|x - q\|, \ \forall q \in S\}.$$

This definition is illustrated in Figure 1. Consider the half-plane of points at least as close to p as to q: $H_{pq} = \{x \in \mathbb{R}^2 : \|x - p\| \leq \|x - q\|\}$. The Voronoi region of p is the intersection of half-planes H_{pq}, for all $q \in S - \{p\}$. It follows that V_p is a convex polygonal region, possibly unbounded, with at most $n - 1$ edges.

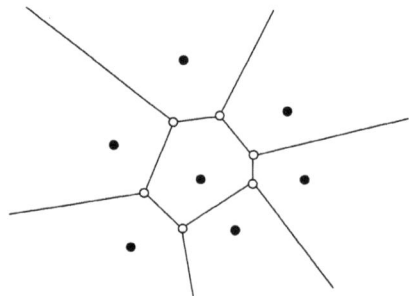

Fig. 1.   Seven points define the same number of
Voronoi regions. One of the regions is bounded
because the defining point is completely surrounded
by the others

Each point $x \in \mathbb{R}^2$ has at least one nearest point in S, so it lies in at least one Voronoi region. It follows that the Voronoi regions cover the entire plane. Two Voronoi regions lie on opposite sides of the perpendicular bisector separating the two generating points. It follows that Voronoi regions do not share interior points, and if a point x belongs to two Voronoi regions then it lies on the bisector of the two generators. The Voronoi regions together with their shared edges and vertices form the *Voronoi diagram* of S.

Delaunay triangulation

We get a dual diagram if we draw a straight *Delaunay edge* connecting points $p, q \in S$ if and only if their Voronoi regions intersect along a common line segment; see Figure 2. In general, the Delaunay edges decompose the convex hull of S into triangular regions, which are referred to as *Delaunay triangles*.

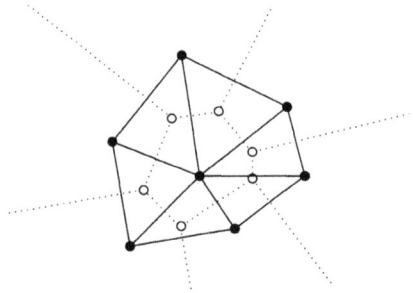

Fig. 2. The Voronoi edges are dotted and
the dual Delaunay edges are solid

To count the Delaunay edges we use some results on *planar graphs* defined by the property that their edges can be drawn in the plane without crossing. It is true that no two Delaunay edges cross each other, but to avoid an argument, we draw each Delaunay edge from one endpoint straight to the midpoint of the shared Voronoi edge and then straight to the other endpoint. Now it is trivial that no two of these edges cross. Using Euler's relation, it can be shown that a planar graph with $n \geq 3$ vertices has at most $3n - 6$ edges and at most $2n - 4$ faces. The same bounds hold for the number of Delaunay edges and triangles. There is a bijection between the Voronoi edges and the Delaunay edges, so $3n - 6$ is also an upper bound on the number of Voronoi edges. Similarly, $2n - 4$ is an upper bound on the number of Voronoi vertices.

Degeneracy

There is an ambiguity in the definition of Delaunay triangulation if four or more Voronoi regions meet at a common point u. One such case is shown in Figure 3. The points generating the four or more regions all have the same distance from u: they lie on a common circle around u. Probabilistically, the chance of picking even just four points on a circle is zero because the circle defined by the first three points has zero measure in $\mathbb{R}^2$. A common way to say the same thing is that four points on a common circle form a *degeneracy* or a *special case*. An arbitrarily small perturbation suffices to remove the degeneracy and to reduce the special to the general case.

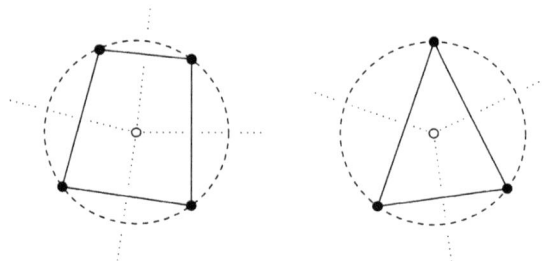

Fig. 3. To the left, four dotted Voronoi edges meet at
a common vertex and the dual Delaunay edges bound
a quadrilateral. To the right, we have the general case,
where only three Voronoi edges meet at a common
vertex and the Delaunay edges bound a triangle

We will often assume *general position*, which is the absence of any degeneracy. This really means that we delay the treatment of degenerate cases to later. The treatment is eventually done by perturbation, which can be actual or conceptual, or by exhaustive case analysis.

Circles and power

For now we assume general position. For a Delaunay triangle, abc, consider the circumcircle, which is the unique circle passing through a, b, and c. Its centre is the corresponding Voronoi vertex, $u = V_a \cap V_b \cap V_c$, and its radius is $\varrho = \|u - a\| = \|u - b\| = \|u - c\|$; see Figure 3. We call the circle *empty* because it encloses no point of S. It turns out that empty circles characterize Delaunay triangles.

Circumcircle Claim.   Let $S \subseteq \mathbb{R}^2$ be finite and in general position, and let $a, b, c \in S$ be three points. Then abc is a Delaunay triangle if and only if the circumcircle of abc is empty.

It is not entirely straightforward to see that this is true, at least not at the moment. Instead of proving the Circumcircle Claim, we focus our attention on a new concept of distance from a circle. The *power* of a point $x \in \mathbb{R}^2$ from a circle U with centre u and radius ϱ is

$$\pi_U(x) \;\; = \;\; \|x - u\|^2 - \varrho^2.$$

If x lies outside the circle, then $\pi_U(x)$ is the square length of a tangent line segment connecting x with U. In any case, the power is positive outside the circle, zero on the circle, and negative inside the circle. We sometimes think of a circle as a weighted point and of the power as a weighted distance to that point. Given two circles, the set of points with equal power from both is a line. Figure 4 illustrates three different arrangements of two circles and their bisectors of points with equal power from both.

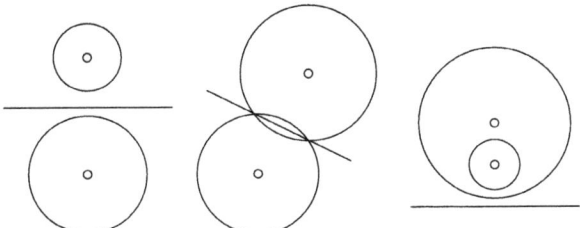

Fig. 4. Three times two circles with bisector.
From left to right: two disjoint and non-nested
circles, two intersecting circles, two nested circles

Acyclicity

We use the notion of power to prove an acyclicity result for Delaunay triangles. Let $x \in \mathbb{R}^2$ be an arbitrary but fixed viewpoint. We say a triangle *abc lies in front of* another triangle *def* if there is a half-line starting at x that first passes through *abc* and then through *def*; see Figure 6. We write *abc* $\prec$ *def* if *abc* lies in front of *def*. The set of Delaunay triangles together with $\prec$ forms a relation. General relations have cycles, which are sequences $\tau_0 \prec \tau_1 \prec \cdots \prec \tau_k \prec \tau_0$. Such cycles can also occur in general triangulations, as illustrated in Figure 5, but they cannot occur if the triangles are defined by empty circumcircles.

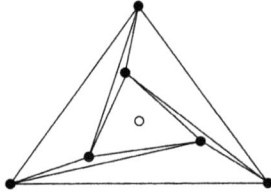

Fig. 5. From the viewpoint in the middle, the three
skinny triangles form a cycle in the in-front relation

Acyclicity Lemma. The in-front relation for the set of Delaunay triangles defined by a finite set $S \subseteq \mathbb{R}^2$ is acyclic.

Proof. We show that *abc* $\prec$ *def* implies that the power of x from the circumcircle of *abc* is less than the power from the circumcircle of *def*. Define

$abc = \tau_0$ and write $\pi_0(x)$ for the power of x from the circumcircle of abc. Similarly define $def = \tau_k$ and $\pi_k(x)$. Because S is finite, we can choose a half-line that starts at x, passes through abc and def, and contains no point of S. It intersects a sequence of Delaunay triangles:

$$abc = \tau_0 \prec \tau_1 \prec \cdots \prec \tau_k = def.$$

For any two consecutive triangles, the bisector of the two circumcircles contains the common edge. Because the third point of τ_{i+1} lies outside the circumcircle of τ_i we have $\pi_i(x) < \pi_{i+1}(x)$, for $0 \le i \le k - 1$. Hence $\pi_0(x) < \pi_k(x)$. The acyclicity of the relation follows because real numbers cannot increase along a cycle.   □

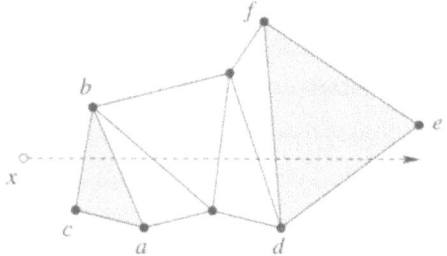

Fig. 6. Triangle abc lies in front of triangle def.
If abc and def belong to a Delaunay triangulation,
then there is a sequence of triangles between them
that all intersect the half-line

Bibliographic notes

Voronoi diagrams are named after the Russian mathematician Georges Voronoi, who published two seminal papers at the beginning of the twentieth century (Voronoi 1907/08). The same concept was discussed about half a century earlier by P. G. L. Dirichlet, and there are unpublished notes by René Descartes suggesting that he was already using Voronoi diagrams in the first half of the seventeenth century. Delaunay triangulations are named after the Russian mathematician Boris Delaunay, who dedicated his paper on empty spheres (Delaunay 1934) to Georges Voronoi. The article by Franz Aurenhammer (1991) offers a nice survey of Voronoi diagrams and their algorithmic applications. The acyclicity of Delaunay triangulations in arbitrary dimensions was proved by Edelsbrunner (1990) and subsequently applied in computer graphics. In particular, the three-dimensional case has been exploited for the visualization of diffuse volumes (Max, Hanrahan and Crawfis 1990, Williams 1992).

3. Edge flipping

This section introduces a local condition for edges, shows it implies a triangulation is Delaunay, and derives an algorithm based on edge flipping. The correctness of the algorithm implies that, among all triangulations of a given point set, the Delaunay triangulation maximizes the smallest angle.

Empty circles

Recall the Circumcircle Claim, which says that three points $a, b, c \in S$ are vertices of a Delaunay triangle if and only if the circle that passes through a, b, c is empty. A Delaunay edge, ab, belongs to one or two Delaunay triangles. In either case, there is a pencil of empty circles passing through a and b. The centres of these circles are the points on the Voronoi edge $V_a \cap V_b$; see Figure 7. What the Circumcircle Claim is for triangles, the Supporting Circle Claim is for edges.

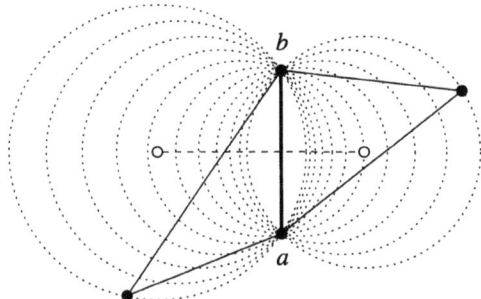

Fig. 7. The Voronoi edge is the dashed line segment of
centres of circles passing through the endpoints of ab

Supporting Circle Claim. Let $S \subseteq \mathbb{R}^2$ be finite and in general position and $a, b \in S$. Then ab is a Delaunay edge if and only if there is an empty circle that passes through a and b.

Delaunay lemma

By a *triangulation* we mean a collection of triangles together with their edges and vertices. A triangulation K *triangulates* S if the triangles decompose the convex hull of S and the set of vertices is S. An edge $ab \in K$ is *locally Delaunay* if

  (i) it belongs to only one triangle and therefore bounds the convex hull of S, or
 (ii) it belongs to two triangles, abc and abd, and d lies outside the circumcircle of abc.

The definition is illustrated in Figure 8. A locally Delaunay edge is not necessarily an edge of the Delaunay triangulation, and it is fairly easy to construct such an example. However, if *every* edge is locally Delaunay then we can show that *all* are Delaunay edges.

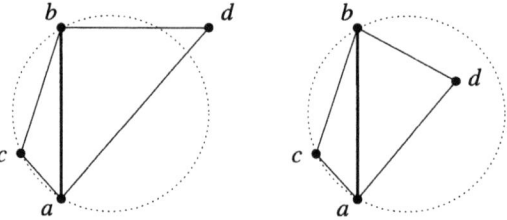

Fig. 8. To the left ab is locally Delaunay
and to the right it is not

Delaunay Lemma. If every edge of K is locally Delaunay then K is the Delaunay triangulation of S.

Proof. Consider a triangle $abc \in K$ and a vertex $p \in K$ different from a, b, c. We show that p lies outside the circumcircle of abc. Because this is then true for every p, the circumcircle of abc is empty, and because this is then true for every triangle abc, K is the Delaunay triangulation of S. Choose a point x inside abc such that the line segment from x to p contains no vertex other than p. Let $abc = \tau_0, \tau_1, \ldots, \tau_k$ be the sequence of triangles that intersect xp, as in Figure 9. We write $\pi_i(p)$ for the power of p to the circumcircle of τ_i, as before. Since the edges along xp are all locally Delaunay, we have $\pi_0(p) > \pi_1(p) > \cdots > \pi_k(p)$. Since p is one of the vertices of the last triangle we have $\pi_k(p) = 0$. Therefore $\pi_0(p) > 0$, which is equivalent to p lying outside the circumcircle of abc. □

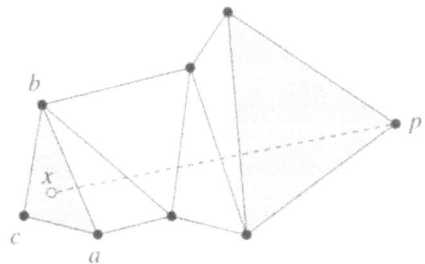

Fig. 9. Sequence of triangles in K that intersect xp

Edge-flip algorithm

If ab belongs to two triangles, abc and abd, whose union is a convex quadrangle, then we can *flip* ab to cd. Formally, this means we remove ab, abc, abd

from the triangulation and we add cd, acd, bcd to the triangulation, as in Figure 10. The picture of a flip looks like a tetrahedron with front and back superimposed. We can use edge flips as elementary operations to convert an

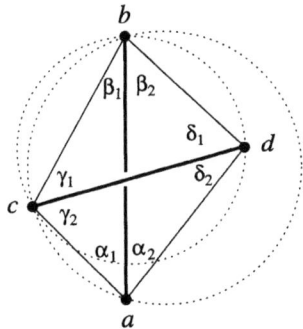

Fig. 10. Flipping ab to cd. If ab is not locally
Delaunay then the union of the two triangles
is convex and cd is locally Delaunay

arbitrary triangulation K to the Delaunay triangulation. The algorithm uses a stack and maintains the invariant that unless an edge is locally Delaunay, it resides on the stack. To avoid duplicates, we mark edges stored on the stack. Initially, all edges are marked and pushed on the stack.

```
while stack is non-empty do
    pop ab from stack and unmark it;
    if ab not locally Delaunay then
        flip ab to cd;
        for xy ∈ {ac, cb, bd, da} do
            if xy not marked then
                mark xy and push it on stack
            endif
        endfor
    endif
endwhile.
```

Let n be the number of points. The amount of memory used by the algorithm is $O(n)$ because there are at most $3n - 6$ edges, and the stack contains at most one copy of each edge. At the time the algorithm terminates every edge is locally Delaunay. By the Delaunay lemma, the triangulation is therefore the Delaunay triangulation of the point set.

Circle and plane

Before proving the algorithm terminates, we interpret a flip as a tetrahedron in three-dimensional space. Let $\hat{a}, \hat{b}, \hat{c}, \hat{d}$ be the vertical projections of points

a, b, c, d in the $x_1 x_2$-plane onto the paraboloid defined as the graph of Π : $x_3 = x_1^2 + x_2^2$; see Figure 11.

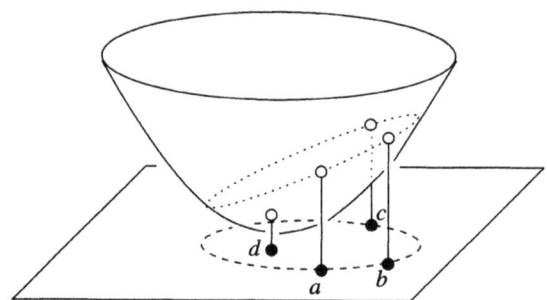

Fig. 11. Points a, b, c lie on the dashed circle in the $x_1 x_2$-plane and d lies inside that circle. The dotted curve is the intersection of the paraboloid with the plane that passes through $\hat{a}, \hat{b}, \hat{c}$. It is an ellipse whose projection is the dashed circle

Lifted Circle Claim.   Point d lies inside the circumcircle of abc if and only if point $\hat{d}$ lies vertically below the plane passing through $\hat{a}, \hat{b}, \hat{c}$.

Proof.   Let U be the circumcircle of abc and H the plane passing through $\hat{a}, \hat{b}, \hat{c}$. We first show that U is the vertical projection of $H \cap \text{gf } \Pi$. Transform the entire space by mapping every point (x_1, x_2, x_3) to $(x_1, x_2, x_3 - x_1^2 - x_2^2)$. Points $\hat{a}, \hat{b}, \hat{c}, \hat{d}$ are mapped back to a, b, c, d and the paraboloid Π becomes the $x_1 x_2$-plane. The plane H becomes a paraboloid that passes through a, b, c. It intersects the $x_1 x_2$-plane in the circumcircle of abc. Plane H partitions gf Π into a patch below H, a curve in H, and a patch above H. The curve in H is projected onto the circumcircle of abc, and the patch below H is projected onto the open disk inside the circle. It follows that $\hat{d}$ belongs to the patch below H if and only if d lies inside the circumcircle of abc.   □

Running time

Flipping ab to cd is like gluing the tetrahedron $\hat{a}\hat{b}\hat{c}\hat{d}$ from below to $\hat{a}\hat{b}\hat{c}$ and $\hat{a}\hat{b}\hat{d}$. The algorithm can be understood as gluing a sequence of tetrahedra. Once we glue $\hat{a}\hat{b}\hat{c}\hat{d}$ we cannot glue another tetrahedron right below $\hat{a}\hat{b}$. In other words, once we flip ab we cannot introduce ab again by some other flip. This implies there are at most as many flips as there are edges connecting n points, namely $\binom{n}{2}$. Each flip takes constant time, hence the total running time is $O(n^2)$.

   There are cases where the algorithm takes $\Theta(n^2)$ flips to change an initial triangulation to the Delaunay triangulation, and one such case is illustrated in Figure 12. Take a convex upper and a concave lower curve and place m points on each, such that the upper points lie to the left of the lower points. The edges connecting the two curves in the initial and the Delaunay triangulation are shown in Figure 12. For each point, count the positions it is away from the middle, and for each edge charge the minimum of the two numbers obtained for its endpoints. In the initial triangulation, the total charge is about m^2, and in the Delaunay triangulation, the total charge is zero. Each flip moves an endpoint by at most one position and therefore decreases the charge by at most one. A lower bound of about m^2 for the number of flips follows.

Fig. 12. To the left we see about one-third of the
edges in the initial triangulation, and to the right
we see the same number of edges in the final
Delaunay triangulation

MaxMin Angle property

A flip substitutes two new triangles for two old triangles. It therefore changes six of the angles. In Figure 10, the new angles are $\gamma_1, \delta_1, \beta_1+\beta_2, \gamma_2, \delta_2, \alpha_1+\alpha_2$ and the old angles are $\alpha_1, \beta_1, \gamma_1 + \gamma_2, \alpha_2, \beta_2, \delta_1 + \delta_2$. We claim that for each of the six new angles there is an old angle that is at least as small. Indeed, $\gamma_1 \geq \alpha_2$ because both angles are opposite the same edge, namely bd, and a lies outside the circle passing through b, c, d. Similarly, $\delta_1 \geq \alpha_1$, $\gamma_2 \geq \beta_2$, $\delta_2 \geq \beta_1$, and for trivial reasons $\beta_1 + \beta_2 \geq \beta_1$ and $\alpha_1 + \alpha_2 \geq \alpha_1$. It follows that a flip does not decrease the smallest angle in a triangulation. Since we can go from any triangulation K of S to the Delaunay triangulation, this implies that the smallest angle in K is no larger than the smallest angle in the Delaunay triangulation.

MaxMin Angle Lemma. Among all triangulations of a finite set $S \subseteq \mathbb{R}^2$, the Delaunay triangulation maximizes the minimum angle.

Figure 13 illustrates the above proof of the MaxMin Angle Lemma by sketching what we call the *flip-graph* of S. Each triangulation is a node, and there is a directed arc from node μ to node ν if there is a flip that changes the triangulation μ to ν. The direction of the arc corresponds to our requirement that the flip substitutes a locally Delaunay edge for one that is not locally Delaunay. The running time analysis implies that the flip-graph is acyclic and that its undirected version is connected. If we allow flips in either direction we can go from any triangulation of S to any other triangulation in less than n^2 flips.

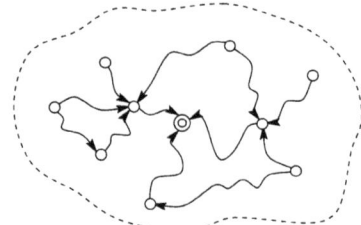

Fig. 13. Sketch of flip-graph. The sink is the
Delaunay triangulation. There is a directed path
from every node to the Delaunay triangulation

Bibliographic notes

A proof of the Delaunay lemma and its generalization to arbitrary finite dimensions is contained in the original paper by Boris Delaunay (1934). The edge-flip algorithm is due to Charles Lawson (1977). The algorithm does not generalize to three or higher dimensions. For planar triangulations, the edge-flip operation is widely used to improve local quality measures; see, *e.g.*, Schumaker (1987). Unfortunately, the algorithm gets caught in local optima for almost all interesting measures. The observation that the Delaunay triangulation maximizes the smallest angle was first made by Robin Sibson (1978). Minimizing the largest angle seems more difficult and the only known polynomial time algorithm uses edge insertions, which are somewhat more powerful than edge flips (Edelsbrunner, Tan and Waupotitsch 1992).

4. Randomized construction

The algorithm in this section constructs Delaunay triangulations incrementally, using edge flips and randomization. After explaining the algorithm, we present a detailed analysis of the expected amount of resources it requires.

Incremental algorithm

We obtain a fast algorithm for constructing Delaunay triangulations if we interleave flipping edges with adding points. Denote the points in $S \subseteq \mathbb{R}^2$ as $p_1, p_2, \ldots, p_n$ and assume general position. When we add a point to the triangulation, it can either lie inside or outside the convex hull of the preceding points. To reduce the outside to the inside case, we start with a triangulation D_0 that consists of a single and sufficiently large triangle xyz. Define $S_i = \{x, y, z, p_1, p_2, \ldots, p_i\}$, and let D_i be the Delaunay triangulation of S_i. The algorithm is a `for`-loop adding the points in sequence. After adding a point, it uses edge flips to satisfy the Delaunay lemma before the next point is added.

```
for i = 1 to n do
    find τ_{i-1} ∈ D_{i-1} containing p_i;
    add p_i by splitting τ_{i-1} into three;
    while ∃ab not locally Delaunay do
        flip ab to other diagonal cd
    endwhile
endfor.
```

The two elementary operations used by the algorithm are shown in Figure 14. Both pictures can be interpreted as the projection of a tetrahedron, though from different angles. For this reason, the addition of a point inside a triangle is sometimes called a 1-to-3 flip, while an edge flip is sometimes also called a 2-to-2 flip.

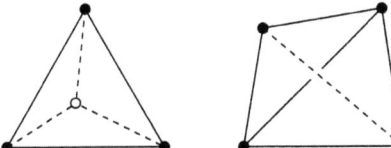

Fig. 14. To the left, the hollow vertex splits the triangle into three. To the right, the dashed diagonal replaces the solid diagonal

Growing star

Note that every new triangle in D_i has p_i as one of its vertices. Indeed, abc is a triangle in D_i if and only if $a, b, c \in S_i$ and the circumcircle is empty of points in S_i. But if p_i is not one of the vertices then $a, b, c \in S_{i-1}$ and if the circumcircle is empty of points in S_i then it is also empty of points in S_{i-1}. So abc is also a triangle in D_{i-1}. This implies that all flips during the insertion of p_i occur right around p_i.

We need some definitions. The *star* of p_i consists of all triangles that contain p_i. The *link* of p_i consists of all edges of triangles in the star that are disjoint from p_i. Both concepts are illustrated in Figure 15. Right after p_i is added, the link consists of three edges, namely the edges of the triangle that contains p_i. These edges are marked and pushed on the stack to start the edge-flipping while-loop. Each flip replaces a link edge by an edge with endpoint p_i. At the same time, it removes one triangle in the star and one outside the star and it adds the two triangles that cover the same quadrangle to the star. The net effect is one more triangle in the star. The number of edge flips is therefore 3 less than the number of edges in the final link, which is the same as 3 less than the degree of p_i in D_i.

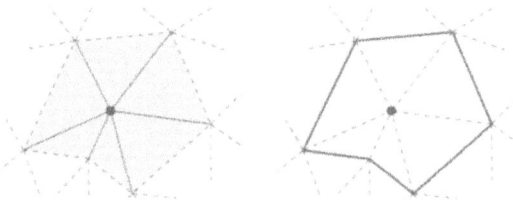

Fig. 15. The star of the solid vertex to the left
and the link of the same vertex to the right

Number of flips

We temporarily ignore the time needed to find the triangles τ_{i-1}. The rest of the time is proportional to the number of flips needed to add $p_1, p_2, \ldots, p_n$. We assume $p_1, p_2, \ldots, p_n$ is a randomly chosen input sequence. Random does not mean arbitrary but rather that every permutation of the n points is equally likely. The expected number of flips is the total number of flips needed to construct the Delaunay triangulation for all $n!$ input permutations divided by $n!$.

Consider inserting the last point, p_n. The sum of degrees of all possible last points is the same as the sum of degrees of all points p_i in D_n. The latter is equal to twice the number of edges and therefore

$$\sum_{i=1}^{n} \deg p_i \ \leq \ 6n.$$

The number of flips needed to add all last points is therefore at most $6n - 3n = 3n$. The total number of flips is

$$F(n) \leq n \cdot F(n-1) + 3n \leq 3n \cdot n!.$$

Indeed, if we assume $F(n-1) \leq 3(n-1) \cdot (n-1)!$ we get $n \cdot F(n-1) + 3n = 3(n-1) \cdot n! + 3n \leq 3n \cdot n!$. The expected number of edge flips needed for n points is therefore at most $3n$.

There is a simple way to say the same thing. The expected number of flips for the last point is at most 3, and therefore the expected number of flips to add any point is at most 3.

The history DAG

We use the evolution of the Delaunay triangulation to find the triangle τ_{i-1} that contains point p_i. Instead of deleting a triangle when it is split or flipped away, we just make it the parent of the new triangles. Figure 16 shows the two operations to the left and the corresponding parent–child relations to the right. Each time we split or flip, we add triangles or nodes to the growing data structure that records the history of the construction. The evolution from D_0 to D_n consists of n splits and an expected number of at most $3n$ flips. The resulting directed acyclic graph, or DAG for short, therefore has an expected size of at most $1 + 3n + 2 \cdot 3n = 9n + 1$ nodes. It has a unique source, the triangle xyz, and its sinks are the triangles in D_n.

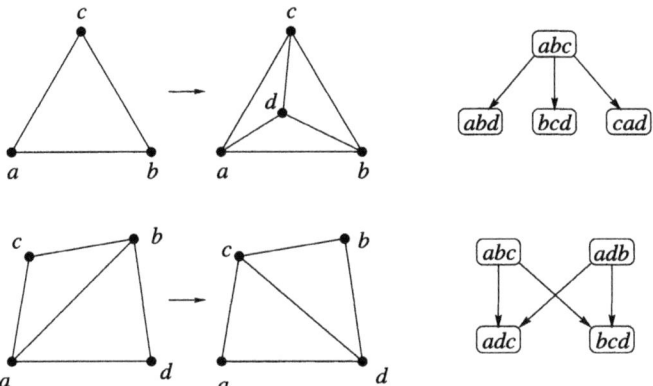

Fig. 16. Splitting a triangle generates a parent with
three children. Flipping an edge generates two parents
sharing the same two children

Searching and charging

Consider adding the point p_i. To find the triangle $\tau_{i-1} \in D_{i-1}$, we search a path of triangles in the history DAG that all contain p_i. The path begins as xyz and ends at τ_{i-1}. The history DAG of D_{i-1} consists of i layers. Layers $0, 1, \ldots, j$ represent the DAG of D_j. Its sinks are the triangles in D_j, and we let $\sigma_j \in D_j$ be the triangle that contains p_i. Triangles $\sigma_0, \sigma_1, \ldots, \sigma_j$ form a not necessarily contiguous subsequence of nodes along the search path. It is quite possible that some of the triangles σ are the same. Let G_j be the set of triangles removed from D_j during the insertion of p_{j+1}, and let H_j be the set of triangles removed from D_j during the hypothetical and independent insertion of p_i into D_j. The two sets are schematically sketched as intervals along the real line representing the Delaunay triangulation in Figure 17. We have $\sigma_j = \sigma_{j+1}$ if G_j and H_j are disjoint. Suppose $\sigma_j \neq \sigma_{j+1}$. Then $X_j = G_j \cap H_j \neq \emptyset$, and all triangles on the portion of the path from σ_j to σ_{j+1} are generated by flips that remove triangles in X_j. The cost for searching with p_i is therefore at most proportional to the sum of card X_j, for j from 0 to $i - 2$.

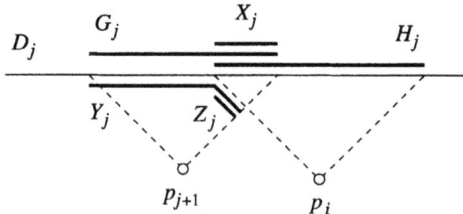

Fig. 17. The intervals represent sets of triangles removed
or added when we insert p_{j+1} and/or p_i to D_j

We write X_j in terms of other sets. These sets represent what happens if we again hypothetically first insert p_i into D_j and then insert p_{j+1} into the Delaunay triangulation of $S_j \cup \{p_i\}$. Let Y_j be the set of triangles removed during the insertion of p_{j+1}, and let $Z_j \subseteq Y_j$ be the subset of triangles that do not belong to D_j. Each triangle in Z_j is created during the insertion of p_i, so p_i must be one of its vertices. We have

$$X_j \;=\; G_j - (Y_j - Z_j).$$

Expectations

We bound the expected search time by bounded the expected total size of the X_j. Write cardinalities using corresponding lower-case letters. Because $Z_j \subseteq Y_j$ and $Y_j - Z_j \subseteq G_j$ we have

$$x_j \;=\; g_j - y_j + z_j.$$

The expected values of g_j and y_{j-1} are the same, because both count triangles removed by inserting a random jth point. Because the expectation of a sum is the sum of expectations, we have

$$\mathrm{E}\left[\sum_{j=0}^{i-2} x_j \right] \;=\; \sum_{j=0}^{i-2} \mathrm{E}[\, g_j \,] - \mathrm{E}[\, y_j \,] + \mathrm{E}[\, z_j \,]$$

$$=\; \mathrm{E}[\, g_0 - g_{i-1} \,] + \sum_{j=0}^{i-2} \mathrm{E}[\, z_j \,].$$

To compute the expected value of z_j, we use the fact that among $j+2$ points, every pair is equally likely to be p_{j+1} and p_i. For example, if p_{j+1} and p_i are not connected by an edge in the Delaunay triangulation of $S_j \cup \{p_{j+1}, p_i\}$ then $Z_j = \emptyset$. In general, a triangle in the Delaunay triangulation of $S_j \cup \{p_i\}$ has probability at most $\frac{3}{j+1}$ of being in the star of p_i. The expected number of triangles removed by inserting p_{j+1} is at most 4. Because the expectation of a product is the product of expectations, we have $\mathrm{E}[\, z_j \,] \le \frac{4 \cdot 3}{j+1}$. The expected length of the search path for p_i is

$$\sum_{j=0}^{i-2} \mathrm{E}[\,x_j\,] \;\leq\; \sum_{j=0}^{i-2} \frac{12}{j+1} \;\leq\; 1 + 12\ln(i-1).$$

The expected total time spent on searching in the history DAG is $\sum \mathrm{E}[\,x_j\,] \leq c \cdot n \log n$.

To summarize, the randomized incremental algorithm constructs the Delaunay triangulation of n points in $\mathbb{R}^2$ in expected time $\mathrm{O}(n \log n)$ and expected amount of memory $\mathrm{O}(n)$.

Bibliographic notes

The randomized incremental algorithm of this section is due to Guibas, Knuth and Sharir (1992). It has been generalized to three and higher dimensions by Edelsbrunner and Shah (1996). All this is based on earlier work on randomized algorithms and in particular on the methods developed by Clarkson and Shor (1989). The arguments used to bound the expected number of flips and the expected search time are examples of the backwards analysis introduced by Raimund Seidel (1993).

5. Symbolic perturbation

The computational technique of symbolically perturbing a geometric input justifies the mathematically convenient assumption of general position. This section describes a particular perturbation known as SoS or Simulation of Simplicity.

Orientation test

Let $a = (\alpha_1, \alpha_2)$, $b = (\beta_1, \beta_2)$, $c = (\gamma_1, \gamma_2)$ be three points in the plane. We consider a, b, c degenerate if they lie on a common line. This includes the case where two or all three points are the same. In the degenerate case, point c is an affine combination of a and b, that is, $c = \lambda_1 a + \lambda_2 b$ with $\lambda_1 + \lambda_2 = 1$. Such λ_1, λ_2 exist if and only if the determinant of

$$\Delta \;=\; \begin{bmatrix} 1 & \alpha_1 & \alpha_2 \\ 1 & \beta_1 & \beta_2 \\ 1 & \gamma_1 & \gamma_2 \end{bmatrix}$$

vanishes. In the non-degenerate case, the sequence a, b, c either forms a left- or a right-turn. We can again use the determinant of Δ to decide which it is.

Orientation Claim.   The sequence a, b, c forms a left-turn if and only if $\det \Delta > 0$, and it forms a right-turn if and only if $\det \Delta < 0$.

Proof. We first check the claim for $a_0 = (0,0)$, $b_0 = (1,0)$, $c_0 = (0,1)$. It is geometrically obvious that a_0, b_0, c_0 form a left-turn, and indeed

$$\det \begin{bmatrix} 1 & 0 & 0 \\ 1 & 1 & 0 \\ 1 & 0 & 1 \end{bmatrix} = 1.$$

We can continuously move a_0, b_0, c_0 to any other left-turn a, b, c without ever having three collinear points. Since the determinant changes continuously with the coordinates, it remains positive during the entire motion and is therefore positive at a, b, c. Symmetry implies that all right-turns have negative determinant.   □

In-circle test

The in-circle test is formulated for four points a, b, c, d in the plane. We consider a, b, c, d degenerate if a, b, c lie on a common line or a, b, c, d lie on a common circle. We already know how to test for points on a common line. To test for points on a common circle, we recall the definition of lifted points, $\hat{a} = (\alpha_1, \alpha_2, \alpha_3)$ with $\alpha_3 = \alpha_1^2 + \alpha_2^2$, etc. Points a, b, c, d lie on a common circle if and only if $\hat{a}, \hat{b}, \hat{c}, \hat{d}$ lie on a common plane in $\mathbb{R}^3$; see Figure 11. In other words, $\hat{d}$ is an affine combination of $\hat{a}, \hat{b}, \hat{c}$, which is equivalent to

$$\Gamma = \begin{bmatrix} 1 & \alpha_1 & \alpha_2 & \alpha_3 \\ 1 & \beta_1 & \beta_2 & \beta_3 \\ 1 & \gamma_1 & \gamma_2 & \gamma_3 \\ 1 & \delta_1 & \delta_2 & \delta_3 \end{bmatrix}$$

having zero determinant. In the non-degenerate case, d either lies inside or outside the circle defined by a, b, c. We can use the determinants of Δ and Γ to decide which it is. Note that permuting a, b, c can change the sign of $\det \Gamma$ without changing the geometric configuration. Since the signs of $\det \Gamma$ and $\det \Delta$ change simultaneously, we can counteract by multiplying the two.

In-circle Claim.   Point d lies inside the circle passing through a, b, c if and only if $\det \Delta \cdot \det \Gamma < 0$, and d lies outside the circle if and only if $\det \Delta \cdot \det \Gamma > 0$.

Proof. We first check the claim for $d_0 = (\frac{1}{2}, \frac{1}{2})$ and $a_0 = (0,0)$, $b_0 = (1,0)$, $c_0 = (0,1)$ as before. Point d_0 lies at the centre and therefore inside the circle passing through a_0, b_0, c_0. The determinant of Δ is 1, and that of Γ is

$$\det \begin{bmatrix} 1 & 0 & 0 & 0 \\ 1 & 1 & 0 & 1 \\ 1 & 0 & 1 & 1 \\ 1 & \frac{1}{2} & \frac{1}{2} & \frac{1}{2} \end{bmatrix} = -\frac{1}{2},$$

so their product is negative. As in the proof of the Orientation Claim, we derive the general result from the special one by continuity. Specifically, every configuration a, b, c, d, where d lies inside the circle of a, b, c, can be obtained from a_0, b_0, c_0, d_0 by continuous motion avoiding all degeneracies. The signs of the two determinants remain the same throughout the motion, and so does their product. This implies the claim for negative products, and symmetry implies the claim for positive products. □

Algebraic framework

Let us now take a more abstract and algebraic view of degeneracy as a geometric phenomenon. For expository reasons, we restrict ourselves to orientation tests in the plane. Let S be a collection of n points, denoted as $p_i = (\phi_{i,1}, \phi_{i,2})$, for $1 \leq i \leq n$. By listing the $2n$ coordinates in a single sequence, we think of S as a single point in $2n$-dimensional space. Specifically, S is mapped to $Z = (\zeta_1, \zeta_2, \zeta_3, \ldots, \zeta_{2n}) \in \mathbb{R}^{2n}$, where $\zeta_{2i-1} = \phi_{i,1}$ and $\zeta_{2i} = \phi_{i,2}$, for $1 \leq i \leq n$. Point Z is degenerate if and only if

$$\det \begin{bmatrix} 1 & \zeta_{2i-1} & \zeta_{2i} \\ 1 & \zeta_{2j-1} & \zeta_{2j} \\ 1 & \zeta_{2k-1} & \zeta_{2k} \end{bmatrix} = 0$$

for some $1 \leq i < j < k \leq n$. The equation identifies a differentiable $(2n-1)$-dimensional manifold in $\mathbb{R}^{2n}$. There are $\binom{n}{3}$ such manifolds, $\mathbb{M}_\ell$, and Z is degenerate if and only if $Z \in \bigcup_\ell \mathbb{M}_\ell$, as sketched in Figure 18. Each manifold

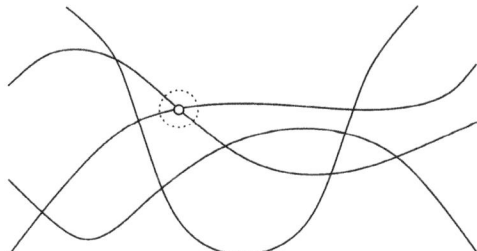

Fig. 18. Schematic picture of the union of $(2n - 1)$-dimensional manifolds in $2n$-dimensional space. The marked point lies on two manifolds and thus has two degenerate subconfigurations. The dotted circle bounds a neighbourhood, and most points in that neighbourhood are non-degenerate

has dimension one less than the ambient space and hence measure zero in $\mathbb{R}^{2n}$. We have a finite union of measure zero sets, which still has measure zero. In other words, most points in an open neighbourhood of $Z \in \mathbb{R}^{2n}$ are non-degenerate. A point nearby Z is often called a perturbation of Z or S.

The result on neighbourhoods thus implies that there are arbitrarily close non-degenerate perturbations of S.

Perturbation

We construct a non-degenerate perturbation of S using positive parameters $\varepsilon_1, \varepsilon_2, \ldots, \varepsilon_{2n}$. These parameters will be chosen anywhere between arbitrarily and sufficiently small, and we may think of them as infinitesimals. They will also be chosen sufficiently different, and we will see shortly what this means. Let $Z \in \mathbb{R}^{2n}$, and for every $\varepsilon > 0$ define

$$Z(\varepsilon) \;=\; (\zeta_1 + \varepsilon_1, \zeta_2 + \varepsilon_2, \ldots, \zeta_{2n} + \varepsilon_{2n}),$$

where $\varepsilon_i = f_i(\varepsilon)$ with $f_i : \mathbb{R} \to \mathbb{R}$ continuous and $f_i(0) = 0$. If the ε_i are sufficiently different, we get the following three properties provided $\varepsilon > 0$ is sufficiently small.

  I. $Z(\varepsilon)$ is non-degenerate.
 II. $Z(\varepsilon)$ retains all non-degenerate properties of Z.
III. The computational overhead for simulating $Z(\varepsilon)$ is negligible.

For example, if $\varepsilon_i = \varepsilon^{2^i}$ then $\varepsilon_1 \gg \varepsilon_2 \gg \cdots \gg \varepsilon_{2n}$ and we can do all computations simply by comparing indices without ever computing a feasible ε. We demonstrate this by explicitly computing the orientation of the points p_i, p_j, p_k after perturbation. By definition, that orientation is the sign of the determinant of

$$\Delta(\varepsilon) \;=\; \begin{bmatrix} 1 & \zeta_{2i-1} + \varepsilon_{2i-1} & \zeta_{2i} + \varepsilon_{2i} \\ 1 & \zeta_{2j-1} + \varepsilon_{2j-1} & \zeta_{2j} + \varepsilon_{2j} \\ 1 & \zeta_{2k-1} + \varepsilon_{2k-1} & \zeta_{2k} + \varepsilon_{2k} \end{bmatrix}.$$

Note that $\Delta(\varepsilon)$ is a polynomial in ε. The terms with smaller power are more significant than those with larger power. We assume $i < j < k$ and list the terms of $\Delta(\varepsilon)$ in the order of decreasing significance, that is,

$$\begin{aligned} \det \Delta(\varepsilon) \;=\; & \det \Delta - \det \Delta_1 \cdot \varepsilon^{2^{2i-1}} \\ & + \det \Delta_2 \cdot \varepsilon^{2^{2i}} + \det \Delta_3 \cdot \varepsilon^{2^{2j-1}} \\ & - 1 \cdot \varepsilon^{2^{2j-1}} \varepsilon^{2^{2i}} \pm \ldots, \end{aligned}$$

where

$$\Delta \;=\; \begin{bmatrix} 1 & \zeta_{2i-1} & \zeta_{2i} \\ 1 & \zeta_{2j-1} & \zeta_{2j} \\ 1 & \zeta_{2k-1} & \zeta_{2k} \end{bmatrix},$$

$$\Delta_1 \;=\; \begin{bmatrix} 1 & \zeta_{2j} \\ 1 & \zeta_{2k} \end{bmatrix},$$

$$\Delta_2 = \begin{bmatrix} 1 & \zeta_{2j-1} \\ 1 & \zeta_{2k-1} \end{bmatrix},$$

$$\Delta_3 = \begin{bmatrix} 1 & \zeta_{2i} \\ 1 & \zeta_{2k} \end{bmatrix}.$$

Property I is satisfied because the fifth term is non-zero, and its influence on the sign of the determinant cannot be cancelled by subsequent terms. Property II is satisfied because the sign of the perturbed determinant is the same as that of the unperturbed one, unless the latter vanishes.

Implementation

In order to show Property III, we give an implementation of the test for $Z(\varepsilon)$. First we sort the indices such that $i < j < k$, and we count the number of transpositions. Then we determine whether the three perturbed points form a left- or a right-turn by computing determinants of the four submatrices listed above.

```
boolean LeftTurn(integer i, j, k):
    assert i < j < k;
    case det Δ ≠ 0: return det Δ > 0;
    case det Δ₁ ≠ 0: return det Δ₁ < 0;
    case det Δ₂ ≠ 0: return det Δ₂ > 0;
    case det Δ₃ ≠ 0: return det Δ₃ > 0;
    otherwise: return FALSE.
```

If the number of transpositions needed to sort i, j, k is odd, then the sorting reverses the sign, and we correct the reversal by reversing the result of the function LeftTurn.

As an important detail we note that signs of determinants need to be computed exactly. With normal floating point arithmetic, this is generally not possible. We must therefore resort to exact arithmetic methods using long integer or other representations of coordinates. These methods are typically more costly than floating point arithmetic, but differences vary widely among different computer hardware. A pragmatic compromise uses floating point arithmetic together with error analysis. After computing the determinant with floating point arithmetic, we check whether the absolute value is large enough for its sign to be guaranteed. Only if that guarantee cannot be obtained do we repeat the computation in exact arithmetic.

Bibliographic notes

The idea of using symbolic perturbation for computational reasons is already present in the work of George Danzig on linear programming (Danzig 1963). It reappeared in computational geometry with the work of four independent groups of authors. Edelsbrunner and Mücke (1990) develop SoS, which is the method described in this section. Yap (1990) studies the class of perturbations obtained with different orderings of infinitesimals. Emiris and Canny (1995) introduce perturbations along straight lines. Michelucci (1995) exploits randomness in the design of perturbations.

Symbolic perturbations as a general computational technique within computational geometry remains a controversial subject. It succeeds in extending partially to completely correct software for some but not all geometric problems. Seidel (1998) addresses this issue, offers a unified view of symbolic perturbation, and discusses limitations of the method. Fortune and Van Wyk (1996) describe a floating point filter that reduces the overhead needed for exact computation.

6. Constrained triangulation

This section studies triangulations in the plane constrained by edges specified as part of the input. We show that there is a unique constrained triangulation that is closest, in some sense, to the (unconstrained) Delaunay triangulation.

Constraining line segments

The preceding sections constructed triangulations for a given set of points. The input now consists of a finite set of points, $S \subseteq \mathbb{R}^2$, together with a finite set of line segments, L, each connecting two points in S. We require that any two line segments are either disjoint or meet at most in a common endpoint. A *constrained triangulation* of S and L is a triangulation of S that contains all line segments of L as edges. Figure 19 illustrates that we can construct a constrained triangulation by adding straight edges connecting points in S as long as they have no interior points in common with previous edges.

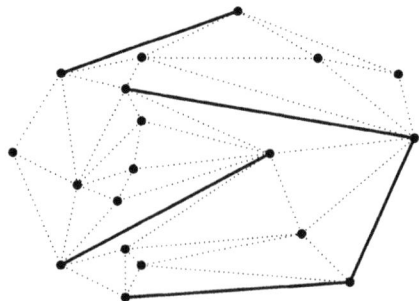

Fig. 19. Given the points and solid edges, we form a
constrained triangulation by adding as many dotted
edges as possible without creating improper intersections

Plane-sweep algorithm

The idea of organizing the actions of the algorithm around a line sweeping over the plane leads to an efficient way of constructing constrained triangulations. We use a vertical line that sweeps over the plane from left to right, as shown in Figure 20. The algorithm uses two data structures. The *schedule*, X, orders events in time. The *cross-section*, Y, stores the line segments in L that currently intersect the sweep-line. The algorithm is defined by the following invariant.

(I) At any moment in time, the partial triangulation contains all edges in L, a maximal number of edges connecting points to the left of the sweep-line, and no other edges.

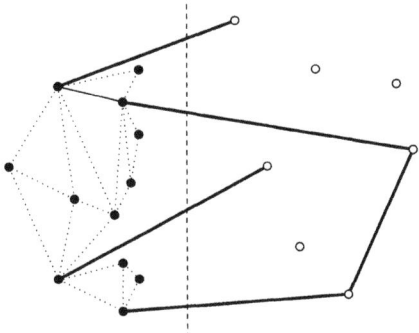

Fig. 20. Snapshot of plane-sweep constructing
a constrained triangulation

Invariant (I) implies that between the left endpoints of two constraining line segments adjacent along the sweep-line we have a convex chain of edges in the partial triangulation. To ensure that new edges can each be added in constant time, the algorithm remembers the rightmost vertex in each chain. If the point p encountered next by the sweep-line falls inside one of the intervals along the sweep-line, the algorithm connects p to the corresponding rightmost vertex. It then proceeds in a clockwise and an anticlockwise order along the convex chain. Each step either adds a new edge or it ends the walk. If p is the right endpoint of a line segment then it separates two intervals along the sweep-line, and the algorithm does the same kind of walking twice, once for each interval.

The schedule is constructed by sorting the points in S from left to right, which can be done in time $O(n \log n)$, where $n = \operatorname{card} S$. The cross-section is maintained as a dictionary, which supports search, insertion, deletion all in time $O(\log n)$. There is a search for each point in S and an insertion–deletion pair for each line segment in L, taking total time $O(n \log n)$. Fewer than $3n$ edges are added to the triangulation, each in constant time. The plane-sweep algorithm thus constructs a constrained triangulation of S and L in time $O(n \log n)$.

Constrained Delaunay triangulations

The triangulations constructed by plane-sweep usually have many small and large angles. We use a notion of visibility between points to introduce a constrained triangulation that avoids small angles to the extent possible.

   Points $x, y \in \mathbb{R}^2$ are *visible* from each other if xy contains no point of S in its interior and it shares no interior point with a constraining line segment. Formally, int $xy \cap S = \emptyset$ and int $xy \cap uv = \emptyset$ for all $uv \in L$. Assume general position. An edge ab, with $a, b \in S$, belongs to the *constrained Delaunay triangulation* of S and L if

(i) $ab \in L$,   or
(ii) a and b are visible from each other and there is a circle passing through a and b such that each point inside this circle is invisible from every point $x \in$ int ab.

We say the circle in (ii) *witnesses* the membership of ab in the constrained Delaunay triangulation. Figure 21 illustrates this definition. Note if $L = \emptyset$ then the constrained Delaunay triangulation of S and L is the Delaunay triangulation of S. More generally, it is however unclear that what we defined is indeed a triangulation. For example, why is it true that no two edges satisfying (i) or (ii) cross?

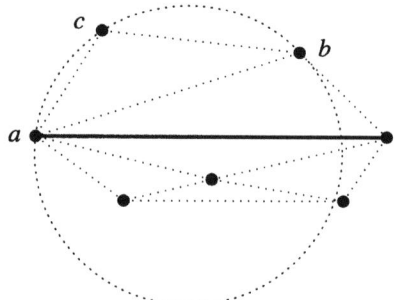

Fig. 21. Constrained Delaunay triangulation for
seven points and one constraining line segment.
The circumcircle of abc encloses only points that
are invisible from all points of int ab

Edge flipping

We introduce a generalized concept of being locally Delaunay, and use it to prove that the above definition makes sense. Let K be any constrained triangulation of S and L. An edge $ab \in K$ is *locally Delaunay* if $ab \in L$, or ab is a convex hull edge, or d lies outside the circumcircle of abc, where $abc, abd \in K$.

Constrained Delaunay Lemma. If every edge of K is locally Delaunay then K is the constrained Delaunay triangulation of S and L.

Proof. We show that every edge in K satisfies (i) or (ii) and therefore belongs to the constrained Delaunay triangulation. The claim follows because every additional edge crosses at least one edge of K and therefore of the constrained Delaunay triangulation.

   Let ab be an edge and p a vertex in K. Assume $ab \notin L$, for else ab belongs to the constrained Delaunay triangulation for trivial reasons. Assume also that ab is not a convex hull edge, for else we can easily find a circle passing through a and b such that p lies outside the circle. Hence, ab belongs to two triangles, and we let abc be the one separated from p by the line passing through ab. We need to prove that if p is visible from a point $x \in \text{int } ab$ then it lies outside the circumcircle of abc. Consider the sequence of edges in K crossing xp. Since x and p are visible from each other, all these edges are not in L. We can therefore apply the argument of the proof of the original Delaunay lemma, which is illustrated in Figure 9. $\square$

   This result suggests we use the edge-flipping algorithm to construct the constrained Delaunay triangulation. The only difference to the original edge-flipping algorithm is that edges in L are not flipped, since they are locally Delaunay by definition. As before, the algorithm halts in time $O(n^2)$ after fewer than $\binom{n}{2}$ flips. The analysis of angle changes during an edge flip presented in Section 3 implies that the MaxMin Angle Lemma also holds in the constrained case.

Constrained MaxMin Angle Lemma. Among all constrained triangulations of S and L, the constrained Delaunay triangulation maximizes the minimum angle.

Extended Voronoi diagrams

Just as for ordinary Delaunay triangulations, every constrained Delaunay triangulation has a dual Voronoi diagram, but in a surface that is more complicated than the Euclidean plane. Imagine $\mathbb{R}^2$ is a sheet of paper, Σ_0, with the points of S and the line segments in L drawn on it. For each $\ell_i \in L$, we cut Σ_0 open along ℓ_i and glue another sheet Σ_i, which is also cut open along ℓ_i. The gluing is done around ℓ_i such that every traveller who crosses ℓ_i switches from Σ_0 to Σ_i and *vice versa*. A cross-section of the particular gluing necessary to achieve that effect is illustrated in Figure 22. It is not possible to do this without self-intersections in $\mathbb{R}^3$, but in $\mathbb{R}^4$ there is already sufficient space to embed the resulting surface. Call Σ_0 the *primary sheet*, and after the gluing is done we have $m = \operatorname{card} L$ *secondary sheets* Σ_i for $1 \le i \le m$. Each secondary sheet is attached to Σ_0, but not connected to any of the other secondary sheets. For each point $x \in \mathbb{R}^2$, we now have $m + 1$ copies $x_i \in \Sigma_i$, one on each sheet.

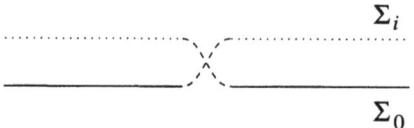

Fig. 22. The gap in Σ_0 represents the cut along ℓ_i.
The secondary sheet Σ_i is glued to Σ_0 so that
each path crossing ℓ_i switches sheets

We know what it means for two points on the primary sheet to be visible from each other. For other pairs we need a more general definition. For $i \ne 0$, points $x_0 \in \Sigma_0$ and $y_i \in \Sigma_i$ are *visible* if xy crosses ℓ_i, and ℓ_i is the first constraining line segment crossed if we traverse xy in the direction from x to y. The *distance* between points x_0 and y_i is

$$d(x_0, y_i) = \begin{cases} \|x - y\|, & \text{if } x_0, y_i \text{ are visible,} \\ \infty, & \text{otherwise.} \end{cases}$$

The new distance function is used to define the *extended Voronoi diagram*, which is illustrated in Figure 23. A circle that witnesses the membership of an edge ab in the constrained Delaunay triangulation has its centre on the primary or on a secondary sheet. In either case, that centre is closer to a and b than to any other point in S. This implies that the Voronoi regions of a and b meet along a non-empty common portion of their boundary. Conversely, every point on an edge of the extended Voronoi diagram is the centre of a circle witnessing the membership of the corresponding edge in the constrained Delaunay triangulation.

Bibliographic notes

The idea of using plane-sweep for solving two-dimensional geometric problems is almost as old as the field of computational geometry itself. It was propagated as a general algorithmic paradigm by Nievergelt and Preparata (1982). Constrained Delaunay triangulations were independently discovered by Lee and Lin (1986) and by Paul Chew (1987). Extended Voronoi diagrams are due to Raimund Seidel (1988), who used them to construct constrained Delaunay triangulations in worst-case time $O(n \log n)$.

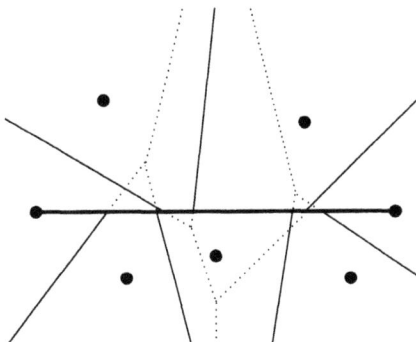

Fig. 23. Extended Voronoi diagram dual to the
constrained Delaunay triangulation in Figure 21.
There is only one secondary sheet glued to the
primary one. The solid Voronoi edges lie in the
primary sheet and the dotted ones in the
secondary sheet

7. Delaunay refinement

This section demonstrates the use of Delaunay triangulations in constructing
triangle meshes in the plane. The idea is to add new vertices until the
triangulation forms a satisfying mesh. Constraining edges are covered by
Delaunay edges, although forcing them into the triangulation as we did in
Section 6 would also be possible.

The meshing problem

The general objective in mesh generation is to decompose a geometric space
into elements. The elements are restricted in type and shape, and the number of elements should not be too big. We discuss a concrete version of the
two-dimensional mesh generation problem.

Input. A polygonal region in the plane, possibly with holes and with constraining edges and vertices inside the region.

Output. A triangulation of the region whose edges cover all input edges and
whose vertices cover all input vertices.

The graph of input vertices and edges is denoted by G, and the output
triangulation is denoted by K. It is convenient to enclose G in a bounding
box and to triangulate everything inside that box. A triangulation of the
input region is obtained by taking a subset of the triangles. Figure 24 shows
input and output for a particular mesh generation problem.

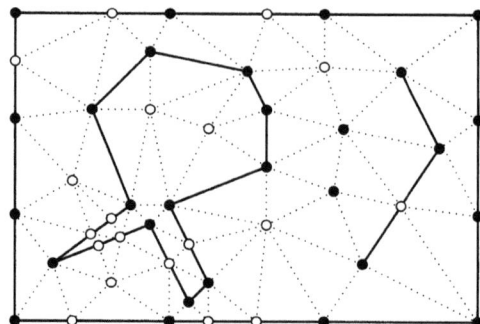

Fig. 24. The solid vertices and edges define the input
graph, and together with the hollow vertices and
dotted edges they define the output triangulation

Triangle quality

The quality of a triangle abc is measured by its smallest angle, θ. Two
alternative choices would be the largest angle and the aspect ratio. We
argue that a good lower bound for the smallest angle implies good bounds
for the other two expressions of quality. The largest angle is at most $\pi - 2\theta$,
so if the smallest angle is bounded away from zero then the largest angle
is bounded away from π. The converse is not true. The aspect ratio is the
length of the longest edge, which we assume is ac, divided by the distance
of b from ac; see Figure 25. Suppose the smallest angle occurs at a. Then
$\|b - x\| = \|b - a\| \cdot \sin\theta$, where x is the orthogonal projection of b onto ac.
The edge ab is at least as long as cb, and therefore $\|b - a\| \geq \|c - a\|/2$. It
follows that

$$\frac{1}{\sin\theta} \leq \frac{\|c - a\|}{\|b - x\|} \leq \frac{2}{\sin\theta}.$$

In words, the aspect ratio is linearly related to one over the smallest angle.
If θ is bounded away from zero then the aspect ratio is bounded from above
by some constant, and *vice versa*.

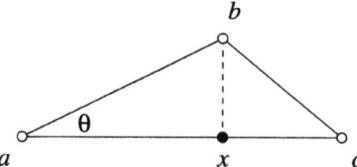

Fig. 25. Triangle with base ac,
height bx, and minimum angle θ

The goal is to construct K so its smallest angle is no less than some
constant, and the number of triangles in K is at most some constant times
the minimum. We see from the example in Figure 24 that a small angle

between two input edges cannot possibly be resolved. A reasonable way to deal with this difficulty is to accept sharp input features as unavoidable and to isolate them so they cause no deterioration of the triangulation nearby. In this section, we assume that there are no sharp input features, and in particular that all input angles are at least $\frac{\pi}{2}$.

Delaunay refinement

We construct K as the Delaunay triangulation of a set of points that includes all input points. Other points are added one by one to resolve input edges that are not covered and triangles that have too small an angle.

(1) Suppose ab is a segment of an edge in G that is not covered by edges of the current Delaunay triangulation. This can only be because some of the vertices lie inside the diameter circle of ab, as in Figure 26. We say these vertices *encroach upon* ab, and we use function SPLIT$_1$ to add the midpoint of ab and to repair the Delaunay triangulation with a series of edge flips.

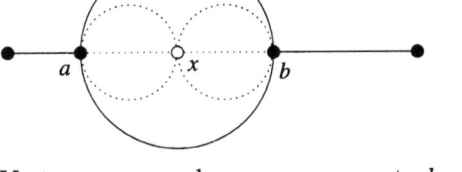

Fig. 26. Vertex p encroaches upon segment ab. After adding the midpoint, we have two smaller diameter circles, both contained in the diameter circle of ab

(2) Suppose a triangle abc in the current Delaunay triangulation K is skinny, that is, it has an angle less than the required lower bound. We use function SPLIT$_2$ to add the circumcentre as a new vertex, such as point x in Figure 27. Since its circumcircle is no longer empty, triangle abc is guaranteed to be removed by one of the edge flips used to repair the Delaunay triangulation.

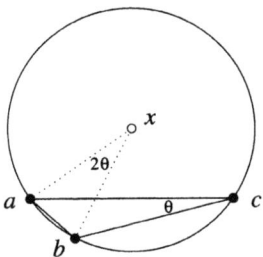

Fig. 27. The angle $\angle axb$ is twice the angle $\angle acb$

Algorithm

The first priority of the algorithm is to cover input edges, and its second priority is to resolve skinny triangles. Before starting the algorithm, we place G inside a rectangular box B. The purpose of the box is to contain the points added by the algorithm and thus prevent the perpetual growth of the meshed region. To be specific, we take B three times the size of the minimum enclosing rectangle of G. Box B has space for nine copies of the rectangle, and we place G inside the centre copy. Each side of B is decomposed into three equally long edges. Refer to Figure 24, where for aesthetic reasons the box is drawn smaller than required but with the right combinatorics. Initially, K is the Delaunay triangulation of the input points, which includes the 12 vertices along the boundary of B.

```
loop
  while ∃ encroached segment ab do
    SPLIT₁(ab)
  endwhile;
  if no skinny triangle left then exit endif;
  let abc ∈ K be skinny and x its circumcentre;
  x encroaches upon segments s₁, s₂, ..., sₖ;
  if k ≥ 1 then SPLIT₁(sᵢ) for all i
        else SPLIT₂(abc)
  endif
forever.
```

The choice of B implies that no circumcentre x will ever lie outside the box. This is because the initial 12 or fewer triangles next to the box boundary have non-obtuse angles opposite to boundary edges. Since the circumcircles of Delaunay triangles are empty, this implies that all circumcentres lie inside B. The algorithm maintains the non-obtuseness of angles opposing input edges and thus limits circumcentres to lie inside B.

Preliminary analysis

The behaviour of the algorithm is expressed by the points it adds as vertices to the mesh. We already know that all points lie on the boundary or inside the box B, which has finite area. If we can prove that no two points are less than a positive constant 2ε apart, then this implies that the algorithm halts after adding finitely many points. To be specific, let w be the width and h the height of B. The area of the box obtained by extending B by ε on each side is $A = (w + 2\varepsilon)(h + 2\varepsilon)$. The number of points inside the box is $n \leq A/\varepsilon^2\pi$. This is because the disks with radius ε centred at the vertices of the mesh have pairwise disjoint interiors, and they are all contained in the extended box. This type of area argument is common in meshing and related to packing, as illustrated in Figure 28. The existence of a positive ε will be established in Section 8. The analysis there will refine the area argument by varying the sizes of disks with their location inside the meshing region.

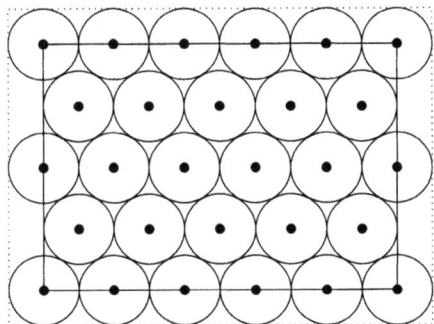

Fig. 28. The centres of the disk are contained in the inner box, and the disks are contained in the box enlarged by the disk radius in all four directions

In terms of running time, the most expensive activity is edge flipping used to repair the Delaunay triangulation. The expected linear bound on the number proved in Section 4 does not apply because points are not added in a random order. The total number of flips is less than $\binom{n}{2}$. This implies an upper bound of $O(n^2)$ on the running time, as long as the cost for adding a new vertex is at most $O(n)$.

Bibliographic notes

The algorithm described in this section is due to Jim Ruppert (1995). Experiments suggest it achieves best results if the skinny triangles are removed in order of non-decreasing smallest angle. A predecessor of Ruppert's algorithm is the version of the Delaunay refinement method by Paul Chew (1989). That algorithm is also described in Chew (1993), where it is generalized to surfaces in three-dimensional space. The main contribution of Ruppert is a detailed analysis of the Delaunay refinement method. The gained insights are powerful enough to permit modifications of the general method that guarantee a close to optimum mesh.

8. Local feature size

This section analyses the Delaunay refinement algorithm of Section 7. It proves an upper bound on the number of triangles generated by the algorithm and an asymptotically matching lower bound on the number of triangles that must be generated.

Local feature size

We understand the Delaunay refinement algorithm through relating its actions to the *local feature size* defined as a map $f : \mathbb{R}^2 \to \mathbb{R}$. For a point $x \in \mathbb{R}^2$, $f(x)$ is the smallest radius r such that the closed disk with centre x and radius r

  (i) contains two vertices of G,
 (ii) intersects one edge of G and contains one vertex of G that is not endpoint of that edge,   or
(iii) intersects two vertex disjoint edges of G.

The three cases are illustrated in Figure 29. Because of (i) we have $f(a) \leq \|a - b\|$ for all vertices $a \neq b$ in G. The local feature size satisfies a one-sided Lipschitz inequality, which implies continuity.

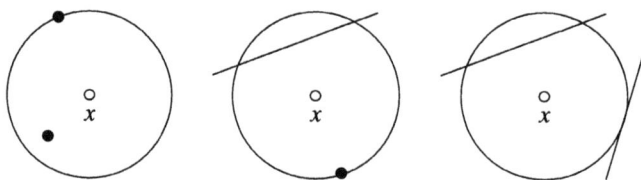

Fig. 29. In each case, the radius of the circle
is the local feature size at x

Lipschitz Condition. $|f(x) - f(y)| \leq \|x - y\|$.

Proof. To get a contradiction, assume there are points x, y with $f(x) < f(y) - \|x - y\|$. The disk with radius $f(x)$ around x is contained in the interior of the disk with radius $f(y)$ around y. We can thus shrink the disk of y while maintaining its non-empty intersection with two disjoint vertices or edges of G. This contradicts the definition of $f(y)$.   □

Constants

The analysis of the algorithm uses two carefully chosen positive constants C_1 and C_2 such that

$$1 + \sqrt{2}C_2 \;\leq\; C_1 \;\leq\; \frac{C_2 - 1}{2 \sin \alpha},$$

where α is the lower bound on angles enforced by the Delaunay refinement algorithm. The constraints that correspond to the two inequalities are bounded by lines, and we have a solution if and only if the slope of the first line is greater than that of the second, $1/\sqrt{2} > 2 \sin \alpha$. Figure 30 illustrates the two constraints for $\alpha < \arcsin \frac{1}{2\sqrt{2}} = 20.7 \ldots °$. The two lines intersect at a point in the positive quadrant, and the coordinates of that point are the smallest constants C_1 and C_2 that satisfy the inequalities.

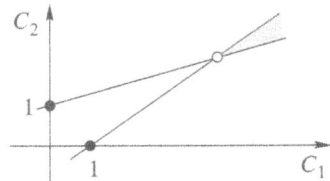

Fig. 30. Each line bounds a half-plane of points
(C_1, C_2) that satisfy one inequality. The shaded
wedge contains all points that satisfy both inequalities

Invariants

The algorithm starts with the vertices of G and generates all other vertices
in sequence. We show that, when a new vertex is added, its distance to
already present vertices is not much smaller than the local feature size.

Invariants.   Let p and x be two vertices such that x was added after p. If
x was added by

(A) SPLIT$_1$ then $\|x - p\| \geq f(x)/C_1$,

(B) SPLIT$_2$ then $\|x - p\| \geq f(x)/C_2$.

Proof.   We first prove (B). In this case, point x is the circumcentre of a
skinny triangle abc. Let $\theta < \alpha$ at c be the smallest angle in abc, as in
Figure 27. Assume that either a and b both belong to G or that a was
added after b. We distinguish three cases depending on how a became to be
a vertex. Let L be the length of ab.

Case 1. a is a vertex of G. Then b is also a vertex of G and $f(a) \leq L$.

Case 2. a was added as the circumcentre of a circle with radius r'. Prior to
the addition of a this circle was empty, and hence $r' \leq L$. By induction,
we have $f(a) \leq r' \cdot C_2$ and therefore $f(a) \leq L \cdot C_2$.

Case 3. a was added as the midpoint of a segment. Then $f(a) \leq L \cdot C_1$,
again by induction.

Since $1 \leq C_2 \leq C_1$, we have $f(a) \leq L \cdot C_1$ in all three cases. Let $r = \|x - a\|$
be the radius of the circumcircle of abc. Using the Lipschitz Condition and
$L = 2r \sin \theta$ from Figure 27 we get

$$
\begin{aligned}
f(x) &\leq & f(a) + r \\
&\leq & L \cdot C_1 + r \\
&\leq & 2r \cdot \sin \theta \cdot C_1 + r.
\end{aligned}
$$

Since $\theta < \alpha$ and $C_2 \geq 1 + 2C_1 \cdot \sin \alpha$ we get

$$r \;\geq\; \frac{f(x)}{1 + 2C_1 \cdot \sin \alpha} \;\geq\; \frac{f(x)}{C_2},$$

as required.

We use a similar argument to prove (A). In this case, x is the midpoint of a segment ab. Let $r = \|x - a\| = \|x - b\|$ be the radius of the smallest circle passing through a and b, and let p be a vertex that encroaches upon ab, as in Figure 26. Consider first the case where p lies on an input edge that shares no endpoint with the input edge of ab. Then $f(x) \leq r$ by condition (iii) of the definition of local feature size. Consider second the case where the splitting of ab is triggered by rejecting the addition of a circumcentre. Let p be this circumcentre and let r' be the radius of its circle. Since p lies inside the diameter circle of ab we have $r' \leq \sqrt{2}r$. Using the Lipschitz Condition and induction we get

$$\begin{aligned} f(x) \;&\leq\; f(p) + r \\ &\leq\; r' \cdot C_2 + r \\ &\leq\; \sqrt{2}r \cdot C_2 + r. \end{aligned}$$

Using $C_1 \geq 1 + \sqrt{2}C_2$ we get

$$r \;\geq\; \frac{f(x)}{1 + \sqrt{2}C_2} \;\geq\; \frac{f(x)}{C_1},$$

as required. $\qquad\square$

Upper bound

Invariants (A) and (B) guarantee that vertices added to the triangulation cannot get arbitrarily close to preceding vertices. We show that this implies that they cannot get close to succeeding vertices either. Recall that K is the final triangulation generated by the Delaunay refinement algorithm.

Smallest Gap Lemma. $\|a - b\| \geq \frac{f(a)}{1 + C_1}$ for all vertices $a, b \in K$.

Proof. If b precedes a then $\|a - b\| \geq f(a)/C_1 \geq f(a)/(1 + C_1)$. Otherwise, we have $\|b - a\| \geq f(b)/C_1$ and therefore

$$f(a) \;\leq\; f(b) + \|a - b\| \;\leq\; \|a - b\| \cdot (1 + C_1),$$

as claimed. $\qquad\square$

Since vertices cannot get arbitrarily close to each other, we can use an area argument to show that the algorithm halts after adding a finite number of vertices. We relate the number of vertices to the integral of $1/f^2(x)$. Recall that B is the bounding box used in the construction of K.

Upper Bound Lemma.   The number of vertices in K is at most some constant times $\int_B \mathrm{d}x/f^2(x)$.

Proof.   For each vertex a of K, let D_a be the disk with centre a and radius $r_a = f(a)/(2+2C_1)$. By the Smallest Gap Lemma, the disks are pairwise disjoint. At least one quarter of each disk lies inside B. Therefore,

$$
\begin{aligned}
\int_B \frac{\mathrm{d}x}{f^2(x)} &\geq \frac{1}{4}\cdot\sum_a \int_{D_a} \frac{\mathrm{d}x}{f^2(x)} \\
&\geq \frac{1}{4}\cdot\sum_a \frac{r_a^2\pi}{(f(a)+r_a)^2} \\
&\geq \frac{1}{4}\cdot\sum_a \frac{\pi}{(3+2C_1)^2}.
\end{aligned}
$$

This is a constant times the number of vertices.   □

Two geometric results

We prepare the lower bound argument with two geometric results on triangles with angles no smaller than some constant $\alpha > 0$. Two edges of such a triangle abc cannot be too different in length, and specifically, $\frac{\|a-c\|}{\|a-b\|} \leq \varrho = 1/\sin\frac{\alpha}{2}$. If we have a chain of triangles connected through shared edges, the length ratio cannot exceed ϱ^t, where t is the number of triangles. Two edges sharing a common vertex are connected by the chain of triangles around that vertex. That chain cannot be longer than $\frac{2\pi}{\alpha}$, simply because we cannot pack more angles into 2π.

Length Ratio Lemma.   The length ratio between two edges sharing a common vertex is at most $\varrho^{2\pi/\alpha}$.

The second result concerns covering a triangle with four disks, one each around the three vertices and the circumcentre. For each vertex we take a disk with radius c_0 times the length of the shortest edge. For the circumcentre we take a disk with radius $1 - c_2$ times the circumradius. For a general triangle, we can keep c_0 fixed and force c_2 as close to zero as we like, just by decreasing the angle. If angles cannot be arbitrarily small, then c_2 can also be bounded away from zero.

Triangle Cover Lemma.   For each constant $c_0 > 0$ there is a constant $c_2 > 0$ such that the four disks cover the triangle.

Proof.   Refer to Figure 31. Let R be the circumradius and ab be the shortest of the three edges. Its length is $\|a - b\| \geq 2R\cdot\sin\frac{\alpha}{2}$. The disk around a covers all points at distance at most $c_0 \cdot \|a - b\|$ from a, and we assume without loss of generality that $c_0 < \frac{1}{2}$. The distance between the circumcentre, z, and the point $y \in ab$ at distance $c_0 \cdot \|a - b\|$ from a is

$$
\begin{aligned}
\|y - z\| &< \sqrt{R^2 - c_0^2 \|a - b\|^2} \\
&\leq \sqrt{R^2 \cdot \left(1 - 4c_0^2 \cdot \sin^2 \frac{\alpha}{2}\right)} \\
&< R \cdot \left(1 - 2c_0^2 \cdot \sin^2 \frac{\alpha}{2}\right).
\end{aligned}
$$

All other points on triangle edges not covered by disks around a, b, c are at most that distance from z. Since c_0 and α are positive constants, $c_2 = 2c_0^2 \cdot \sin^2 \frac{\alpha}{2}$ is also a positive constant.   □

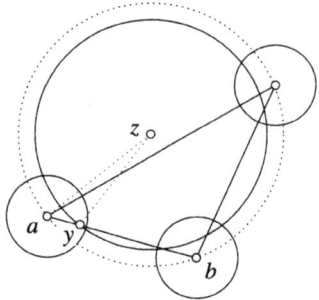

Fig. 31. The disks constructed for a triangle
and its three vertices cover the triangle

Lower bound

The reason for picking the disk of radius $(1 - c_2)R$ around the circumcentre is that for a point x inside this disk the local feature size cannot be arbitrarily small. In particular, it cannot be smaller than the distance from the circumcircle times the cosine of half the smallest angle, $f(x) \geq c_2 R \cdot \cos \frac{\alpha}{2}$. To get a similar result for disks around vertices, let L be the length of the shortest edge incident to a vertex a. The local feature size of a is at least $L \cdot \sin \alpha$. By choosing $c_0 = \frac{\sin \alpha}{2}$ we get $f(a) \geq 2c_0 L$ and therefore $f(x) \geq f(a) - \|a - x\| \geq c_0 L$ for every point x inside the disk with radius $c_0 L$ around a.

We use these observations to show that any algorithm that constructs triangles with angles no smaller than some constant $\alpha > 0$ generates at least some constant times the integral of $1/f^2(x)$ many vertices. It follows that the algorithm in Section 7 constructs meshes with asymptotically minimum size.

Lower Bound Lemma.   If K is a triangle mesh of G with all angles larger than α, then the number of vertices is at least some constant times $\int_B \mathrm{d}x / f^2(x)$.

Proof. Around each vertex $a \in K$ draw a disk with radius equal to $\frac{\sin \alpha}{2}$ times the length of the shortest incident edge. Let $c_0 = \frac{\sin \alpha}{2} \varrho^{\pi/\alpha}$ and use the Triangle Cover Lemma to pick a matching constant $c_2 > 0$. For each triangle $abc \in K$ draw the disk with radius $1 - c_2$ times the circumradius around the circumcentre. Each triangle is covered by its four disks, which implies that the mesh is covered by the collection of disks.

For each disk D_i in the collection, let f_i be the minimum local feature size at any point $x \in D_i$. By what we said earlier, that minimum is at least some constant fraction of the radius of D_i, $f_i \geq r_i/C$. Given that the disks cover the mesh we have

$$\int_B \frac{\mathrm{d}x}{f^2(x)} \;\leq\; \sum_i \int_{D_i} \frac{\mathrm{d}x}{f^2(x)}$$

$$\leq\; \sum_i \frac{r_i^2 \pi}{f_i^2}$$

$$\leq\; \sum_i C^2 \pi.$$

The number of triangles is less than twice the number of vertices, which we denote as n. Hence,

$$n \;\geq\; \sum_i \frac{1}{3} \;\geq\; \frac{1}{3C^2\pi} \int_B \frac{\mathrm{d}x}{f^2(x)},$$

as claimed. $\square$

Bibliographic notes

The idea of using the local feature size function in the analysis of the Delaunay refinement algorithm is due to Jim Ruppert. The details of the analysis left out in the journal publication Ruppert (1995) can be found in the technical report Ruppert (1992). Bern, Eppstein and Gilbert (1994) show that the same technical result (constant minimum angle and constant times minimum number of triangles) can also be achieved using quad-trees. Experimentally, the approach with Delaunay triangulations seems to generate meshes with fewer and nicer triangles. One reason for the better performance might be the absence of any directional bias from Delaunay triangulations.

9. Lifting and polarity

The Delaunay tetrahedrization of a finite set of points in $\mathbb{R}^3$ is dual to the Voronoi diagram of the same set. This section introduces both concepts and shows how they can be obtained as projections of the boundary of convex polyhedra.

Voronoi diagrams

The *Voronoi region* of a point p in a finite collection $S \subseteq \mathbb{R}^3$ is the set of points at least as close to p as to any other point in S,

$$V_p = \{x \in \mathbb{R}^3 : \|x - p\| \leq \|x - q\|, \ \forall q \in S\}.$$

Each inequality defines a closed half-space, and V_p is the intersection of a finite collection of such half-spaces. In other words, V_p is a convex polyhedron, maybe like the one shown in Figure 32. In the generic case, every vertex of V_p belongs to only three facets and three edges of the polyhedron. If V_p is bounded then it is the convex hull of its vertices. It is also possible that V_p is unbounded. This is the case if and only if there is a plane through p with all points of S on or on one side of the plane.

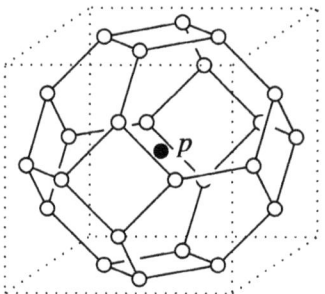

Fig. 32. The Voronoi polyhedron of a point in a
body-centred cube lattice. The relevant neighbours of
the cube centre p are the eight corners of the cube and
the centres of the six adjacent cubes

The Voronoi regions together with their shared facets, edges, vertices form the *Voronoi diagram* of S. A point x that belongs to k Voronoi regions is equally far from the k generating points. It follows that the k points lie on a common sphere. If the points are in general position then $k \leq 4$. A Voronoi vertex x belongs to at least four Voronoi regions, and assuming general position it belongs to exactly four regions.

Delaunay tetrahedrization

We obtain the *Delaunay tetrahedrization* by taking the dual of the Voronoi diagram. The Delaunay vertices are the points in S. The Delaunay edges connect generators of Voronoi regions that share a common facet. The Delaunay facets connect generators of Voronoi regions that share a common edge. Assuming general position, each edge is shared by three Voronoi regions and the Delaunay facets are triangles. The Delaunay polyhedra connect generators of Voronoi regions that share a common vertex. Assuming general position, each vertex is shared by four Voronoi regions and the Delaunay polyhedra are tetrahedra. Consider point p in Figure 32. Its Voronoi polyhedron has 14 facets, 36 edges, and 24 vertices. It follows that p belongs to 14 Delaunay edges, 36 Delaunay triangles, and 24 Delaunay tetrahedra, as illustrated in Figure 33.

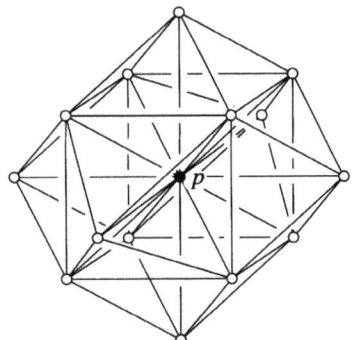

Fig. 33. The Delaunay neighbourhood of a point
in a body-centred cube lattice

Assuming general position of the points in S, the Delaunay tetrahedrization is a collection of simplices. To prove that it is a simplicial complex, we still need to show that the simplices avoid improper intersections. We do this by introducing geometric transformations that relate Voronoi diagrams and Delaunay tetrahedrizations in $\mathbb{R}^3$ with boundary complexes of convex polyhedra in $\mathbb{R}^4$.

Distance maps

The square distance from $p \in S$ is the map $\pi_p : \mathbb{R}^3 \to \mathbb{R}$ defined by $\pi_p(x) = \|x - p\|^2$. Its graph is a paraboloid of revolution in $\mathbb{R}^4$. We simplify notation by supressing the difference between a function and its graph. Figure 34 illustrates this idea in one lower dimension. Take the collection of all square distance functions defined by points in S. The pointwise minimum is the map $\pi_S : \mathbb{R}^3 \to \mathbb{R}$ defined by

$$\pi_S(x) \;=\; \min\{\pi_p(x) : p \in S\}.$$

Its graph is the lower envelope of the collection of paraboloids. By definition of Voronoi region, $\pi_S(x) = \pi_p(x)$ if and only if $x \in V_p$. We can therefore think of V_p as the projection of the portion of the lower envelope contributed by the paraboloid π_p.

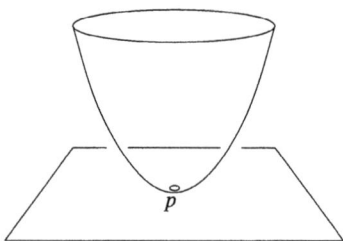

Fig. 34. The graph of the square
distance function of a point p in the plane

Linearization

All square distance functions have the same quadratic term, which is $\|x\|^2$. If we subtract that term we get linear functions, namely

$$
\begin{aligned}
f_p(x) &= \pi_p(x) - \|x\|^2 \\
&= (x-p)^T \cdot (x-p) - x^T \cdot x \\
&= -2p^T \cdot x + \|p\|^2.
\end{aligned}
$$

The graph of f_p is a hyperplane in $\mathbb{R}^4$. The same transformation warps the hyperplane $x_4 = 0$ to the upside-down paraboloid Π defined as the graph of the map defined by $\Pi(x) = -\|x\|^2$. Figure 35 shows the result of the transformation applied to the plane and paraboloid in Figure 34. We can apply the transformation to the entire collection of paraboloids at once. Each point in $\mathbb{R}^4$ travels vertically, that is, parallel to the x_4-axis. The travelled distance is the square distance to the x_4-axis. Paraboloids go to hyperplanes, intersections of paraboloids go to intersections of hyperplanes, and the lower envelope of the paraboloids goes to the lower envelope of the hyperplanes.

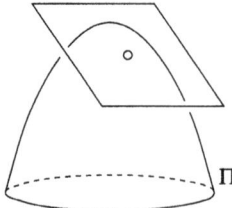

Fig. 35. The plane in Figure 34
becomes an upside-down paraboloid,
and the paraboloid becomes a plane

Replace each hyperplane by the closed half-space bounded from above by the hyperplane. The intersection of the half-spaces is a convex polyhedron F in $\mathbb{R}^4$, and the lower envelope of the hyperplanes is the boundary of F. It is a complex of convex faces of dimension 3, 2, 1, 0. Since the transformation moves points vertically, the projection onto $x_4 = 0$ of the lower envelope of paraboloids and the lower envelope of hyperplanes are the same. In particular, the projection of each three-dimensional face of F is a Voronoi region, and the projection of the entire boundary complex is the Voronoi diagram.

Polarity

We still need to describe what all this has to do with the Delaunay tetrahedrization of S. Instead of addressing this question directly, we first study the relationship between non-vertical hyperplanes and their polar points in $\mathbb{R}^4$.

A non-vertical hyperplane is the graph of a linear function $f : \mathbb{R}^3 \to \mathbb{R}$, which can generally be defined by a point $p \in \mathbb{R}^3$ and a scalar $c \in \mathbb{R}$, that is,

$$f(x) = -2p^T \cdot x + \|p\|^2 - c.$$

The hyperplane parallel to f and tangent to Π is defined by the equation $-2p^T \cdot x + \|p\|^2$. The vertical distance between the two hyperplanes is $|c|$. The *polar point* of f is $g = f^* = (p, -\|p\|^2 + c)$. The vertical distance between g and f is $2|c|$, and the parallel tangent hyperplane lies right in the middle between g and f. Furthermore, the vertical line through g also passes through the point where the tangent hyperplane touches Π. It follows that $g \in \Pi$ if and only if f is tangent to Π. Figure 36 shows a few examples of hyperplanes and their polar points in $\mathbb{R}^2$. Since hyperplanes are non-vertical, the points lying *above, on, below* are unambiguously defined. Let f_1, f_2 be two non-vertical hyperplanes and g_1, g_2 their polar points.

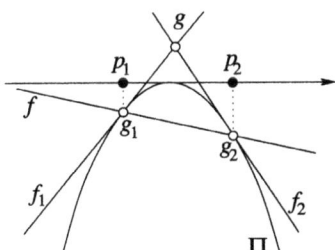

Fig. 36. Points g_1, g_2, g are polar to the lines (hyperplanes) f_1, f_2, f. Lines f_1, f_2 are warped images of the distance square functions of the points p_1, p_2 on the real line

Order Reversal Claim. Point g_1 lies above, on, below hyperplane f_2 if and only if point g_2 lies above, on, below hyperplane f_1.

Proof. Let $g_i = (p_i, -\|p_i\|^2 + c_i)$ for $i = 1, 2$. The algebraic expression for g_1 above f_2 is

$$-\|p_1\|^2 + c_1 \; > \; -2p_2^T \cdot p_1 + \|p_2\|^2 - c_2.$$

We move terms left and right and use the fact that vector products are commutative to get

$$-\|p_2\|^2 + c_2 \; > \; -2p_1^T \cdot p_2 + \|p_1\|^2 - c_1.$$

This is the algebraic expression for g_2 above f_1. The arguments for point g_1 lying on and below hyperplane f_2 are the same. □

Polar polyhedron

We are now ready to construct the Delaunay tetrahedrization as the projection of the boundary complex of a convex polyhedron in $\mathbb{R}^4$. For each point $p \in S$, let $g_p = (p, -\|p\|^2)$ be the polar point of the corresponding hyperplane. All points g_p lie on the upside-down paraboloid Π, as shown in Figure 37. For a non-vertical hyperplane f, we consider the closed half-space bounded from above by f. Let G be the intersection of all such half-spaces that contain all points g_p. G is a convex polyhedron in $\mathbb{R}^4$. Its boundary consists of the upper portion of the convex hull boundary plus the silhouette extended to infinity in the $-x_4$ direction. The Order Reversal Claim implies the following correspondence between G and F. A hyperplane *supports* G if it has non-empty intersection with the boundary and empty intersection with the interior.

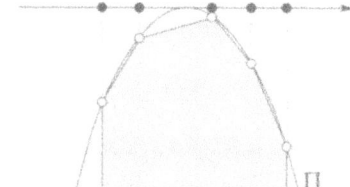

Fig. 37. The boundary complex of the shaded
polyhedron projects onto the Delaunay
tetrahedrization of the set of solid points

Support Claim. A hyperplane f supports G if and only if the polar point $g = f^*$ lies in the boundary of F.

Imagine exploring G by rolling the supporting hyperplane along its boundary. The dual image of this picture is the polar point moving inside the boundary of F. For each k-dimensional face of G we get a $(3-k)$-dimensional face of F and *vice versa*. An exception is the set of vertical faces of G, which do not correspond to any faces of F, except possibly to faces stipulated at infinity. The relationship between the two boundary complexes is the same as that between the Delaunay tetrahedrization and the Voronoi diagram. The isomorphism between the boundary complex of F and the Voronoi diagram implies the isomorphism between the boundary complex of G (excluding vertical faces) and the Delaunay tetrahedrization. Since the vertices of G project onto points in S, it follows that the boundary complex of G projects onto the Delaunay tetrahedrization of S. This finally implies that there are no improper intersections between Delaunay simplices. The Delaunay tetrahedrization of a set S of finitely many points in general position is indeed a simplicial complex.

Bibliographic notes

Voronoi diagrams and Delaunay triangulation are named after Georges Voronoi (1907/08) and Boris Delaunay (1934). The concepts themselves are older and can be traced back to prominent mathematicians of earlier centuries, including Friedrich Gauß and René Descartes. The connection to convex polytopes has also been known for a long time. The combinatorial theory of convex polytopes is a well developed field within mathematics. We refer to the texts by Branko Grünbaum (1967) and by Günter Ziegler (1995) for excellent sources of the accumulated knowledge in that subject.

10. Weighted distance

The correspondence between Voronoi diagrams and convex polyhedra hints at a generalization of Voronoi and Delaunay diagrams forming a richer class of objects. This section describes this generalization using points with real weights. Within this larger class of diagrams we find a symmetry between Voronoi and Delaunay diagrams absent in the smaller class of unweighted diagrams.

Commuting diagram

Figure 38 illustrates the correspondence between Voronoi diagrams and Delaunay tetrahedrizations in $\mathbb{R}^3$ and convex polyhedra in $\mathbb{R}^4$, as worked out in Section 9. V and D are dual to each other. F is obtained from V through linearization of distance functions, and V is formed by the projections of the boundary complex of F. F and G are polar to each other. G is the convex hull of the points projected onto Π (extended to infinity along the $-x_4$-direction), and D is the projection of the boundary complex of G.

We call G an *inscribed* polyhedron because each vertex lies on the upside-down paraboloid Π. Similarly, we call F a *circumscribed* polyhedron because each hyperplane spanned by a 3-face is tangent to the Π. Being inscribed or

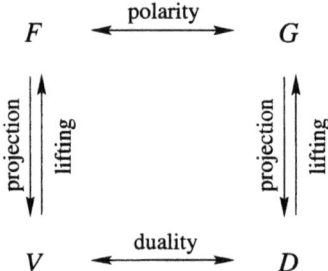

Fig. 38. Relationship between Voronoi diagram, V,
Delaunay tetrahedrization, D,
and convex polyhedra, F and G

circumscribed is a rather special property. We use weights to generalize the concepts of Voronoi diagrams and Delaunay tetrahedrization in a way that effectively frees the polyhedra from being inscribed or circumscribed. For technical reasons, we still require that every vertical line intersects F in a half-line and G either in a half-line or the empty set. This is an insubstantial although sometimes inconvenient restriction.

Weighted points

We prepare the definition of weighted Delaunay tetrahedrization by introducing points with real weights. It is convenient to write the weight of a point as the square of a non-negative real or a non-negative multiple of the imaginary unit. We think of the weighted point $\hat{p} = (p, P^2) \in \mathbb{R}^3 \times \mathbb{R}$ as the sphere with centre $p \in \mathbb{R}^3$ and radius P. The *power* or *weighted distance function* of $\hat{p}$ is the map $\pi_{\hat{p}} : \mathbb{R}^3 \to \mathbb{R}$ defined by

$$\pi_{\hat{p}}(x) \;=\; \|x - p\|^2 - P^2.$$

It is positive for points x outside the sphere, zero for points on the sphere, and negative for points inside the sphere. The various cases permit intuitive geometric interpretations of weighted distance. For example, for positive P^2 and x outside the sphere, it is the square length of a tangent line segment connecting x with a point on the sphere. This is illustrated in Figure 39. What is it if x lies inside the sphere? In Section 2, we have seen that the set of points with equal weighted distance from two circles is a line. Similarly, the set of points with equal weighted distance from two spheres in $\mathbb{R}^3$ is a plane. If the two spheres intersect then the plane passes through the intersection circle, and if the two spheres are disjoint and lie side by side then the plane separates the two spheres.

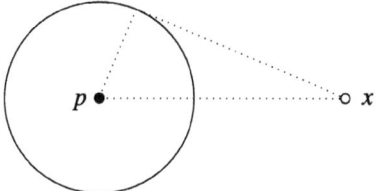

Fig. 39. The segment px, the tangent segment
from x to the circle, and the connecting radius
form a right-angled triangle

Orthogonality

Given two spheres or weighted points $\hat{p} = (p, P^2)$ and $\hat{q} = (q, Q^2)$, we
generalize weighted distance to the symmetric form

$$\pi_{\hat{p},\hat{q}} = \|p - q\| - P^2 - Q^2.$$

For $Q^2 = 0$, this is the weighted distance from q to $\hat{p}$, and for $P^2 = 0$, this
is the weighted distance from p to $\hat{q}$. We call $\hat{p}$ and $\hat{q}$ *orthogonal* if $\pi_{\hat{p},\hat{q}} = 0$.
Indeed, if $P^2, Q^2 > 0$ then $\pi_{\hat{p},\hat{q}} = 0$ if and only if the two spheres meet
in a circle and the two tangent planes at every point of this circle form a
right angle. Orthogonality is the key concept in generalizing Delaunay to
weighted Delaunay tetrahedrizations. We call $\hat{p}$ and $\hat{q}$ *further than orthogonal* if $\pi_{\hat{p},\hat{q}} > 0$.

   Let us contemplate for a brief moment how weights affect the lifting process. The graph of the weighted distance function is a paraboloid whose
zero-set, $\pi_{\hat{p}}^{-1}(0)$, is the sphere $\hat{p}$. We can linearize as before and get a hyperplane defined by

$$
\begin{aligned}
f_{\hat{p}}(x) &= \pi_{\hat{p}}(x) - \|x\|^2 \\
&= -2p^T \cdot x + \|p\|^2 - P^2.
\end{aligned}
$$

We can also polarize and get

$$g_{\hat{p}} = (p, -\|p\|^2 + P^2).$$

Orthogonality between two spheres now translates to a point-hyperplane
incidence.

Orthogonality Claim.   Spheres $\hat{p}$ and $\hat{q}$ are orthogonal if and only if
point $g_{\hat{p}}$ lies on the hyperplane $f_{\hat{q}}$.

Proof.   The algebraic expression for $g_{\hat{p}} \in f_{\hat{q}}$ is

$$-2q^T \cdot p + \|q\|^2 - Q^2 = -\|p\|^2 + P^2.$$

This is equivalent to

$$(p - q)^T \cdot (p - q) - P^2 - Q^2 \;=\; 0,$$

which is equivalent to $\pi_{\hat{p},\hat{q}} = 0$.   □

Weighted Delaunay tetrahedrization

Let S be a finite set of spheres. Depending on the application, we think of an element of S as a point in $\mathbb{R}^3$ or a weighted point in $\mathbb{R}^3 \times \mathbb{R}$. The weighted distance can be used to construct the *weighted Voronoi diagram*, and the *weighted Delaunay tetrahedrization* is dual to that diagram, as usual. Instead of going through the technical formalism of the construction, which is pretty much the same as for unweighted points, we illustrate the concept in Figure 40. For unweighted points, a tetrahedron belongs to the Delaunay tetrahedrization if and only if the circumsphere passing through the four vertices is empty. For weighted points, the circumsphere is replaced by the *orthosphere*, which is the unique sphere orthogonal to all four spheres whose centres are the vertices of the tetrahedron. Its centre is the Voronoi vertex shared by the four Voronoi regions, and its weight is the common weighted distance of that vertex from the four spheres. We summarize by generalizing the Circumcircle Claim of Section 2 to three dimensions and to the weighted case.

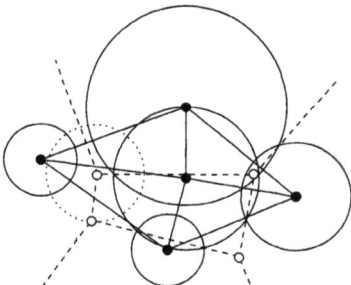

Fig. 40. Dashed weighted Voronoi diagram and solid
weighted Delaunay triangulation of five weighted
points in the plane. Each Voronoi vertex is the centre
of a circle orthogonal to the generating circles of the
regions that meet at that vertex. Only one such
circle is shown

Orthosphere Claim. A tetrahedron belongs to the weighted Delaunay tetrahedrization if and only if the orthosphere of the four spheres is further than orthogonal from all other sphere in the set.

A sphere in S is *redundant* if its Voronoi region is empty. By definition, the centre of a sphere is a vertex of the weighted Delaunay triangulation if and only if it is non-redundant. All extreme points are non-redundant, which implies that the underlying space is the convex hull of S, as in the unweighted case.

Local convexity

Recall the Delaunay lemma of Section 3, which states that a triangulation of a finite set in $\mathbb{R}^2$ is the Delaunay triangulation if and only if every one of its edges is locally Delaunay. This result generalizes to three (and higher) dimensions and to the weighted case. For the purpose of this discussion, we define a *tetrahedrization* of S as a simplicial complex K whose underlying space is conv S and whose vertex set is a subset of S. A triangle abc in K is *locally convex* if

 (i) it belongs to only one tetrahedron and therefore bounds the convex hull of S,   or

 (ii) it belongs to two tetrahedra, $abcd$ and $abce$, and $\hat{e}$ is further than orthogonal from the orthosphere of $abcd$.

If all triangles in K are locally convex, then after lifting we get the boundary complex of a convex polyhedron. This is consistent with the right side of the commuting diagram in Figure 38. However, to be sure this polyhedron is G, we also require that no lifted point lies vertically below the boundary.

Local Convexity Lemma. If Vert K contains all non-redundant weighted points and every triangle is locally convex, then K is the weighted Delaunay tetrahedrization of S.

The proof is rather similar to that of the Delaunay lemma in Section 3 and does not need to be repeated. Similarly, we can extend the Acyclicity Lemma of Section 2 to three (and higher) dimensions and to the weighted case. Details should be clear and are omitted.

Bibliographic notes

Weighted Voronoi diagrams are possibly as old as unweighted ones. Some of the earliest references appear in the context of quadratic forms, which arise in the study of the geometry of numbers (Gruber and Lekkerkerker 1987). These forms are naturally related to weighted as opposed to unweighted diagrams. Examples of such work are the papers by Dirichlet (1850) and Voronoi (1907/08). Weighted Delaunay triangulations and their generalizations to three and higher dimensions seem less natural and have a shorter history. Nevertheless, they have already acquired at least three different names, namely regular triangulations (Billera and Sturmfels 1992) and coherent triangulations (Gelfand, Kapranov and Zelevinsky 1994) besides the one used in this paper.

11. Flipping

The goal of this section is to generalize the idea of edge flipping to three and higher dimensions. We begin with two classic theorems in convex geometry. Helly's theorem talks about the intersection structure of convex sets. It can be proved using Radon's theorem, which talks about partitions of finite point sets and is directly related to flips in d dimensions. We then define flips and discuss structural issues that arise in $\mathbb{R}^3$.

Radon's theorem

This is a result on $n \geq d + 2$ points in $\mathbb{R}^d$. The case of $n = 4$ points in $\mathbb{R}^2$ is related to edge flipping in the plane.

Radon's Theorem. Every collection S of $n \geq d + 2$ points in $\mathbb{R}^d$ has a partition $S = A \,\dot\cup\, B$ with $\operatorname{conv} A \cap \operatorname{conv} B \neq \emptyset$.

Proof. Since there are more than $d+1$ points, they are affinely dependent. Hence there are coefficients λ_i, not all zero, with $\sum \lambda_i p_i = 0$ and $\sum \lambda_i = 0$. Let I be the set of indices i with $\lambda_i > 0$, and let J contain all other indices. Note that $c = \sum_{i \in I} \lambda_i = -\sum_{j \in J} \lambda_j > 0$, and also

$$ x = \frac{1}{c} \cdot \sum_{i \in I} \lambda_i p_i = -\frac{1}{c} \cdot \sum_{j \in J} \lambda_j p_j. $$

Let A be the collection of points p_i with $i \in I$ and let B contain all other points. Point x is a convex combination of the points in A as well as of the points in B. Equivalently, $x \in \operatorname{conv} A \cap \operatorname{conv} B$. □

A $(d{+}1)$-dimensional simplex has $d+2$ vertices and a face for every subset of the vertices. If we project its boundary complex onto $\mathbb{R}^d$ we get a simplex for every subset of at most $d+1$ vertices. By Radon's theorem, at least two of these simplices have an improper intersection. This intersection comes from projecting the two sides of the simplex boundary on top of each other.

Helly's theorem

This is a result on $n \geq d{+}2$ convex sets in $\mathbb{R}^d$. For $d = 1$ it states that if every pair of a collection of $n \geq 2$ closed intervals has a non-empty intersection then the entire collection has a non-empty common intersection. This is true because the premise implies that the rightmost left endpoint is to the left or equal to the leftmost right endpoint. The interval between these two endpoints belongs to every interval in the collection.

Helly's Theorem. If every $d + 1$ sets in a collection of $n \geq d + 2$ closed convex sets in $\mathbb{R}^d$ have a non-empty common intersection, then the entire collection has a non-empty intersection.

Proof. Assume inductively that the claim holds for $n - 1$ closed convex sets. For each C_i in the collection of n sets, let p_i be a point in the common intersection of the other $n - 1$ sets. Let S be the collection of points p_i. By Radon's theorem, there is a partition $S = A \overset{.}{\cup} B$ and a point $x \in \operatorname{conv} A \cap \operatorname{conv} B$. By construction, $\operatorname{conv} A$ is contained in all sets C_j with $p_j \in B$, and symmetrically, $\operatorname{conv} B$ is contained in all sets C_i with $p_i \in A$. Hence, x is contained in every set of the collection.   □

Flipside of a simplex

Consider the case $d = 2$. The projection of a 3-simplex (tetrahedron) onto $\mathbb{R}^2$ is either a convex quadrangle or a triangle. In the former case the two diagonals cross, and in the latter case one vertex lies in the triangle spanned by the other three. Both cases are illustrated in Figure 41. The direction of projection defines an *upper* and a *lower* side of the tetrahedron boundary, and the two sides meet along the *silhouette*. Let $\alpha = \operatorname{conv} A$ and $\beta = \operatorname{conv} B$ be the two faces whose projections have an improper intersection. They lie on opposite sides, and we assume that α belongs to the upper and β to the lower side. The quadrangle case defines an edge flip, which replaces the projection of the upper by the projection of the lower side, or *vice versa*. We also call this a *2-to-2 flip* because it replaces 2 old by 2 new triangles. The triangle case defines a new type of flip, which we refer to as a *1-to-3* or a *3-to-1 flip* depending on whether a new vertex is added or an old vertex is removed.

Fig. 41. The two generic projections
of a tetrahedron onto the plane

How do these considerations generalize to the case $d = 3$? As illustrated in Figure 42, the projection of a 4-simplex onto $\mathbb{R}^3$ is either a double pyramid or a tetrahedron. In the double pyramid case, α is an edge and β is a triangle. There are three tetrahedra that share α and they form the upper side of the 4-simplex. The remaining two tetrahedra share β and form the lower side. The *3-to-2 flip* replaces the projection of the upper side by the projection of the lower side, and the *2-to-3 flip* does it the other way round. In the tetrahedron case, α is one vertex and β is the tetrahedron spanned by the other four vertices. The *1-to-4 flip* adds α, effectively replacing β by four tetrahedra, and the *4-to-1 flip* removes α.

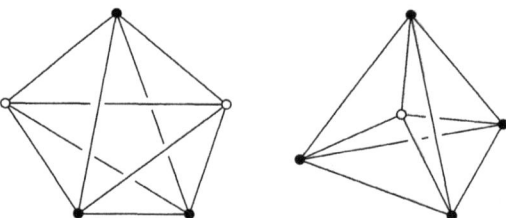

Fig. 42. The two generic projections
of a 4-simplex onto 3-dimensional space

Transformability

In using flips to construct a Delaunay tetrahedrization in $\mathbb{R}^3$, we encounter cases where we would like to flip but we cannot. This happens only for 2-to-3 flips. Let *abcd* and *bcde* share the triangle *bcd*. If the edge *ae* crosses *bcd* we can replace *abcd, bcde* by *baec, caed, daeb*, which is a 2-to-3 flip. However, if the edge *ae* misses *bcd*, as illustrated in Figure 43 where *ae* passes behind *bd*, we cannot add *ae* because it might cross other triangles in the current tetrahedrization. In this case, the union of the two tetrahedra is non-convex. Assume without loss of generality that *bd* is the non-convex edge. There are two cases. If *bd* belongs to only three tetrahedra then the third one is *abde*, and we can replace *abdc, cbde, ebda* by *bace, aced*. This is a 3-to-2 flip. However, if *bd* belongs to four or more tetrahedra then we are stuck and cannot remove the triangle *bcd*. This is the *non-transformable* case.

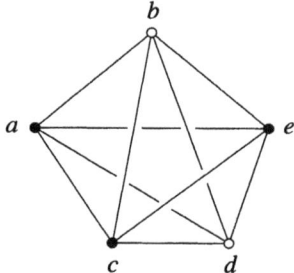

Fig. 43. The edge *ae* does not pass through
the triangle *bcd* but rather behind the edge *bd*

The reason for studying flips is of course the interest in an algorithm that constructs a weighted Delaunay tetrahedrization by flipping. The occurrence of non-transformable cases does not imply that all hope is lost. It might still be possible to flip elsewhere in a way that resolves non-transformable cases by changing their local neighbourhood. But this requires further analysis.

Bibliographic notes

Radon's theorem is a by-product of the attempt by Johann Radon (1921) to prove Helly's theorem, communicated to him by Eduard Helly (1923). The two theorems are equivalent and form a cornerstone of modern convex geometry. Helly was missing as a prisoner of war in Russia, so Radon published his theorem and proof. After returning from Russia, Helly published his theorem and his own proof, which is inductive in the size of the collection *and* the dimension. Years later, Helly generalized his theorem to a topological setting where convexity is replaced by requirements of connectivity (Helly 1930). The concept of an edge flip was generalized to three and higher dimensions by Lawson (1986) without, however, realizing the connection to Radon's theorem.

12. Incremental algorithm

This section generalizes the algorithm of Section 4 to three dimensions and to the weighted case. The algorithm is incremental and adds a point in a sequence of flips. We describe the algorithm, prove its correctness, and discuss its running time.

Algorithm

Let S be a finite set of weighted points in $\mathbb{R}^3$. We denote the points by $\hat{p}_1, \hat{p}_2, \ldots, \hat{p}_n$ and assume they are in general position. To reduce the number of cases, we let $wxyz$ be a sufficiently large tetrahedron. In particular, we assume $wxyz$ contains all points of S in its interior. Define $S_i = \{w, x, y, z, \hat{p}_1, \hat{p}_2, \ldots, \hat{p}_i\}$ for $0 \leq i \leq n$, and let D_i be the weighted Delaunay tetrahedrization of S_i. The algorithm starts with D_0 and adds the weighted points in order. Adding $\hat{p}_1$ is done in a sequence of flips.

```
for i = 1 to n do
    find pqrs ∈ D_{i-1} that contains p_i;
    if p̂_i is non-redundant among p̂, q̂, r̂, ŝ then
        add p̂_i with a 1-to-4 flip
    endif;
    while ∃ triangle bcd not locally convex do
        flip bcd
    endwhile
endfor.
```

The algorithm maintains a tetrahedrization, which we denote as K. Sometimes, K is a weighted Delaunay tetrahedrization of a subset of the points, but often it is not. Consider flipping the triangle bcd in K. Let $abcd$ and $bcde$ be the two tetrahedra that share bcd. If their union is convex, then flipping *bcd* means a 2-to-3 flip that replaces *bcd* by edge *ae* together with triangles *aeb*, *aec*, *aed*. Otherwise, we consider the subcomplex induced by *a*, *b*, *c*, *d*, *e*. It consists of the simplices in K spanned by subsets of the five points. If the underlying space of the induced subcomplex is non-convex then *bcd* cannot be flipped. If the underlying space is convex then it is either a double pyramid or a tetrahedron. In the former case, flipping means a 3-to-2 flip. In the latter case, flipping means a 4-to-1 flip, which effectively removes a vertex. The various types of flips are illustrated in Figure 44.

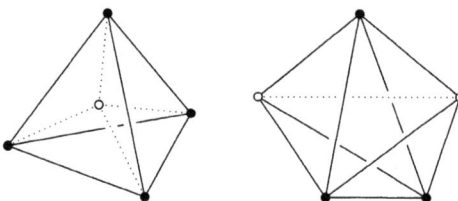

Fig. 44. To the left, a 1-to-4 or a 4-to-1 flip depending
on whether the hollow vertex is added or removed.
To the right, a 2-to-3 or a 3-to-2 flip depending on
whether the dotted edge is added or removed

Stack of triangles

Flipping is done in a sequence controlled by a stack. At any moment, the stack contains all triangles in the link of p_i that are not locally convex. It may also contain other triangles in the link, but it contains each triangle at most once. Initially, the stack consists of the four triangles of *pqrs*. Flipping continues until the stack is empty.

```
while stack is non-empty do
    pop bcd from stack;
    if bcd ∈ K and bcd is not locally convex
              and bcd is transformable then
        apply a 2-to-3, 3-to-2, or 4-to-1 flip;
        push new link triangles on stack
    endif
endwhile.
```

Why can we restrict our attention to triangles in the link of p_i? Outside the link, K is equal to D_{i-1}, hence all triangles are locally convex. A triangle inside the link connects p_i with an edge *cd* in the link. Let xp_icd and p_icdy be the two tetrahedra sharing p_icd. If their union is convex, we can remove p_icd by a 2-to-3 flip. This creates a new tetrahedron *acde* not incident to p_i, which contradicts that D_{i-1} is a weighted Delaunay tetrahedrization.

If their union is non-convex, the triangles xcd and cdy in the link are also not locally convex.

Correctness

Let K be the tetrahedrization at some moment in time after adding $\hat{p}_i$ when it is not yet the weighted Delaunay tetrahedrization of S_i. It suffices to show that K has at least one link triangle that is not locally convex and transformable. To get a contradiction, we suppose all triangles that are not locally convex are non-transformable. Let L be the set of tetrahedra in $K - \operatorname{St} p_i$ that have at least one triangle in the link. These tetrahedra form a spiky sphere around p_i, not unlike the spiky circle in Figure 45. Let $L' \subseteq L$ contain all tetrahedra whose triangles in the link are not locally convex. By assumption, $L' \neq \emptyset$. For each tetrahedron in L, consider the orthosphere $\hat{z}$ and the weighted distance $\pi_{\hat{p}_i, \hat{z}}$. Let $abcd \in L$ be the tetrahedron whose orthosphere minimizes that function. We have $abcd \in L'$, or equivalently $\pi_{\hat{p}_i, \hat{z}} < 0$, for else the triangle bcd in the link would be locally convex, and so would every other link triangle.

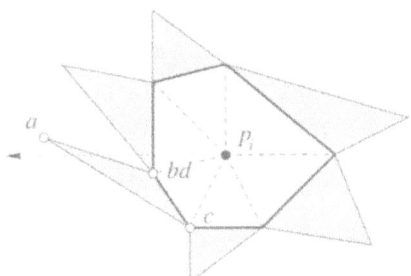

Fig. 45. The bold edges belong to the link of p_i
and the shaded triangles belong to L

We argue that bcd is transformable. To get a contradiction assume it is not. Let bd be a non-convex edge of the union of $abcd$ and $bcdp_i$, and let $abdx$ be the tetrahedron on the other side of abd. If bd is the only non-convex edge then $x \neq p_i$, for else bcd would be transformable. Otherwise, there is another non-convex edge, say bc. Let $abcy$ be the tetrahedron on the other side of abc. If $x = y = p_i$ we again have a contradiction because this would imply that bcd is transformable. We may therefore assume that $x \neq p_i$. Equivalently, abd is not in the link of $\hat{p}_i$. Consider a half-line that starts at p_i and passes through an interior point of abd. After crossing the link, the half-line goes through a tetrahedron of L before it encounters $abcd$. This is illustrated in Figure 45. Outside the link, we have a genuine weighted Delaunay tetrahedrization, namely a portion of D_{i-1}. For tetrahedra in D_{i-1}, the weighted distance of $\hat{p}_i$ from their orthospheres increases along the half-line, which contradicts the minimality assumption in the choice of *abcd*. This finally proves that flipping continues until D_i is reached.

Number of flips

To upper-bound the number of flips in the worst case, we interpret that algorithm as gluing 4-simplices to a three-dimensional surface consisting of tetrahedra in $\mathbb{R}^4$. Each flip corresponds to a 4-simplex. It either removes or introduces one or four edges. Once an edge is removed it cannot be introduced again. This implies that the total number of flips is less than $2\binom{n}{2} < n^2$. Modulo implementation details, we thus have an algorithm that constructs the Delaunay tetrahedrization of n points in $\mathbb{R}^3$ in $O(n^2)$ time. The size of the final Delaunay tetrahedrization is therefore at most some constant times n^2.

There are sets of n points in $\mathbb{R}^3$ with at least some constant times n^2 Delaunay tetrahedra. Take, for example, two skew lines and place $\frac{n}{2}$ un-weighted points on each line, as shown in Figure 46. Consider two contiguous point on one line together with two contiguous points on the other line. The sphere passing through the four points is empty, which implies that the four points span a Delaunay tetrahedron. The total number of such tetrahedra is roughly $\frac{n^2}{4}$. However, for point sets that seem to occur in practice, the number of Delaunay tetrahedra is typically less than some constant times n. Examples of such sets are dense packing of spheres common in molecular modelling, and well-spaced sets as produced by three-dimensional mesh generation software.

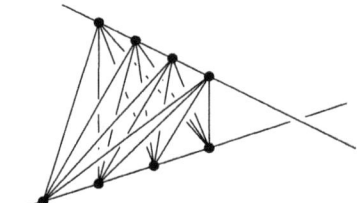

Fig. 46. A tetrahedral mesh whose edge skeleton
contains a complete bipartite graph

Expected running time

It is a good idea to first compute a random permutation of the points so that the construction proceeds in a random order. However, because the size of the tetrahedrization can vary between linearly and quadratically many simplices, the analysis is more involved than in two dimensions. We cannot even claim that the expected running time is at most $\log_2 n$ times the size of the final tetrahedrization. Indeed, this is false because there exist point sets with linear size Delaunay tetrahedrizations that reach quadratic intermediate size with positive constant probability. Nevertheless, such a claim holds if we further relativize the statement by drawing points from a fixed distribution. Suppose the expected size of the Delaunay tetrahedrization of k points chosen randomly from the distribution is $O(f(k))$. If $f(k) = \Omega(k^{1+\varepsilon})$, for some constant $\varepsilon > 0$, then the expected running time is $O(f(n))$, and otherwise it is $O(f(n) \log n)$. The argument is similar to the one presented in Section 4 and details are omitted.

Bibliographic notes

Algorithms that construct a Delaunay tetrahedrization in $\mathbb{R}^3$ through flips have first been considered by Barry Joe. In the paper Joe (1989) he gives an example where the non-transformable cases form a deadlock situation and flipping does not lead to the Delaunay tetrahedrization. In Joe (1991) he shows that flipping succeeds if the points are added one at a time. The proof of Joe's result in this section is taken from Edelsbrunner and Shah (1996), where the same is shown for weighted Delaunay tetrahedrization in $\mathbb{R}^d$.

13. Meshing polyhedra

In this paper, meshing a spatial domain means decomposing a polyhedron into tetrahedra that form a simplicial complex. This section introduces polyhedra and studies the problem of how many tetrahedra are needed to mesh them.

Polyhedra and faces

A *polyhedron* is the union of convex polyhedra, $P = \bigcup_{i \in I} \bigcap H_i$, where I is a finite index set and each H_i is a finite set of closed half-spaces. For example the polyhedron in Figure 47 can be specified as the union of four convex polyhedra. As we can see, faces are not necessarily simply connected. We use a definition that permits faces even to be disconnected.

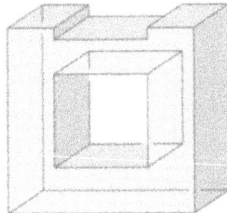

Fig. 47. A non-convex polyhedron

Let b be the open ball with unit radius centred at the origin of $\mathbb{R}^3$. For a point x we consider a sufficiently small neighbourhood, $N_\varepsilon(x) = (x + \varepsilon \cdot b) \cap P$. The *face figure* of x is the enlarged version of this neighbourhood within the polyhedron, $x + \bigcup_{\lambda>0} \lambda \cdot (N_\varepsilon(x) - x)$. A *face* of P is the closure of a maximal collection of points with identical face figure. To distinguish the faces of P from the edges and triangles of the Delaunay tetrahedrization to be constructed, we call 1- and 2-faces of P *segments* and *facets*. Observe that the polyhedron in Figure 47 has 24 vertices, 30 segments, 11 facets, and two 3-faces, namely the inside with face figure $\mathbb{R}^3$ and the outside with empty face figure. Six of the segments and three of the facets are non-connected. Two of the facets are connected but not simply connected, namely the front and the back facets.

Tetrahedrizations

A *tetrahedrization* of P is a simplicial complex K whose simplices decompose P. Since simplicial complexes are finite by definition, only bounded polyhedra have tetrahedrizations. A tetrahedrization of P triangulates every facet and every segment by a subcomplex each. Every vertex of P is necessarily also a vertex of K.

We will see shortly that every bounded polyhedron has a tetrahedrization. Interestingly, there are polyhedra whose tetrahedrizations have necessarily more vertices than the polyhedra. The smallest such example is the Schönhardt polyhedron shown in Figure 48. It can be obtained from a triangular prism by a slight rotation of one triangular facet relative to the other. The six vertices of the polyhedron span $\binom{6}{4} = 15$ tetrahedra, which we classify into three types exemplified by $abcA$, $abAB$, $bcCA$. All three tetrahedra share bA as an edge. But this edge lies outside the Schönhardt polyhedron, which implies that none of the 15 tetrahedra is contained in the polyhedron. The Schönhardt polyhedron can therefore not be tetrahedrized using tetrahedra spanned by its vertices. There are of course other tetrahedrizations. The simplest uses a vertex z in the centre and cones from z to the 6 vertices, 12 edges, 8 triangles in the boundary.

Fencing off

We give a constructive proof that every polyhedron P has a tetrahedrization. For simplicity we assume that P is everywhere three-dimensional. Equivalently, P is the closure of its interior, $P = \text{cl int } P$. It is convenient to place P in space such that no facet lies in a vertical plane and no segment is contained in a vertical line. Call two points $x, y \in P$ *vertically visible* if x, y lie on a common vertical line and the edge xy is contained in P. The *fence* of a segment consists of all points $x \in P$ vertically visible from some

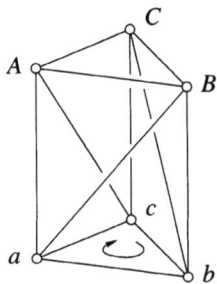

Fig. 48. The Schönhardt polyhedron.
The edges aB, bC, cA are non-convex

point y of the segment. The tetrahedrization is constructed in three steps, the first of which is illustrated in Figure 49.

Step 1. Erect the fence of each segment. The fences decompose P into vertical cylinders, each bounded by a top and a bottom facet and a circle of fence pieces called *walls*.

Step 2. Triangulate the bottom facet of every cylinder and erect fences from the new segments, effectively decomposing P into triangular cylinders.

Step 3. Decompose each wall into triangles and finally tetrahedrize each cylinder by constructing cones from an interior point to the boundary.

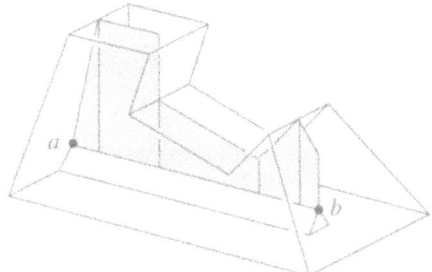

Fig. 49. The fence of the segment ab consists
of five walls, each a triangle or a quadrangle

Upper bound

We analyse the tetrahedrization obtained by erecting fences and prove that the final number of tetrahedra is at most some constant times the square of the number of segments.

Upper Bound Claim. The three steps tetrahedrize a bounded polyhedron with m segments using fewer than $28m^2$ tetrahedra.

Proof. Fences erected in **Step 1** may meet in vertical edges. Each inter-section corresponds to a crossing between vertical projections of segments. The total number of crossings is at most $\binom{m}{2}$. Each segment creates a fence, and each crossing involving this segment may cut one wall of the fence into two. The total number of walls is therefore no more than $m + 2\binom{m}{2} = m^2$. A cylinder bounded by k walls is decomposed into $k-2$ triangular cylinders separated from each other by $k-3$ new walls. **Step 2** thus increases the total number of walls to less than $3m^2$. The total number of cylinders at this stage is less than $2m^2$. Each wall is a triangle or a quadrangle, and it may be divided into two by the piece of the segment that defines it. **Step 2** there-fore triagulates each wall using four or fewer triangles, and it tetrahedrizes each cylinder using 14 or fewer tetrahedra. The final tetrahedrization thus contains fewer than $28m^2$ tetrahedra.   □

Saddle surface

We prepare a matching lower bound by studying the *hyperbolic paraboloid* specified by the equation $x_3 = x_1 \cdot x_2$. Figure 50 illustrates the paraboloid by showing its intersection with the vertical planes $\pm x_1 \pm x_2 = 1$. A general

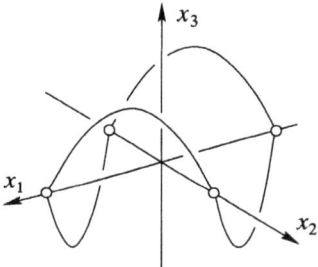

Fig. 50. Hyperbolic paraboloid indicated
through its intersection with vertical walls

line in the x_1x_2-plane is specified by $ax_1 + bx_2 + c = 0$. To determine the intersection of the paraboloid with the vertical plane through that line, we can either express x_1 in terms of x_2 or *vice versa*,

$$x_3 = -\frac{b}{a}x_2^2 - \frac{c}{a}x_2,$$
$$x_3 = -\frac{a}{b}x_2^2 - \frac{c}{b}x_2.$$

For $a \cdot b \neq 0$ we get a parabola. For $a = 0$ we get a line for every value of $\frac{c}{b}$, and we sample this family at integer values. Similarly, we sample the 1-parameter family of lines we get for $b = 0$ at integer values of $\frac{c}{a}$. Figure 51 shows a small portion of the two families in top view. If two points x and y lie on the paraboloid then the segment between them lies on the surface if and only if the vertical projections of x, y onto the $x_1 x_2$-plane line on a common horizontal or vertical line. If the line has positive slope then the segment lies below the surface, and if the line has negative slope then it lies above the surface.

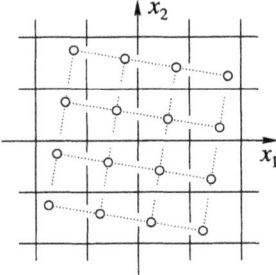

Fig. 51. Top view of hyperbolic paraboloid. We see
samples of the two ruling families of lines and dotted
edges connecting points sampled on the surface

Lower bound construction

We build a polyhedron Q out of a cube by cutting deep wedges, each close to a line of the two ruling families. The construction is illustrated in Figure 52. Assuming we have n cuts from the top and n from the bottom, we have $m = 14n + 8$ segments forming the polyhedron.

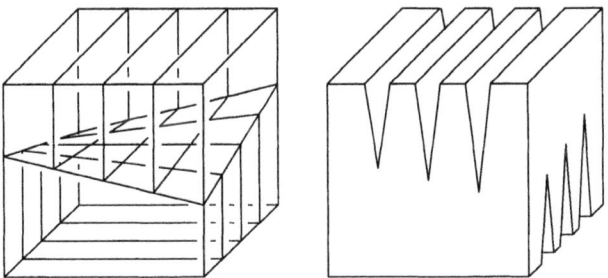

Fig. 52. Polyhedron Q with two families of cuts
almost meeting along the saddle surface

Lower Bound Claim.   Every tetrahedrization of Q consists of at least $(n + 1)^2$ tetrahedra.

Proof.   Consider the checkerboard produced by the $2n + 4$ lines on the saddle surface that mark the ends of the $2n$ cuts and the intersection with the boundary of the cube. Choose a point in each square of the checkerboard producing the slightly tilted square grid pattern of Figure 51. The edges connecting any two points intersect at least one of the wedges, provided the sharp ends of the wedges reach sufficiently close to the saddle surface. It follows that in any tetrahedrization of Q, the $(n+1)^2$ points lie inside pairwise different tetrahedra.   □

Bibliographic notes

The definition of a polyhedron as the union of intersections of closed half-spaces is taken from Hadwiger (1957). The definition of a face is taken from Edelsbrunner (1995) and should be contrasted with that suggestion in Grünbaum and Shephard (1994). The Schönhardt polyhedron was named after E. Schönhardt who described the polyhedron in 1928 (Schönhardt 1928). The same construction was mentioned 17 years earlier in a paper by Lennes (1911). Ruppert and Seidel (1992) build on this construction, and show that deciding whether or not a polyhedron can be tetrahedrized without adding new vertices is NP-complete. The quadratic upper and lower bounds for tetrahedrizing polyhedra are taken from a paper by Bernard Chazelle (1984).

14. Tetrahedral shape

This section looks at the various shapes tetrahedra can assume. For the time being, good shape quality is defined as having a small circumradius over shortest edge length ratio. We will see later that meshes of tetrahedra with small ratio also have nice combinatorial properties, such as constant size vertex stars.

Classifying tetrahedra

The classification of tetrahedra into shape types is a fuzzy undertaking. We normalize by scaling tetrahedra to unit diameter. A normalized tetrahedron has small volume either because its vertices are close to a line, or, if that is not the case, its vertices are close to a plane. In the first case, the tetrahedron is *skinny*, and we distinguish five types depending on how its vertices cluster along the line. Up to symmetry, the possibilities are 1-1-1-1, 1-1-2, 1-2-1, 1-3, 2-2, as shown from left to right in Figure 53. A *flat* tetrahedron has small volume but is not skinny. We have four types depending on whether two vertices are close to each other, three vertices lie close to a line, the orthogonal projection of the tetrahedron onto the close plane is a triangle, or the projection is a quadrangle. All four types are shown from left to right in Figure 54.

Circumradius over shortest edge length

A tetrahedron $abcd$ has a unique circumsphere. Let $R = R(abcd)$ be that radius and $L = L(abcd)$ the length of the shortest edge. We measure the

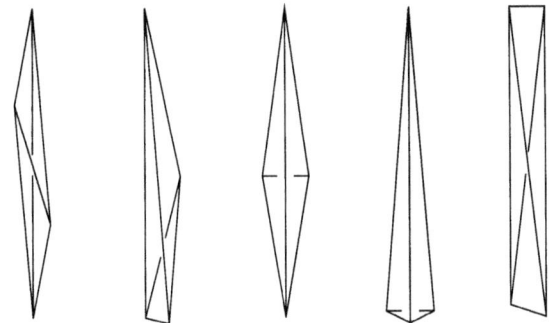

Fig. 53. Five fuzzy types of skinny tetrahedra

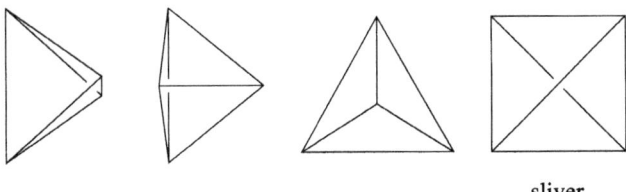

sliver

Fig. 54. Four fuzzy types of flat tetrahedra

quality of the tetrahedron shape by taking the ratio, that is,

$$\varrho \;=\; \varrho(abcd) \;=\; \frac{R}{L}.$$

We also define ϱ for triangles, taking the radius of the circumcircle over the length of the shortest edge. Observe that the ratio of a tetrahedron is always larger than or equal to the ratio of each of its triangles.

A triangle abc minimizes the ratio if and only if it is equilateral, in which case the circumcentre is also the barycentre,

$$y \;=\; \frac{1}{3} \cdot (a+b+c) \;=\; \frac{2}{3} \cdot x + \frac{1}{3} \cdot c,$$

where $x = \frac{1}{2} \cdot (a+b)$. Normalization implies that the three edges have length 1. The ratio is therefore equal to the circumradius, which is

$$\|c - y\| \;=\; \frac{2}{3} \cdot \|c - x\| \;=\; \frac{2}{3} \cdot \sqrt{1 - \frac{1}{4}}$$

$$\;=\; \frac{\sqrt{3}}{3} \;=\; 0.577\ldots$$

A tetrahedron $abcd$ minimizes the ratio if and only if it is regular, in which case the circumcentre is again the barycentre,

$$z \;=\; \frac{1}{4} \cdot (a+b+c+d) \;=\; \frac{3}{4} \cdot y + \frac{1}{4} \cdot d.$$

Normalization implies that the six edges have length 1. The ratio is therefore equal to the circumradius, which is

$$\|d - z\| = \frac{3}{4} \cdot \|d - y\| = \frac{3}{4} \cdot \sqrt{1 - \frac{3}{9}}$$

$$= \frac{\sqrt{6}}{4} = 0.612\ldots$$

Both calculations are illustrated in Figure 55.

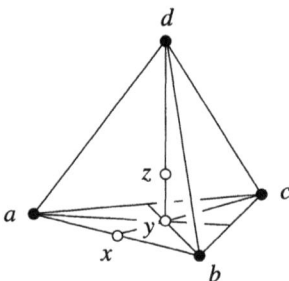

Fig. 55. A regular tetrahedron and the barycentres
of an edge, a triangle, the tetrahedron

A *skinny* triangle has small area. It has either a short edge or a large circumradius. In either case, its ratio is large. A skinny tetrahedron has skinny triangles, hence its ratio is large. A flat triangle that is not a sliver has either a short edge or a large circumradius and thus a large ratio. The only remaining small volume tetrahedron is the sliver, and it can have ϱ as small as $\frac{\sqrt{2}}{2} = 0.707\ldots$ or even a tiny amount smaller.

Ratio property

A mesh of tetrahedra has the *ratio property for ϱ_0* if $\varrho \le \varrho_0$ for all tetrahedra. We assume that every triangle in the mesh is the face of a tetrahedron in the mesh. It follows that $\varrho \le \varrho_0$ also for every triangle. We prove two elementary facts about edge lengths in a mesh K that has the ratio property for a constant ϱ_0.

Claim A. If abc is a triangle in K then

$$\frac{1}{2\varrho_0} \cdot \|a - b\| \le \|a - c\| \le 2\varrho_0 \cdot \|a - b\|.$$

Proof. The length of an edge is at most twice the circumradius, $\|a - b\| \le 2Y$. By assumption, $\|a - b\| \ge Y/\varrho_0$. The same inequalities hold for $\|a - c\|$, which implies the claim. $\square$

Next we show that, if K has the ratio property and it is a Delaunay tetrahedrization, then edges that share a common endpoint and form a small angle cannot have very different lengths. For this to hold, it is not necessary that the two edges belong to a common triangle. Define

$$\eta_0 = \arctan 2 \left(\varrho_0 - \sqrt{\varrho_0^2 - 1/4} \right).$$

Since ϱ_0 is a constant, so is η_0.

Claim B. If the angle between ab and ap is less than η_0 then

$$\frac{1}{2} \cdot \|a - b\| \;<\; \|a - p\| \;<\; 2 \cdot \|a - b\|.$$

Proof. Consider the circumsphere of a tetrahedron that contains ab as an edge, and let $\hat{y} = (y, Y^2)$ be the circle in which the plane passing through a, b, p intersects the sphere. We use Figure 56 as an illustration throughout

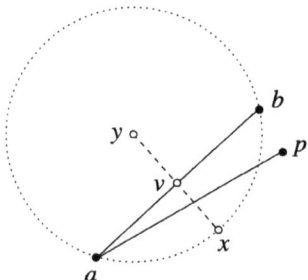

Fig. 56. Section through a circumsphere
of a Delaunay tetrahedron with edge ab

the proof. Let v be the midpoint of ab, and let x be the point on the circle such that y, v, x lie in this sequence on a common line. We have $Y \leq \varrho_0 \cdot \|a - b\|$ by assumption. The distance between x and v is

$$
\begin{aligned}
\|x - v\| \;&=\; Y - \sqrt{Y^2 - \|a - b\|^2/4} \\
&\geq\; \left(\varrho_0 - \sqrt{\varrho_0^2 - 1/4} \right) \cdot \|a - b\|,
\end{aligned}
$$

because the difference between Y and $\sqrt{Y^2 - C}$ decreases with increasing Y. The angle between ab and ax is

$$
\begin{aligned}
\angle bax \;&=\; \arctan \frac{2\|x - v\|}{\|a - b\|} \\
&\geq\; \arctan 2 \left(\varrho_0 - \sqrt{\varrho_0^2 - 1/4} \right) \\
&=\; \eta_0.
\end{aligned}
$$

The claimed lower bound follows because the circle forces ap to be at least as long as ax, which is longer than half of ab. The claimed upper bound on the length of ap follows by a symmetric argument that reverses the roles of b and p. $\square$

Length variation

We use Claims A and B to show that the length variation of edges with a common endpoint a in K is bounded by some constant. As before, we assume K has the ratio property and is a Delaunay tetrahedrization. Define $m_0 = 2/(1 - \cos\frac{\eta_0}{4})$ and $\nu_0 = 2^{2m_0-1} \cdot \varrho_0^{m_0-1}$. Since ϱ_0 and η_0 are constants, so are m_0 and ν_0.

Length Variation Lemma. If ab, ap are edges in K then

$$\frac{1}{\nu_0} \cdot \|a - b\| \; < \; \|a - p\| \; < \; \nu_0 \cdot \|a - b\|.$$

Proof. Let Σ be the sphere of directions around a. We form a maximal packing of circular caps, each with angle $\eta_0/4$. This means if y is the centre and x a boundary point of a cap then $4\angle xay = \eta_0$. The area of each cap is $(1 - \cos\frac{\eta_0}{4})/2$ times the area of Σ, which implies that there are at most m_0 caps.

By increasing the caps to radius $\eta_0/2$ we change the maximal packing into a covering of Σ. For each edge ab in the star of a, let $b' \in \Sigma$ be the radial projection of b. Similarly, for each triangle abc consider the arc on Σ that is the radial projection of bc. The points and arcs form a planar graph. Let ab be the longest and ap the shortest edge in the star of a. We walk in the graph from b' to p'. This path leads from cap to cap, and we record the sequence ignoring detours that return to previously visited caps. The sequence consists of at most m_0 caps. Let us track the edge length during the walk. As long as we stay within a cap, Claim B implies the length decreases by less than a factor $\frac{1}{2}$. If we step from one cap to the next, Claim A implies the length decreases by at most a factor $\frac{1}{2\varrho_0}$. Hence $\|a - p\| > \frac{1}{\nu_0} \cdot \|a - b\|$. The upper bound follows by a symmetric argument that exchanges b and p. $\square$

Constant degree

A straightforward volume argument together with the Length Variation Lemma implies that each vertex in K belongs to at most some constant number of edges. Define $\delta_0 = (2\nu_0^2 + 1)^3$, which is a constant.

Degree Lemma. Every vertex a in K belongs to at most δ_0 edges.

Proof. Let ab be the longest and ap the shortest edge in the star of a. Assume without loss of generality that $\|a - p\| = 1$. Let c be a neighbour of a and let d be a neighbour of c. We have $\|a - c\| \geq 1$ by assumption and $\|c - d\| \geq \frac{1}{\nu_0}$ by the Length Variation Lemma. For each neighbour c of a let Γ_c be the open ball with centre c and radius $\frac{1}{2\nu_0}$. The balls are pairwise disjoint and fit inside the ball Γ with centre a and radius $\|a - b\| + \frac{1}{2\nu_0}$. The volume of Γ is

$$\operatorname{vol}\Gamma = \frac{4\pi}{3}\left(\|a - b\| + \frac{1}{2\nu_0}\right)^3$$
$$\leq \frac{4\pi}{3}\left(\frac{2\nu_0^2 + 1}{2\nu_0}\right)^3$$
$$= (2\nu_0^2 + 1)^3 \cdot \operatorname{vol}\Gamma_c.$$

In words, at most $\delta_0 = (2\nu_0^2 + 1)^3$ neighbour balls fit into Γ. This implies that δ_0 is an upper bound on the number of neighbours of a.   □

The constant δ_0 in the Degree Lemma is miserably large. The main reason is that the constant ν_0 in the Length Variation Lemma is miserably large. It would be nice to find a possibly more direct proof of that lemma and bring the constant down to reasonable size.

Bibliographic notes

The idea of measuring the quality of a tetrahedron by its circumradius over shortest edge length ratio is due to Miller and co-authors (Miller, Talmor, Teng and Walkington 1995). The proofs of the Length Variation and Degree Lemmas are taken from the same source. Further results on meshes of tetrahedra that have the ratio property can be found in the doctoral thesis by Talmor (1997).

15. Delaunay refinement

This section generalizes the Delaunay refinement algorithm of Section 7 from two to three dimensions. The additional dimension complicates matters. In particular, special care must be taken to avoid infinite loops bouncing back and forth between refining segments and facets of the input polyhedron.

Refinement algorithm

For technical reasons, we restrict ourselves to bounded polyhedra P without interior angles smaller than $\frac{\pi}{2}$. The condition applies to angles between two segments, between a segment and a facet, and between two facets. The polyhedron in Figure 47 satisfies the condition, but the polyhedron in Figure 48 does not. The goal is to construct a Delaunay tetrahedrization D

with a subcomplex $K \subseteq D$ that subdivides P and has the ratio property for a constant ϱ_0. The first step of the algorithm computes D as the Delaunay tetrahedrization of the set of vertices of P. Unless we are lucky, there will be segments that are not covered by edges of D, and there will be facets that are not covered by triangles of D. To recover these segments and facets, we add new points and update the Delaunay tetrahedrization using the incremental algorithm of Section 12. The points are added using the three rules given below.

We need some definitions. A segment of P is decomposed into *subsegments* by vertices of the Delaunay tetrahedrization that lie on the segment, and a facet is decomposed into (triangular) *subfacets* by the Delaunay triangulation of the vertices on the facet and its boundary. A vertex *encroaches upon* a subsegment if it is enclosed by the diameter sphere of that subsegment, and it *encroaches upon* a subfacet if it is enclosed by the equator sphere of that subfacet. Both spheres are the smallest that pass through all vertices of the subsegment and the subfacet.

Rule 1. If a subsegment is encroached upon, we split it by adding the midpoint as a new vertex to the Delaunay tetrahedrization. The new subsegments may or may not be encroached upon, and splitting continues until none of the subsegments is encroached upon.

Rule 2. If a subfacet is encroached upon, we split it by adding the circumcentre x as a new vertex to the Delaunay tetrahedrization. However, if x encroaches upon one or more subsegments then we do not add x and instead split the subsegments.

Rule 3. If a tetrahedron inside P has circumradius over shortest edge length ratio $R/L > \varrho_0$ then we split the tetrahedron by adding the circumcentre x as a new vertex to the Delaunay tetrahedrization. However, if x encroaches upon any subsegments or subfacets, we do not add x and instead split the subsegments and subfacets.

Rule 1 takes priority over **Rule 2**, and **Rule 2** takes priority over **Rule 3**. At the time we add a point on a facet, the prioritization guarantees that the boundary segments of the facet are subdivided by edges of the Delaunay tetrahedrization. Similarly, at the time we add a point in the interior of P, the boundary of P is subdivided by triangles in the Delaunay tetrahedrization. A point considered for addition to the Delaunay tetrahedrization has a *type*, which is the number of the rule that considers it or equivalently the dimension of the simplex it splits. Points of type 1 split subsegments and are always added once they are considered. Points of type 2 and 3 may be added or rejected.

Local density

Just as in two dimensions, the *local feature size* is crucial to understanding the Delaunay refinement algorithm. It is the function $f : \mathbb{R}^3 \to \mathbb{R}$ with $f(x)$ the radius of the smallest closed ball with centre x that intersects at least two disjoint faces of P. Note that f is bounded away from zero by some positive constant. It is easy to show that f satisfies the Lipschitz condition

$$f(x) \leq f(y) + \|x - y\|.$$

This implies that f is continuous over $\mathbb{R}^3$, but more than that, the condition says that f varies only slowly with x.

The local feature size is related to the *insertion radius* r_x of a point x, which is the length of the shortest Delaunay edge with endpoint x immediately after adding x. If x is a vertex of P then r_x is the distance to the nearest other vertex of P. If x is type 1 or 2 then r_x is the distance to the nearest encroaching vertex. If that encroaching vertex does not exist because it was rejected, then r_x is either half the length of the subsegment if x is type 1, or it is the circumradius of the subfacet if x is type 2. Finally, r_x is the circumradius of the tetrahedron it splits if x is type 3. We also define the insertion radius for a point that is considered for addition but rejected, because it encroaches upon subsegments or subfacets. This is done by hypothetically adding the point and taking the length of the shortest edge in the hypothetical star.

Radii and parents

Points are added in a sequence, and for each new point there are predecessors that we can make responsible for the addition. If x is type 1 or 2 then we define the responsible *parent* $p = p_x$ as the encroaching point that triggers the event. The point p may be a Delaunay vertex or a rejected circumcentre. If there are several encroaching points then p is the one closest to x. If x is type 3 then p is the most recently added endpoint of the shortest edge of the tetrahedron x splits.

Radius Claim. Let x be a vertex of D and p its parent, if it exists. Then $r_x \geq f(x)$ or $r_x \geq c \cdot r_p$, where $c = 1/\sqrt{2}$ if x is type 1 or 2 and $c = \varrho_0$ if x is type 3.

Proof. If x is a vertex of P then $f(x)$ is less than or equal to the distance to the nearest other vertex. This distance is $r_x \geq f(x)$. For the rest of the proof assume x is not a vertex of P. It therefore has a parent $p = p_x$. First consider the case where p is a vertex of P. If x is type 1 or 2, it lies in a segment or facet of P, and p is not contained in that segment or facet. Hence $r_x = \|x - p\| \geq f(x)$. If x is type 3 then the tetrahedron split by x has at least two vertices in P. Hence $r_x = \|x - p\| \geq f(x)$ as before. Secondly, consider the case where p is not a vertex of P. If x is type 1 or 2 then p was rejected for triggering the insertion of x. Since p encroaches upon the subsegment or subfacet split by x, its distance to the closest vertex of that subsegment or subfacet is at most $\sqrt{2}$ times the distance of x from that same vertex. Hence $r_x \geq r_p/\sqrt{2}$. Finally, if x is type 3 then $r_p \leq L$, where L is the length of the shortest edge of the tetrahedron split by x. The algorithm splits that tetrahedron only if $R > L\varrho_0$. Hence $r_x = R > L\varrho_0 \geq \varrho_0 r_p$.   □

Termination

The Radius Claim limits how quickly the insertion radius can decrease. We aim at choosing the only independent constant, which is ϱ_0, such that the insertion radii are bounded from below by a positive constant. Once this is achieved, we can prove termination of the algorithm using a standard packing argument. Figure 57 illustrates the possible parent–child relations between the three types of points added by the algorithm. We follow an arc of the digraph whenever the insertion radius of a point x is less than $f(x)$. The arc is labelled by the smallest possible factor relating the insertion radius of x to that of its parent. Note that there is no arc from type 1 to type 2 and there are no loops from type 1 back to type 1 and from type 2 back to type 2. This is because the angle constraint on the input polyhedron prevents parent–child relations for points on segments and facets with non-empty intersection. If there is a relation between points on segments and facets with empty intersection then $r_x \geq f(x)$ and there is no need to follow an arc in the digraph.

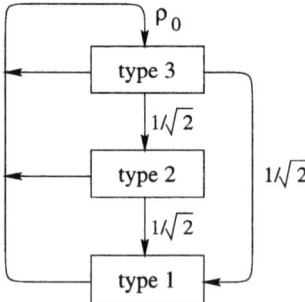

Fig. 57. The directed arcs indicate possible parent–child relations, and their labels give the worst case factors relating insertion radii

Observe that every cycle in the digraph contains the arc labelled ϱ_0 leading into type 3. We choose $\varrho_0 \geq 2$ to guarantee that the products of arc labels for all cycles are 1 or larger. The smallest product of any path in the digraph is therefore $\frac{1}{2}$. In cases where r_x is not at least $f(x)$, there exist ancestors q

with $r_x \geq r_q/2$ and $r_q \geq f(q)$. Since $f(q)$ is bounded away from zero by some positive constant, we conclude that the insertion radii cannot get arbitrarily small. It follows that the Delaunay refinement algorithm terminates. For $\varrho_0 < 2$ there are cases where the algorithm does not terminate.

Graded meshes

With little additional effort we can show that for ϱ_0 strictly larger than 2, insertion radii are directly related to local feature size, and not just indirectly through chains of ancestors. We begin with a relation between the local feature size over insertion radius ratio of a vertex and of its parent.

Ratio Claim.   Let x be a Delaunay vertex with parent p and assume $r_x \geq c \cdot r_p$. Then

$$\frac{f(x)}{r_x} \leq 1 + \frac{f(p)}{c \cdot r_p}.$$

Proof.   We have $r_x = \|x - p\|$ if p is a Delaunay vertex and $r_x \geq \|x - p\|$ if p is a rejected midpoint or circumcentre. Starting with the Lipschitz condition we get

$$\begin{aligned} f(x) &\leq f(p) + \|x - p\| \\ &\leq \frac{f(p)}{c \cdot r_p} \cdot r_x + r_x, \end{aligned}$$

and the result follows after dividing by r_x. $\square$

To prepare the next step we assume $\varrho_0 > 2$ and define constants

$$\begin{aligned} C_1 &= \frac{(3 + \sqrt{2}) \cdot \varrho_0}{\varrho_0 - 2}, \\ C_2 &= \frac{(1 + \sqrt{2}) \cdot \varrho_0 + \sqrt{2}}{\varrho_0 - 2}, \\ C_3 &= \frac{\varrho_0 + 1 + \sqrt{2}}{\varrho_0 - 2}. \end{aligned}$$

Note that $C_1 > C_2 > C_3 > 1$.

Invariant.   If x is a type i vertex in the Delaunay tetrahedrization, for $1 \leq i \leq 3$, then $r_x \geq f(x)/C_i$.

Proof.   If the parent p of x is a vertex of the input polyhedron P then $r_x \geq f(x)$ and we are done. Otherwise, assume inductively that the claimed inequality holds for vertex p. We finish the proof by case analysis. If x is type 3 then $c = \varrho_0$ and $r_x \geq \varrho_0 \cdot r_p$ by the Radius Claim. By induction we get $f(p) \leq C_1 r_p$, no matter what type p is. Using the Ratio Claim we get

$$\frac{f(x)}{r_x} \leq 1 + \frac{C_1}{\varrho_0} = C_3.$$

If x is type 2 then $c = \frac{1}{\sqrt{2}}$. We have $r_x \geq f(x)$ unless p is type 3, and therefore $f(p) \leq C_3 r_p$ by inductive assumption. Then $r_x \geq r_p/\sqrt{2}$ by the Radius Claim, and

$$\frac{f(x)}{r_x} \leq 1 + \sqrt{2} \cdot C_3 = C_2$$

by the Ratio Claim. If x is type 1 then $c = \frac{1}{\sqrt{2}}$. We have $r_x \geq f(x)$ unless p is type 2 or 3, and therefore $f(x) \leq C_2 r_p$ by inductive assumption. Then $r_x \leq 1 + r_p/\sqrt{2}$ by the Radius Claim, and

$$\frac{f(x)}{r_x} \leq 1 + \sqrt{2} \cdot C_2 = C_1$$

by the Ratio Claim. $\qquad\qquad\qquad\qquad\qquad\qquad\qquad\qquad\qquad\qquad\square$

Because C_1 is the largest of the three constants, we can simplify the Invariant to $r_x \geq f(x)/C_1$ for every Delaunay vertex x. From this we conclude

$$\|x - y\| \geq \frac{f(x)}{1 + C_1}$$

for any two vertices x, y in the Delaunay tetrahedrization, using the argument in the proof of the Smallest Gap Lemma in Section 8.

Bibliographic notes

The bulk of the material in this section is taken from a paper by Jonathan Shewchuk (1998). In that paper, the assumed input is a so-called piecewise linear complex as defined by Miller et al. (1996). This is a 3-face of a polyhedron together with its faces, which is slightly more general than a three-dimensional polyhedron.

16. Sliver exudation

The sliver is the only type of small volume tetrahedron whose circumradius over shortest edge length ratio does not grow with decreasing volume. Experimental studies indicate that slivers frequently exist right between other well-shaped tetrahedra inside Delaunay tetrahedrizations. This section explains how point weights can be used to remove slivers.

Periodic meshes

Suppose S is a finite set of points in $\mathbb{R}^3$ whose Delaunay tetrahedrization has the ratio property for a constant ϱ_0. The goal is to prove that there are weights we can assign to the points such that the weighted Delaunay tetrahedrization is free of slivers. This cannot be true in full generality, for if S consists of only four points forming a sliver then no weight assignment can make that sliver disappear. We avoid this and similar boundary effects by replacing the finite by a periodic set $S = P + \mathbb{Z}^3$, where P is a finite set of points in the half-open unit cube $[0,1)^3$ and $\mathbb{Z}^3$ is the three-dimensional integer grid. The periodic set S contains all points $p + \mathbf{v}$, where $p \in P$ and $\mathbf{v}$ is an integer vector. Like S, the Delaunay tetrahedrization D of S is periodic. Specifically, for every tetrahedron $\tau \in D$, the shifted copies $\tau + \mathbb{Z}^3$ are also in D. This idea is illustrated for a periodic set generated by four points in the half-open unit square in Figure 58.

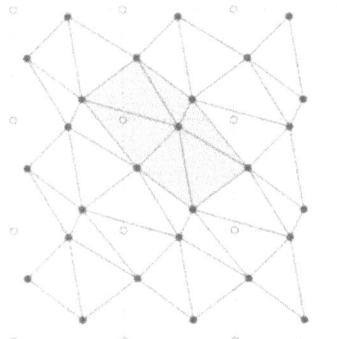

Fig. 58. Periodic tiling of the plane.
The shaded triangles form a domain whose
shifted copies tile the entire plane

Weight assignment

A *weight assignment* is a function $\omega : P \to \mathbb{R}$. The resulting set of spheres is denoted as $S_\omega = \{(a, \omega(p)) : p \in P, a \in p + \mathbb{Z}^3\}$. Depending on ω, a point p may or may not be a vertex of the weighted Delaunay triangulation of S_ω, which we denote as D_ω. Let $N(p)$ be the minimum distance to any other point in S. To prevent points from becoming redundant, we limit ourselves to *mild* weight assignments that satisfy $0 \leq \omega(p) < \frac{1}{3}N(p)$ for all $p \in P$. Every sphere in S_ω has a real radius and every pair is disjoint and not nested. It follows that none of the points is redundant. Another benefit of a mild weight assignment is that it does not drastically change the shape of triangles and tetrahedra. In particular, D_ω has the ratio property for a constant ϱ_1 that only depends on ϱ_0. It follows that the area of each triangle is bounded from below by some constant times the square of its circumcircle. The same is not true for volumes of tetrahedra, which is why eliminating slivers is difficult.

A crucial step towards eliminating slivers is a generalization of the Degree Lemma of Section 14. Let K be the set of simplices that occur in weighted Delaunay tetrahedrizations for mild weight assignments of S. In other words, $K = \bigcup_\omega D_\omega$, which is a three-dimensional simplicial complex but not necessarily geometrically realized in $\mathbb{R}^3$. The vertex set of K is Vert $K = S$, and the *degree* of a vertex is the number of edges in K that share the vertex.

Weighted Degree Lemma. There exists a constant δ_1 depending only on ϱ_0 such that the degree of every vertex in K is at most δ_1.

The proof is fairly tedious and partially a repeat of the proofs of the Length Variation and Degree Lemmas of Section 14. It is therefore omitted.

Slicing orthogonal spheres

We need an elementary fact about spheres (a, A^2) and (z, Z^2) that are orthogonal, that is, $\|a - z\|^2 = A^2 + Z^2$. A plane intersects the two spheres in two circles, which may have real or imaginary radii.

Slicing Lemma. A plane passing through a intersects the two spheres in two orthogonal circles.

Proof. Let $(x, X^2), (y, Y^2)$ be the circles where the plane intersects the two spheres. We have $x = a$, $X^2 = A^2$, and $Y^2 = Z^2 - \|z - y\|^2$. Hence

$$
\begin{aligned}
\|x - y\|^2 &= \|x - z\|^2 - \|z - y\|^2 \\
&= (A^2 + Z^2) - (Z^2 - Y^2) \\
&= X^2 + Y^2.
\end{aligned}
$$

In words, the two circles are also orthogonal.   □

As an application of the Slicing Lemma consider three spheres and the plane that passes through their centres, as in Figure 59. The plane intersects the three spheres in three circles, and there is a unique circle orthogonal to all three. The Slicing Lemma implies that every sphere orthogonal to all three spheres intersects the plane in this same circle.

Variation of orthoradius

Another crucial step towards eliminating slivers is the stability analysis of their orthospheres. We will see that a small weight change can increase the size of the orthosphere dramatically. This is useful because a tetrahedron in

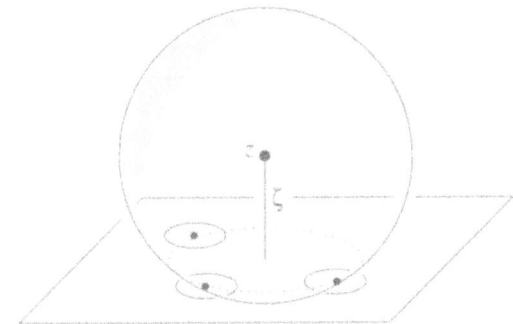

Fig. 59. Slice through three spheres and another
sphere orthogonal to the first three

D_ω cannot have a large orthosphere, for else that orthosphere would be closer than orthogonal to some weighted point. We later exploit this observation and change weights to increase orthospheres of slivers.

Let us analyse how the radius of the orthosphere of four spheres changes as we manipulate the weight of one of the sphere. Let (y, Y^2) be the smallest sphere orthogonal to the first three spheres, let (p, P^2) be the fourth sphere, and let (z, Z^2) be the orthosphere of all four spheres, as illustrated in Figure 60. Let ζ and ϕ be the distances of z and p from the plane h that passes through the centres of the first three spheres. With varying P^2, the centre of the orthosphere moves along the line that meets h orthogonally at y. The distance of z from h is a function of the weight of p, $\zeta : \mathbb{R} \to \mathbb{R}$.

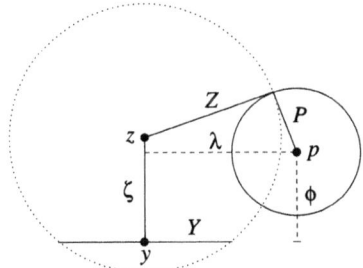

Fig. 60. The orthocentre z moves downward
as the weight of p increases

Distance Variation Lemma. $\zeta(P^2) = \zeta(0) - \frac{P^2}{2\phi}$.

Proof. Let λ be the distance from p to the line along which z moves. We have $Z^2 + P^2 = (\zeta(P^2) - \phi)^2 + \lambda^2$. The weight of the orthosphere is

$Z^2 = \zeta(P^2)^2 + Y^2$. Hence

$$\begin{aligned}\zeta(P^2)^2 &= Z^2 - Y^2 \\ &= (\zeta(P^2) - \phi)^2 + \lambda^2 - P^2 - Y^2.\end{aligned}$$

After cancelling $\zeta(P^2)^2$ we get

$$\zeta(P^2) = \frac{\phi^2 + \lambda^2 - Y^2}{2\phi} - \frac{P^2}{2\phi}.$$

The first term on the right-hand side is $\zeta(0)$. $\Box$

The term $P^2/2\phi$ is the displacement of the orthocentre that occurs as we change the weight of p from 0 to P^2. For slivers, the value of ϕ is small which implies that the displacement is large.

Sliver theorem

We finally show that there is a mild weight assignment that removes all slivers. The proof is constructive and assigns weights in sequence to the points in P. To quantify the property of being a sliver, we define $\xi = V/L$, where V is the volume and L is the length of the shortest edge of the tetrahedron. Only slivers can have bounded R/L as well as small ξ. Note that the volume of the tetrahedron indicated in Figure 60 is one-third the area of the base triangle times ϕ. As mentioned above, the area of the base triangle is some positive constant fraction Y^2. Similarly, L is some positive constant fraction of Y, which implies that ξ is some positive constant fraction of $Y\phi$.

Sliver Theorem.   There are constants $\varrho_1, \xi_0 > 0$ and a mild weight assignment ω, such that the weighted Delaunay tetrahedrization has the ratio property for ϱ_1 and $\xi > \xi_0$ for all its tetrahedra.

Proof. We focus on proving $\xi > \xi_0$ for all tetrahedra in D_ω. Assume without loss of generality that the distance from a point p to its nearest neighbour in S is $N(p) = 1$. The weight assigned to p can be anywhere in the interval $[0, \frac{1}{3}]$. According to the Weighted Degree Lemma, there is only a constant number of tetrahedra that can possibly be in the star of p. Each such tetrahedron can exist in D_ω only if its orthosphere is not too big. In other words, the tetrahedron can only exist if $\omega(p)$ is chosen inside some subinterval of $[0, \frac{1}{3}]$. The Distance Variation Lemma implies that the length of this subinterval decreases linearly with ϕ and therefore linearly with ξ. We can choose ξ_0 small enough such that the constant number of subintervals cannot possibly cover $[0, \frac{1}{3}]$. By the pigeonhole principle, there is a value $\omega(p) \in [0, \frac{1}{3}]$ that excludes all slivers from the star of p. $\Box$

Removing slivers

The proof of the Sliver Theorem suggests an algorithm that assigns weights to individual points in an arbitrary sequence. For each point $p \in P$, the algorithm considers the interval of possible weights and the subintervals in which tetrahedra in K can occur in the weighted Delaunay tetrahedrization. We could consider all tetrahedra in the star of p in K, but it is more convenient to consider only the subset in the 1-parameter family of weighted Delaunay tetrahedrizations generated by continuously increasing the weight of p from 0 through $\frac{1}{3}N(p)$. For each such tetrahedron, we get the ξ value and a subinterval during which it exists in D_ω. Figure 61 draws each tetrahedron as a horizontal line segment in the $\omega\xi$-plane. The lower envelope of the line segments is the function that maps the weight of p to the worst ξ value of any tetrahedron in its star. The algorithm finds the weight where that function has a maximum and assigns it to p. Since there is only a constant number of tetrahedra to be considered, this can be accomplished in constant time. The overall running time of the algorithm is therefore $O(n)$, where $n = \operatorname{card} P$.

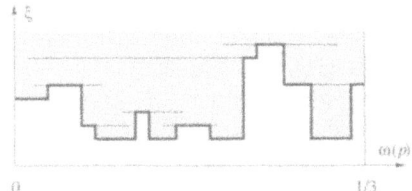

Fig. 61. Each tetrahedron in the star
is represented by a horizontal line segment

A source of possible worry is that, after we have fixed the weight of p we may modify the weight of a neighbour q of p. Modifying the weight of q may change the star of p. However, all new tetrahedra in the star of p also belong to the star of q and thus cannot have arbitrarily small ξ values. We thus do not have to reconsider p, and $O(n)$ time indeed suffices. The Sliver Theorem guarantees the algorithm is successful as quantified by the positive constant ξ_0. While the algorithm does not find the globally optimum weight assignment, it finds the optimum for each point individually, assuming fixed weights of other points. It might therefore achieve a minimum ξ value that is much better than the rather pessimistic estimate for ξ_0 guaranteed by the Sliver Theorem.

Bibliographic notes

The material of this section is taken from the sliver exudation paper by Cheng et al. (1999). The occurrence of slivers as a menace in three-dimensional Delaunay tetrahedrizations was reported by Cavendish, Field and

Frey (1985). Besides the sliver exudation method described in this section, there are two other methods that provably remove slivers. The first by Chew (1997) adds points and uses randomness to avoid creating new slivers. The second by Edelsbrunner et al. (1999) moves points and relies on the ratio property of the Delaunay tetrahedrization, as in the weight assignment method of this section.

REFERENCES

F. Aurenhammer (1991), Voronoi diagrams: a study of a fundamental geometric data structure, *ACM Comput. Surveys* **23**, 345–405.

M. Bern and D. Eppstein (1992), Mesh generation and optimal triangulations, in *Computing in Euclidean Geometry*, Vol. 1 (D.-Z. Du and F. K. Hwang, eds), World Scientific, Singapore, pp. 23–90.

M. Bern, D. Eppstein and J. Gilbert (1994), Provably good mesh generation, *J. Comput. Syst. Sci.* **48**, 384–409.

L. J. Billera and B. Sturmfels (1992), Fiber polytopes, *Ann. Math.* **135**, 527–549.

J. C. Cavendish, D. A. Field and W. H. Frey (1985), An approach to automatic three-dimensional finite element mesh generation, *Internat. J. Numer. Methods Engrg* **21**, 329–347.

B. Chazelle (1984), Convex partitions of polyhedra: a lower bound and worst case algorithm, *SIAM J. Comput.* **13**, 488–507.

S.-W. Cheng, T. K. Dey, H. Edelsbrunner, M. A. Facello and S.-H. Teng (1999), Sliver exudation, in *Proc. 15th Ann. Sympos. Comput. Geom., 1999*, ACM, pp. 1–14.

L. P. Chew (1987), Constrained Delaunay triangulations, in *Proc. 3rd Ann. Sympos. Comput. Geom., 1987*, ACM, pp. 215–222.

L. P. Chew (1989), Guaranteed-quality triangular meshes, Report TR-98-983, Comput. Sci. Dept., Cornell University, Ithaca, NY.

L. P. Chew (1993), Guaranteed-quality mesh generation for curved surfaces, in *Proc. 9th Ann. Sympos. Comput. Geom., 1993*, ACM, pp. 274–280.

L. P. Chew (1997), Guaranteed-quality Delaunay meshing in 3D, in *Proc. 13th Ann. Sympos. Comput. Geom., 1997*, ACM, pp. 391–393.

K. L. Clarkson and P. W. Shor (1989), Applications of random sampling in computational geometry, *Discrete Comput. Geom.* **4**, 387–421.

G. B. Danzig (1963), *Linear Programming and Extensions*, Princeton University Press, Princeton, NJ.

M. de Berg, M. van Kreveld, M. Overmars and O. Schwarzkopf (1997), *Computational Geometry: Algorithms and Applications*, Springer, Berlin.

B. Delaunay (1934), Sur la sphère vide, *Izv. Akad. Nauk SSSR, Otdelenie Matematicheskii i Estestvennyka Nauk* **7**, 793–800.

P. G. L. Dirichlet (1850), Über die Reduktion der positiven quadratischen Formen mit drei unbestimmten ganzen Zahlen, *J. Reine Angew. Math.* **40**, 209–227.

H. Edelsbrunner (1987), *Algorithms in Combinatorial Geometry*, Springer, Heidelberg.

H. Edelsbrunner (1990), An acyclicity theorem for cell complexes in d dimensions, *Combinatorica* **10**, 251–260.

H. Edelsbrunner (1995), Algebraic decomposition of non-convex polyhedra, in *Proc. 36th Ann. IEEE Sympos. Found. Comput. Sci. 1995*, pp. 248–257.

H. Edelsbrunner, X.-Y. Li, G. L. Miller, A. Stathopoulos, D. Talmor, S.-H. Teng, A. Üngör and N. Walkington (1999), Smoothing cleans up slivers, Manuscript.

H. Edelsbrunner and E. P. Mücke (1990), Simulation of simplicity: a technique to cope with degenerate cases in geometric algorithms, *ACM Trans. Graphics* **9**, 66–104.

H. Edelsbrunner and N. R. Shah (1996), Incremental topological flipping works for regular triangulations, *Algorithmica* **15**, 223–241.

H. Edelsbrunner, T. S. Tan and R. Waupotitsch (1992), An $O(n^2 \log n)$ time algorithm for the minmax angle triangulation, *SIAM J. Sci. Stat. Comput.* **13**, 994–1008.

I. Emiris and J. Canny (1995), A general approach to removing geometric degeneracies, *SIAM J. Comput.* **24**, 650–664.

P. Erdős (1979), Combinatorial problems in geometry and number theory, *Proc. Sympos. Pure Math.* **34**, 149–162.

S. Fortune and C. J. Van Wyk (1996), Static analysis yields efficient exact integer arithmetic for computational geometry, *ACM Trans. Graphics* **15**, 223–248.

I. M. Gelfand, M. M. Kapranov and A. V. Zelevinsky (1994), *Discriminants, Resultants and Multidimensional Determinants*, Birkhäuser, Boston.

J. E. Goodman and J. O'Rourke, eds (1997) *Handbook of Discrete and Computational Geometry*, CRC Press, Boca Raton, FL.

P. M. Gruber and C. G. Lekkerkerker (1987), *Geometry of Numbers*, 2nd edn, North-Holland, Amsterdam.

B. Grünbaum (1967), *Convex Polytopes*, Wiley, London.

B. Grünbaum and G. C. Shephard (1994), A new look at Euler's theorem for polyhedra, *Amer. Math. Monthly* **101**, 109–128.

L. J. Guibas, D. E. Knuth and M. Sharir (1992), Randomized incremental construction of Delaunay and Voronoi diagrams, *Algorithmica* **7**, 381–413.

H. Hadwiger (1957), *Vorlesungen über Inhalt, Oberfläche und Isoperimetrie*, Springer, Berlin.

E. Helly (1923), Über Mengen konvexer Körper mit gemeinschaftlichen Punkten, *Jahresber. Deutsch. Math.-Verein.* **32**, 175–176.

E. Helly (1930), Über Systeme von abgeschlossenen Mengen mit gemeinschaftlichen Punkten, *Monatsh. Math. Physik* **37**, 281–302.

B. Joe (1989), Three-dimensional triangulations from local transformations, *SIAM J. Sci. Statist. Comput.* **10**, 718–741.

B. Joe (1991), Construction of three-dimensional Delaunay triangulations from local transformations, *Comput. Aided Geom. Design* **8**, 123–142.

R. Klein (1997), *Algorithmische Geometrie*, Addison-Wesley, Bonn.

C. L. Lawson (1977), Software for C^1 surface interpolation, in *Mathematical Software III*, Academic Press, New York, pp. 161–194.

C. L. Lawson (1986), Properties of n-dimensional triangulations, *Computer Aided Geometric Design* **3**, 231–246.

D. T. Lee and A. K. Lin (1986), Generalized Delaunay triangulations for planar graphs, *Discrete Comput. Geom.* **1**, 201–217.

N. J. Lennes (1911), Theorems on the simple finite polygon and polyhedron, *Amer. J. Math.* **33**, 37–62.

N. Max, P. Hanrahan and R. Crawfis (1990), Area and volume coherence for efficient visualization of 3D scalar functions, in *San Diego Workshop on Volume Visualization*, published in *Computer Graphics* **24**, 27–33.

D. Michelucci (1995), An ε-arithmetic for removing degeneracies, in *Proc. IEEE Sympos. Comput. Arithmetic, 1995*.

G. L. Miller, D. Talmor, S.-H. Teng and N. Walkington (1995), A Delaunay based numerical method for three dimensions: generation, formulation, and partition, in *Proc. 27th Ann. ACM Sympos. Theory Comput., 1995*, pp. 683–692.

G. L. Miller, D. Talmor, S.-H. Teng, N. Walkington and H. Wang (1996), Control volume meshes using sphere packing: generation, refinement and coarsening, in *Proc. 5th Internat. Meshing Roundtable, 1996*, pp. 47–61.

K. Mulmuley (1994), *Computational Geometry: An Introduction Through Randomized Algorithms*, Prentice-Hall, Englewood Cliffs, NJ.

J. Nievergelt and F. P. Preparata (1982), Plane-sweep algorithms for intersecting geometric figures, *Comm. ACM* **25**, 739–747.

A. Okabe, B. Boots and K. Sugihara (1992), *Spatial Tessellations: Concepts and Applications of Voronoi Diagrams*, Wiley, Chichester.

J. O'Rourke (1987), *Art Gallery Theorems and Algorithms*, Oxford University Press, New York.

J. O'Rourke (1994), *Computational Geometry in C*, Cambridge University Press, Cambridge.

J. Pach and P. K. Agarwal (1995), *Combinatorial Geometry*, Wiley-Interscience, New York.

F. P. Preparata and M. I. Shamos (1985), *Computational Geometry: An Introduction*, Springer, New York.

J. Radon (1921), Mengen konvexer Körper, die einen gemeinschaftlichen Punkt enthalten, *Math. Ann.* **83**, 113–115.

J. Ruppert (1992), A new and simple algorithm for quality 2-dimensional mesh generation, Report UCB/CSD 92/694, Comput. Sci. Div., University of California, Berkeley, CA.

J. Ruppert (1995), A Delaunay refinement algorithm for quality 2-dimensional mesh generation, *J. Algorithms* **18**, 548–585.

J. Ruppert and R. Seidel (1992), On the difficulty of triangulating three-dimensional non-convex polyhedra, *Discrete Comput. Geom.* **7**, 227–254.

E. Schönhardt (1928), Über die Zerlegung von Dreieckspolyedern in Tetraeder, *Math. Ann.* **98**, 309–312.

L. L. Schumaker (1987), Triangulation methods, in *Topics in Multivariate Approximation*, (C. K. Choi, L. L. Schumaker and F. I. Utreras, eds), Academic Press, pp. 219–232.

R. Seidel (1988), Constrained Delaunay triangulations and Voronoi diagrams with obstacles, in *1978–1988 Ten Years IIG*, pp. 178–191.

R. Seidel (1993), Backwards analysis of randomized geometric algorithms, in *New Trends in Discrete and Computational Geometry*, (J. Pach, ed.), Springer, Berlin, pp. 37–67.

R. Seidel (1998), The nature and meaning of perturbations in geometric computing, *Discrete Comput. Geom.* **19**, 1–18.

M. I. Shamos (1975), Geometric complexity, in *Proc. 7th Ann. ACM Sympos. Theory Comput., 1975*, pp. 224–233.

M. I. Shamos and D. Hoey (1975), Closest-point problems, in *Proc. 16th Ann. IEEE Sympos. Found. Comput. Sci., 1975*, pp. 151–162.

M. I. Shamos and D. Hoey (1976), Geometric intersection problems, in *Proc. 17th Ann. IEEE Sympos. Found. Comput. Sci., 1976*, pp. 208–215.

J. R. Shewchuk (1998), Tetrahedral mesh generation by Delaunay refinement, in *Proc. 14th Ann. Sympos. Comput. Geom., 1998*, ACM, pp. 86–95.

R. Sibson (1978), Locally equiangular triangulations, *Comput. J.* **21**, 243–245.

D. Talmor (1997), Well-spaced points for numerical methods, Report CMU-CS-97-164, Dept. Comput. Sci., Carnegie-Mellon University, Pittsburgh, PA.

G. Voronoi (1907/08), Nouvelles applications des paramètres continus à la théorie des formes quadratiques, *J. Reine Angew. Math.* **133** (1907), 97–178, and **134** (1908), 198–287.

P. L. Williams (1992), Visibility ordering meshed polyhedra, *ACM Trans. Graphics* **11**, 103–126.

C. K. Yap (1990), Symbolic treatment of geometric degeneracies, *J. Symbolic Comput.* **10**, 349–370.

G. M. Ziegler (1995), *Lectures on Polytopes*, Springer, New York.

Acta Numerica (2000), pp. 215–365

Lie-group methods

Arieh Iserles

Department of Applied Mathematics and Theoretical Physics,
University of Cambridge, England
E-mail: `a.iserles@damtp.cam.ac.uk`

Hans Z. Munthe-Kaas

Department of Computer Science,
University of Bergen, Norway
E-mail: `hans@ii.uib.no`

Syvert P. Nørsett

Institute of Mathematics,
Norwegian University of Science and Technology,
Trondheim, Norway
E-mail: `norsett@math.ntnu.no`

Antonella Zanna

Department of Computer Science,
University of Bergen, Norway
E-mail: `anto@ii.uib.no`

Many differential equations of practical interest evolve on Lie groups or on manifolds acted upon by Lie groups. The retention of Lie-group structure under discretization is often vital in the recovery of qualitatively correct geometry and dynamics and in the minimization of numerical error. Having introduced requisite elements of differential geometry, this paper surveys the novel theory of numerical integrators that respect Lie-group structure, highlighting theory, algorithmic issues and a number of applications.

CONTENTS

1. Numerical analysts in Plato's temple

'Ageometretos medeis eisito': let nobody enter who does not understand geometry. These were the words written at the entrance to Plato's Temple of the Muses. Are numerical analysts welcome in Plato's temple?

Historically, the answer is negative. Computational mathematics is all about rendering mathematical phenomena in an algorithmic form, amenable to sufficiently precise, affordable and robust number crunching. A mathematical phenomenon can be approached in one of two ways: either by exploring its *qualitative* features (which, to a large extent, are synonymous with geometry or, at the very least, can be formulated in geometric terminology) or by approximating its *quantitative* character. Although only purists reside completely at either end of the spectrum, it is fair to point out that numerical analysis, by its very 'rules of engagement', is what 'quantitative mathematics' is all about. Ask a numerical analyst 'How good is the solution?' and the likely answer will address itself to a subtly different question: 'How small is the magnitude of the error?'

In principle, the emphasis on quantitative aspects in mathematical computing has served it well. It is hard to imagine modern technological civilization without the multitude of silent computer programs in the background, flying the aeroplanes, predicting the weather, making sense of CAT scans, controlling robots, identifying fingerprints, keeping reactions from running away, and modelling the behaviour of stock markets. This is the success story of numerical analysis, of this 'quantitative number crunching', and nothing should be allowed to obscure it. So, perhaps if we are doing so well everywhere else, we might cede Plato's temple to our 'purer' brothers and sisters: let them engage in sterile intellectual discourse while we change the world!

The main contention of this review and of the emerging discipline of *geometric integration* is that this approach, although tempting, is at best incomplete, at worst badly misguided. The history is not just a heroic tale of numerical algorithms fleshing out mathematical concepts as numbers and graphs. Progress has always occurred along parallel, intertwined tracks: *both* better theoretical understanding of qualitative attributes of a mathematical construct *and* its better computation. The aeroplane-flying, weather-predicting and CAT-scanning programs can do their job only because they deliver an answer that explains in a satisfactory manner qualitative features, as well as producing the 'right' numbers! Indeed, an artificial dichotomy of quantitative and qualitative aspects of mathematical research is in our opinion misleading and it ill serves mathematical and applied communities alike.

On the one hand, computation tells pure mathematics *what to prove*. Phenomena are often initially identified when observed under discretization and subsequently subjected to the full rigour of mathematical analysis. A familiar case in point is the discovery of solitons in the solution of the Korteweg–de Vries equation by Zabusky and Kruskal (1965), an event which launched a whole new mathematical discipline; other examples abound. Indeed, we are so used to relying on the computer as a laboratory of pure mathematics that it is difficult to imagine the heroic work of Gaston Julia (1918) on the geometry of fractals while bearing in mind that he had no access to computers, and was never able to calculate easily a sequence of rational iterations or to see a fractal on a computer monitor!

On the other hand, qualitative analysis tells computation, quite literally, *what to compute*. Every seasoned numerical analyst knows that the procedure of 'discretize everything in sight and throw it on a computer' works only with toy problems. The more we know about the qualitative behaviour of the underlying mathematical construct, the more we can identify the right computational approach, concentrate resources at the right place, focus on features that influence more the quantitative behaviour and, by the conclusion of the computation, have well-founded expectation that the graph on the computer monitor corresponds to a genuine solution of the problem in hand.

Moreover, consumers of numerical calculations are not interested just in numbers, graphs and impressive visualization. Very often it is the qualitative features, most conveniently phrased in the language of geometry, that draw genuine interest in applications: periodicity, chaoticity, conservation of energy or angular momentum, reduction to lower-dimensional manifolds, symmetry, reversibility, ...

The contention of this review is not just that the contribution of geometry to computation in the special case of time-evolving systems of differential equations is absolutely crucial, but that the terminology of differential geometry, and in particular Lie groups, creates the right backdrop to this process. The following example will help to elucidate this point, while serving as a convenient introduction to the theme of this paper.

Let us denote by S_N the set of all $N \times N$ real symmetric matrices and consider the solution of the *isospectral flow*

$$Y' = B(Y)Y - YB(Y), \quad t \geq 0, \qquad Y(0) = Y_0 \in S_N, \qquad (1.1)$$

where the (sufficiently smooth) function B maps S_N to $N \times N$ real skew-symmetric matrices. The solution itself remains in S_N for all $t \geq 0$. Such flows occur in a variety of applications. Perhaps the earliest (and the best known) is the *Toda lattice* of material points subjected to nearest-neighbour interaction. It was demonstrated by Flaschka that, in the case of an exponential interaction potential, the underlying Hamiltonian system can be rendered in the form (1.1), where Y is tridiagonal and B maps

$$\begin{bmatrix} \alpha_1 & \beta_1 & 0 & \cdots & 0 \\ \beta_1 & \alpha_2 & \beta_2 & & \vdots \\ 0 & \ddots & \ddots & \ddots & 0 \\ \vdots & \ddots & \beta_{N-2} & \alpha_{N-1} & \beta_{N-1} \\ 0 & \cdots & 0 & \beta_{N-1} & \alpha_N \end{bmatrix} \text{ to } \begin{bmatrix} 0 & \beta_1 & 0 & \cdots & 0 \\ -\beta_1 & 0 & \beta_2 & & \vdots \\ 0 & \ddots & \ddots & \ddots & 0 \\ \vdots & \ddots & -\beta_{N-2} & 0 & \beta_{N-1} \\ 0 & \cdots & 0 & -\beta_{N-1} & 0 \end{bmatrix}$$

(Toda 1981). Another important application of (1.1) is to *Lax pairs* in fluid dynamics, whence S_N needs to be replaced by a suitable function space, B is a differential operator and the outcome is a partial differential equation of a hyperbolic type (Toda 1981). Before we mention another application of isospectral flows, we need to single out their most remarkable qualitative feature which, coincidentally, explains their name: as the time evolves, the *eigenvalues* of $Y(t)$ stay put! Upon a moment's reflection, this renders such flows interesting in the context of numerical algebra. Indeed, the classical QR algorithm is intimately related to sampling the solution of (1.1) at unit intervals (Deift, Nanda and Tomei 1983). Many other iterative algorithms can be phrased in this terminology and, perhaps more importantly, many interesting algorithms rely on this construct in the first place. Pride of place belongs here to methods for the *inverse eigenvalue problem*: seeking a matrix of a given structure that possesses a specified set of eigenvalues (or singular values). Such problems are important in a wide range of applications, ranging from the theory of vibrations to control theory, tomography, system identification, geophysics, all the way to particle physics. Isospectral flows are a common denominator to perhaps the most powerful approach toward the design of practical algorithms for the inverse eigenvalue problem, which has been pioneered in the main by Chu (1998). Suppose that we are seeking a matrix in a class $\mathcal{T} \subset S_N$ with the eigenvalues $\boldsymbol{\eta} \in \mathbb{R}^N$. Often it is possible to design a matrix function B so that attractive fixed points of (1.1) lie in $\mathcal{T}$. In that case, letting $Y(0) = \mathrm{diag}\,\boldsymbol{\eta}$ results in a flow that converges to the solution of the inverse eigenvalue problem.

As an example of such a procedure we mention the inverse eigenvalue problem for *Toeplitz matrices*. Thus, $\mathcal{T}$ consists of symmetric $N \times N$ Toeplitz matrices:

$$ X \in \mathcal{T} \qquad \Leftrightarrow \qquad x_{k,l} = t_{|k-l|}, \quad k,l = 1, 2, \ldots, N, $$

where $t_0, t_1, \ldots, t_{N-1}$ are arbitrary real numbers. Such problems are important in the design of control systems but, remarkably, even the very existence of a solution has until very recently been an open problem, which has been answered by Landau (1994) in a beautiful, yet non-constructive, existence proof. Following Chu (1993) and Trench (1997), we let

$$ b_{k,l}(Y) = \begin{cases} y_{k,l-1} - y_{k+1,l}, & 1 \le k < l \le N, \\ 0, & 1 \le k = l \le N, \\ y_{k+1,l} - y_{k,l-1}, & 1 \le l < k \le N \end{cases} $$

be a *Toeplitz annihilator*. Note that B is indeed skew symmetric and that $B(Y) = \boldsymbol{O}$ for $Y \in \mathcal{T}$: thus, a solution is a fixed point.

While remarking that many important questions with regard to the convergence of the above algorithm are still wide open, we should draw the reader's attention to a crucial observation. For numerical purposes, sooner or later we must replace (1.1) by a computational time-stepping scheme. Will such a scheme respect the eigenstructure of Y? This is not simply an optional extra since the whole point of the exercise is to evaluate an answer in $\mathcal{T}$. Yet, as proved in Calvo, Iserles and Zanna (1997), the most popular numerical methods, multistep and Runge–Kutta schemes, do not respect isospectral structure and they fail to converge to the correct element of $\mathcal{T}$: the error on the eigenvalues, of the same order of magnitude as the error in the numerical trajectory itself, is unacceptable.

An alternative, proposed by Calvo et al. (1997), is to observe that all the elements of the *isospectral manifold*

$$ \mathcal{I}(\boldsymbol{\eta}) = \{ X \in S_N \ : \ \sigma(X) = \boldsymbol{\eta} \}, $$

where $\sigma(X)$ denotes the spectrum of X, can be written in the form $X = QY_0Q^{\mathrm{T}}$, where $Y_0 = \mathrm{diag}\,\boldsymbol{\eta}$ is our initial condition and $Q \in \mathrm{SO}(N)$, the set of all $N \times N$ real orthogonal matrices with unit determinant. The main idea is to seek, in place of (1.1), a differential equation that is satisfied by $Q(t)$

in the representation

$$Y(t) = Q(t)Y_0Q(t)^{\mathrm{T}}, \quad t \ge 0. \tag{1.2}$$

It is easy to ascertain that this equation has the form

$$Q' = B(QY_0Q^{\mathrm{T}})Q, \quad t \ge 0, \qquad Q(0) = I \tag{1.3}$$

and that, provided we can solve it while retaining orthogonality, we can easily recover the solution of the original isospectral flow.

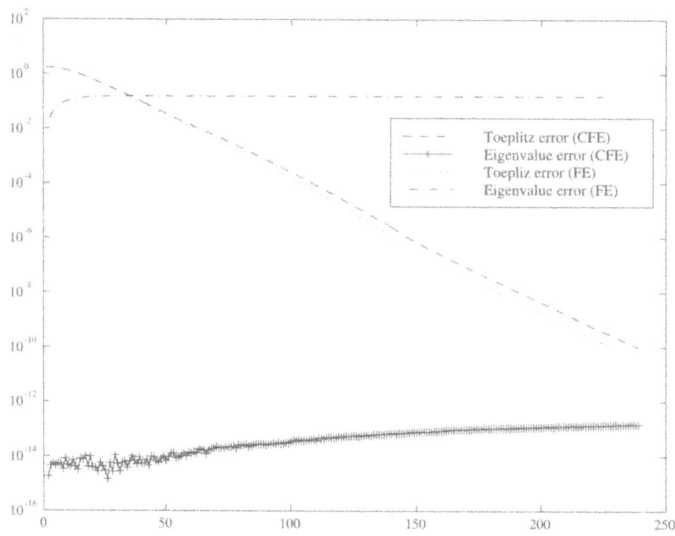

Fig. 1.1. Toeplitz error and error on the eigenvalues versus number of iterations for a 5×5 symmetric inverse eigenvalue problem when solved with the Forward Euler scheme (FE) and the Cayley-based Forward Euler (CFE). Although both methods converge asymptotically to a Toeplitz matrix, the error on the eigenvalues of CFE stays within machine accuracy while the error of FE is completely determined by the choice of integration step-size

To illustrate our point, let us consider a simple numerical experiment. We choose $\boldsymbol{\eta} = [1, 2, 3, 4, 5]^{\mathrm{T}}$, and solve the symmetric inverse eigenvalue problem (1.1), where the matrix function B is the Toeplitz annihilator introduced above. We consider first the standard Forward Euler (FE) scheme,

$$Y_{n+1} = Y_n + h[B(Y_n), Y_n], \qquad n \in \mathbb{Z}^+,$$

with initial condition $Y_0 = \operatorname{diag}(\boldsymbol{\eta})$ and step-size $h = \frac{1}{10}$. The Toeplitz error, $\|B(Y_n)\|_2$, is plotted in Figure 1.1 as a dotted line. Clearly, as n tends to infinity, the Toeplitz error becomes progressively smaller and Y_n tends to a Toeplitz matrix, as the theory predicted. Do the eigenvalues stay put?

The answer is negative, since the FE scheme is not isospectral, as proven in Calvo et al. (1997). Therefore the error on the eigenvalues after the first step is of the same order of the error of FE and is carried along the whole integration (dash-dotted line in Figure 1.1).

Consider next the iteration

$$Q_{n+1} = \left(I - \tfrac{1}{2}hB(Y_n)\right)^{-1}\left(I + \tfrac{1}{2}hB(Y_n)\right),$$
$$Y_{n+1} = Q_{n+1}Y_nQ_{n+1}^{\mathrm{T}}, \qquad n \in \mathbb{Z}^+,$$

which is equivalent to solving (1.3) with a modified version of the Forward Euler scheme based on the Cayley expansion (note that Q_{n+1} is orthogonal) in tandem with a similarity transformation for the update Y_{n+1}. This scheme, to which we will refer as CFE (Cayley-type Forward Euler), is explicit, has the same order of accuracy and requires only slightly more computations than the more classical FE. However, unlike FE, it is isospectral *by design* and preserves the eigenvalues to machine accuracy. In conclusion, FE tends to a Toeplitz matrix with the wrong eigenvalues, while CFE, a simple modification of FE that instead preserves the qualitative features of the flow, tends to a Toeplitz matrix with the right eigenvalues. The Toeplitz and the eigenvalue error of CFE are displayed in Figure 1.1 and correspond to the dashed and 'plus' lines respectively.

The set $\mathrm{SO}(N)$ in which the matrix Q of the above example evolves is an instance of a *Lie group*, a concept that will be described and debated in great detail in Section 2, while (1.2) and (1.3) are special instances of a *group action* and a *Lie-group equation*.

Let us comment briefly on the contents of this review. In Section 2 we have assembled the common mathematical denominator underlying this paper: elements of differential geometry, Lie groups and algebras, homogeneous spaces and differential equations evolving on such objects. Section 3 is devoted to Runge–Kutta–Munthe-Kaas schemes, the most natural approach to Lie-group solvers in our setting. In Section 4 we describe expansions, originally due to Magnus and to Fer, which can be converted into interesting computational tools. Section 5 is concerned with a make-or-break issue for many Lie-group methods, multivariate quadrature of multilinear forms over polytopes. We demonstrate there that some very technical tools from Lie-algebra theory can be used to a great effect in reducing the numerical cost. Lie-group methods are typically based on local imposition of a convenient coordinate system in the group. In Section 6 we debate less conventional choices of the coordinate map, which are suitable for important Lie groups and equations of practical interest. The theme of Section 7 is time-symmetric methods that, by design, exhibit many favourable features, while the concern of Section 8 is the practical approximation of a matrix exponential from a Lie algebra, so that the result lies in the right Lie group.

Stability issues are addressed in Section 9, while Section 10 reviews practical issues of implementation and error control and introduces the *DiffMan* package. A sample of the many applications of Lie-group methods is presented in Section 11. Finally, in Appendix A we list practical Lie-group methods, while Appendix B displays useful explicit formulae for integration in SO(3), perhaps the single Lie group with greatest relevance to problems in science and engineering.

The purpose of this survey is not to cover the entire corpus of Lie-group methods but to present a unified introduction to a young discipline that is likely to undergo many exciting further developments. We have omitted many interesting methods and papers, with due apologies to their authors, to keep our narrative more focused and clear. Only the future can tell which methods and techniques will survive.

It is vital throughout the paper to distinguish what exactly is the type of objects under consideration. We will often be mixing in our formulae elements of Lie groups and Lie algebras, scalars, matrices and vectors. To assist the reader, we have adhered to a consistent naming convention.

- Elements in a Lie algebra are denoted by the Roman letters $a, b, \ldots, h$ and $A, B, \ldots, H$ and by the Greek letters $\alpha, \beta, \ldots, \xi$ and Δ, Θ, Ξ.

- Elements in a Lie group are denoted by the Roman letters $p, q, \ldots, z$ and $P, Q, \ldots, Z$ and by the remaining Greek letters: $\pi, \rho, \ldots, \omega$ and $\Upsilon, \Phi, \Psi, \Omega$.

- Elements in an abstract construct (*e.g.*, an abstract Lie group) are denoted mostly with lower-case letters. However, as soon as we are concerned with specific representation, we reserve lower-case letters for scalars, upper case for matrices and lower-case, boldfaced letters for vectors.

- Special elements deserve special names. Thus, I is the identity in a Lie group, while O is the zero of a Lie algebra. A generic Lie group will be denoted by $\mathcal{G}$ and a generic Lie algebra by $\mathfrak{g}$. In general, we reserve Gothic font for Lie algebras.

- As in all naming conventions, we make obvious compromises with standard mathematical practice and common sense. Proper names remain unchanged: thus, $\sin t$ is the familiar sine function, not an element of a group, $\{B_k\}_{k \in \mathbb{Z}^+}$ are Bernoulli numbers, Φ is the set of roots of a Lie algebra, h is the step-size of a time-stepping algorithm and so on. We also employ a plethora of integration variables, summation indices, constants etc. Occasionally variables evolve in structures which are neither Lie groups nor Lie algebras, *e.g.*, the isospectral manifold $\mathcal{I}(\boldsymbol{\eta})$. In all these cases, which should be obvious from the context, we use *ad hoc* notation.

2. Theory and background

Lie groups and Lie algebras are mathematical objects which originated in the seminal work of Sophus Lie (1842–1899) on solving differential equations by quadrature, using symmetry methods. Originally these concepts were quite concrete, related to flows of differential equations on $\mathbb{R}^N$. Early in the twentieth century an abstract view of Lie-group theory emerged, commencing from the work of Elie Cartan on the classification of Lie algebras. The advantage of abstract formulation is that it simplifies mathematical analysis, and hence this presentation has become dominant throughout the mathematical literature. However, the abstract theory concentrates on understanding mathematical structures rather than exposing applications in solving differential equations. Hence it is not at all clear to most applied mathematicians that Lie groups are useful objects in applied and computational mathematics, and it might be difficult to inspire the motivation to learn an abstract theory.

We believe that the original idea of arriving at Lie algebras via continuous actions on a domain should be an excellent starting point for computationally oriented mathematicians, and in fact for many applications it is important to keep this view in focus. In this presentation we will commence from this perspective and gradually move towards somewhat more abstract formulations. Eventually, in Section 2.5 we will return to concrete matrix formulation, concentrating on the numerical solution of matrix differential equations of the form

$$Y' = A(t, Y)Y, \quad t \geq 0, \quad Y(0) = Y_0,$$

where Y and $A(t, Y)$ are $N \times N$ matrices. It turns out that all our solution techniques can be derived in this concrete matrix setting and without major modifications they can be applied to more general situations. All the algorithms of this paper will be derived within the matrix framework, and hence the theory from this point and up to Section 2.5 might be read in a relaxed manner, without the need to master all the details at first reading.

Numerical integration of ordinary differential equations (ODEs) is traditionally concerned with solving initial value problems evolving on $\mathbb{R}^N$,

$$\boldsymbol{y}' = \boldsymbol{f}(t, \boldsymbol{y}), \quad t \geq 0, \quad \boldsymbol{y}(0) = \boldsymbol{y}_0, \quad \boldsymbol{y}(t) \in \mathbb{R}^N,$$

where $\boldsymbol{f}$ is a vector field on $\mathbb{R}^+ \times \mathbb{R}^N$. Well-known numerical integrators, such as Runge–Kutta and multistep methods, advance a time-stepping procedure by adding vectors in $\mathbb{R}^N$,

$$\boldsymbol{y}_{n+1} = \boldsymbol{y}_n + h\boldsymbol{a}_n,$$

where $\boldsymbol{a}_n = \boldsymbol{a}_n(h, \boldsymbol{y}_n, \dots)$ is computed by the given numerical method and h is the time-step. One might say that *classical integrators are formulated* *using a set of 'basic motions' given by translations on $\mathbb{R}^N$ to advance the numerical solution.*

A major motivation for Lie-group methods is the possibility of replacing the domain $\mathbb{R}^N$ with more general configuration spaces and replacing translations on $\mathbb{R}^N$ by more general families of 'basic motions' on the domain. For example, if $\boldsymbol{y}(t)$ is a vector known to evolve on a sphere, one might consider rotations $\boldsymbol{y}_{n+1} = Q_n \boldsymbol{y}_n$, where Q_n is an orthogonal matrix, as basic motions. We have already encountered another example, isospectral flows (1.1). In that case $Y(t)$ evolves on the isospectral manifold $\mathcal{I}(\sigma(Y_0))$ and this is the right configuration space. Moreover, the natural 'basic motions' are (1.2) and they rest upon the fact that any two elements in $\mathcal{I}(\sigma(Y_0))$ are similar via an orthogonal matrix. A generalization of the last example no longer requires that Y_0 be symmetric and that the matrix function B be skew symmetric: any nonsingular initial value and sufficiently smooth matrix function will do. The flow remains isospectral but the elements of the configuration space are no longer orthogonally similar. The representation (1.2) is valid, however, if we allow $Q(t)$ to range across all possible nonsingular real matrices and replace Q^{T} with Q^{-1}.

An important reason why a manifold, rather than the entire $\mathbb{R}^N$, is a suitable configuration space is that it often expresses crucial geometric attributes of the underlying differential system, for instance conservation laws, symmetries or symplectic structure. As will be seen later, an added bonus of this approach is that it frequently leads to interesting numerical advantages, in particular to slower error accumulation.

In seeking abstractions and generalizations of classical numerical methods it is important to bear in mind abstractions in pure-mathematical treatment of differential equations. However, the transition from pure to computational mathematics is not straightforward. Whenever a pure mathematician says 'There exists an animal such that ... ,' an applied mathematician must add questions like 'Can we compute this animal efficiently?' and 'How can we represent it in software?'

2.1. Vector fields and flows on manifolds

A cornerstone of all abstract mathematical presentations of differential equations is the concept of *differential manifolds* as the definition of domains on which differential equations evolve. A good general introduction to manifolds and to differential geometry are the books by Abraham and Marsden (1978) and Guillemin and Pollack (1974).

Intuitively one should think of a d-dimensional manifold as being a smooth domain which in a (small) neighbourhood of any point 'looks like' $\mathbb{R}^d$, but typically looks different globally. It is known that any d-dimensional manifold can be *represented* as a d-dimensional surface embedded in $\mathbb{R}^N$ for some

$N \geq d$. This is a conceptually very useful view of manifolds. It is, however, also important to know that all geometric properties of manifolds exist independently of any particular representation.[1]

It is possible to discuss these properties in a *coordinate-free* language, which is independent of any particular representation. Although this abstract presentation is mathematically very elegant and it provides vital clues toward writing good software in an object-oriented language (Engø, Marthinsen and Munthe-Kaas 1999), it requires quite a bit of work to present and comprehend. The advantage of the abstract language is that it focuses on the essential structures. However, in most of the applications we will deal with, the manifolds exist naturally as surfaces embedded in $\mathbb{R}^N$, and furthermore it is fully possible to understand and use the numerical techniques we will discuss in this paper without knowing the abstract theory of manifolds. We have therefore decided to base our discussion on the following very concrete definition.

Definition 2.1. A d-dimensional manifold $\mathcal{M}$ is a d-dimensional smooth surface $\mathcal{M} \subset \mathbb{R}^N$ for some $N \geq d$.

It should be made crystal clear that all the numerical techniques we are about to present rely solely on those properties of $\mathcal{M}$ that exist independently of any particular embedding in $\mathbb{R}^N$. We believe that a reader with knowledge of coordinate-free presentations will have no difficulty whatsoever in translating the algorithms and results to a more general setting.

Example 2.1. It is easy to construct examples of manifolds in a number of ways. An example of an abstract definition is the specification of an *atlas* of local coordinate charts. However, given our focus on 'concrete' manifolds, we present examples already embedded in $\mathbb{R}^N$.

- Any smooth surface will do and a few familiar examples are displayed in Figure 2.1. Smoothness is important: the torus

$$\{(\cos\psi + \rho\cos\theta, \sin\psi + \rho\sin\theta, \rho\cos\theta) \; : \; 0 \leq \psi, \theta \leq 2\pi\}$$

  is a manifold for $\rho \in (0,1)$, but not when $\rho = 1$, because of a singularity at the origin.

- An important representation of a manifold is as a smooth subset of solutions of a smooth algebraic equation, $g(\boldsymbol{x}) = 0$. Thus, for example, $g(\boldsymbol{x}) = \|\boldsymbol{x}\|_2^2 - 1$ defines a unit sphere.

---

[1] Existence independently of representations should in fact be taken as the very *definition* of a geometric property.

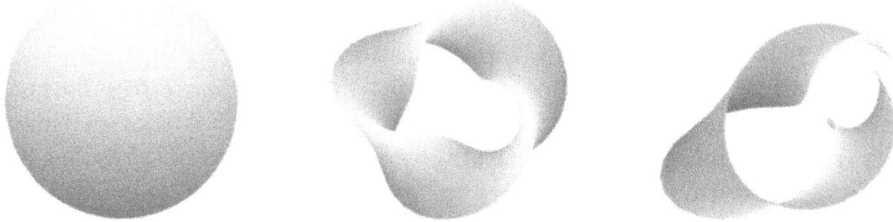

Fig. 2.1. Examples of manifolds embedded in $\mathbb{R}^3$: a sphere, a doubly twisted
Möbius-strip-like torus and a twisted ribbon

- The algebraic-equation representation is of direct relevance to geometric integration, since conservation laws and integrals of differential systems are nothing else but algebraic equations constant along the solution trajectory. Thus, for example, a Hamiltonian system with Hamiltonian energy $H(\boldsymbol{p}, \boldsymbol{q})$ evolves on the manifold $\{H(\boldsymbol{p}, \boldsymbol{q}) = H(\boldsymbol{p}(0), \boldsymbol{q}(0)) : \boldsymbol{p}, \boldsymbol{q} \in \mathbb{R}^N\}$.

- The set $\mathrm{O}(N)$ of all $N \times N$ orthogonal matrices is a manifold, since $X \in \mathrm{O}(N)$ is equivalent to $g(X) = \|X^\mathrm{T} X - \boldsymbol{I}\|_2^2 = 0$. So is $\mathrm{SL}(N)$, the set of all $N \times N$ matrices with unit determinant, since $g(X) = \det X - 1$ is a smooth function.

- A *Stiefel manifold* is the set of all real $M \times N$ matrices X such that $X^\mathrm{T} X = \boldsymbol{I}$. Typically $M > N$ – such matrices are 'long and skinny' – and $X X^\mathrm{T} \neq \boldsymbol{I}$.

- A *Grassmann manifold* is a Stiefel manifold, equivalenced by $\mathrm{O}(M)$. In other words, we identify X_1, X_2 satisfying $X_1^\mathrm{T} X_1 = X_2^\mathrm{T} X_2 = \boldsymbol{I}$ if there exists an orthogonal $M \times M$ matrix Q such that $X_1 = Q X_2$. An alternative interpretation of a Grassmann manifold is as the set of all N-dimensional subspaces of $\mathbb{R}^M$.

The single most important property of a manifold is the existence of *tangents* to the manifold in any point $p \in \mathcal{M}$. If we think of the manifold as a surface in $\mathbb{R}^N$, then a tangent at p can be defined as a vector $\boldsymbol{a}$ such that $\mathrm{dist}(p + \varepsilon \boldsymbol{a}, \mathcal{M}) = \mathcal{O}(\varepsilon^2)$. This construction of a tangent is relying on the embedding of $\mathcal{M}$ in $\mathbb{R}^N$, on the linear-space structure of $\mathbb{R}^N$ and even on the metric structure of $\mathbb{R}^N$. An alternative way to define tangents is by differentiating a curve. This approach has the advantage of making no use of the embedding of $\mathcal{M}$ in $\mathbb{R}^N$ and hence also makes sense on general manifolds.

Definition 2.2. Let $\mathcal{M}$ be a d-dimensional manifold and suppose that $\rho(t) \in \mathcal{M}$ is a smooth curve such that $\rho(0) = p$. A *tangent vector at p is* defined as

$$a = \frac{d\rho(t)}{dt}\bigg|_{t=0}.$$

The set of all tangents at p is called the *tangent space at p* and is denoted by $T\mathcal{M}|_p$. It has the structure of a d-dimensional linear space: if $a, b \in T\mathcal{M}|_p$ then $a + b \in T\mathcal{M}|_p$ and $\alpha a \in T\mathcal{M}|_p$ for any real α. The collection of all tangent spaces at all points $p \in \mathcal{M}$ is called the *tangent bundle* of $\mathcal{M}$ and is denoted by $T\mathcal{M} = \bigcup_{p \in \mathcal{M}} T\mathcal{M}|_p$.

Note that whereas it is fine to add tangents *based at the same point*, there is in general no rule for adding tangent vectors based at different points. Thus, to specify a tangent completely, we need to provide both the basepoint p and the tangent itself. Hence $T\mathcal{M}$ is a $2d$-dimensional space, with elements (p, a) consisting of every possible tangent a for any possible basepoint p.

Definition 2.3. A (tangent) *vector field on* $\mathcal{M}$ is a smooth function $F :$ $\mathcal{M} \to T\mathcal{M}$ such that $F(p) \in T\mathcal{M}|_p$ for all $p \in \mathcal{M}$. The collection of all vector fields on $\mathcal{M}$ is denoted by $\mathfrak{X}(\mathcal{M})$.

Addition and scalar multiplication of vector fields are defined pointwise in a natural way by $(F + G)(p) = F(p) + G(p)$ and $(\alpha F)(p) = \alpha(F(p))$. If $F, G \in \mathfrak{X}(\mathcal{M})$, then $F + G \in \mathfrak{X}(\mathcal{M})$ and $\alpha F \in \mathfrak{X}(\mathcal{M})$ for all real α.

Definition 2.4. Let F be a tangent vector field on $\mathcal{M}$. By a *differential equation (evolving) on* $\mathcal{M}$ we mean a differential equation of the form

$$y' = F(y), \quad t \geq 0, \quad y(0) \in \mathcal{M}, \qquad (2.1)$$

where $F \in \mathfrak{X}(\mathcal{M})$. Whenever convenient, we allow F in (2.1) to be a function of time, $F = F(t, y)$. The *flow* of F is the solution operator $\Psi_{t,F} : \mathcal{M} \to \mathcal{M}$ such that

$$y(t) = \Psi_{t,F}(y_0)$$

solves (2.1).

Note that we can find the vector field F from $\Psi_{t,F}$ by differentiation:

$$F(y) = \frac{d}{dt}\Psi_{t,F}(y)\bigg|_{t=0}.$$

F is often called the *infinitesimal generator* of the flow $\Psi_{t,F}$.

By reparametrizing time (or scaling the vector field) we can see that the flow operator satisfies the identity

$$\Psi_{\alpha,F} = \Psi_{1,\alpha F}. \qquad (2.2)$$

The task of computing the flow of a given vector field is often called the *exponentiation of the vector field*. We will occasionally employ the notation

$$\Psi_{1,F} \equiv \exp(F) \qquad \Leftrightarrow \qquad \Psi_{t,F} \equiv \exp(tF).$$

Computation of a flow is a particular example of an *exponential map*. We will return to a more general definition of exponential maps later. The notation Ψ will be used when we want to emphasize that we are discussing flows of vector fields, and exp whenever the map can be conveniently given the more general interpretation.

Recall our goal of integrating (2.1) numerically, using a chosen set of 'basic motions' to advance the numerical solution. If the analytical solution evolves on $\mathcal{M}$ it is natural to choose a set of basic motions that are everywhere tangent to $\mathcal{M}$, which will produce a numerical solution also evolving on $\mathcal{M}$. It is also useful to consider these basic motions as flows of a finite or infinite collection of vector fields, $B_1, B_2, \ldots$. If we want an efficient solver, we must be able to compute the flow, that is, exponentiate these B_is, efficiently.

From the standpoint of geometry, numerical integration of ODEs is concerned with the task of approximating the exponential of a general vector field F by exponentials originating in a family of simpler vector fields $B_1, B_2, \ldots$.

Example 2.2. Let $\mathcal{M} = \mathbb{R}^N$ and let $T_{\boldsymbol{a}}$ stand for the constant vector field $T_{\boldsymbol{a}}(\boldsymbol{y}) = \boldsymbol{a}$ for some vector $\boldsymbol{a} \in \mathbb{R}^N$. The flow of $T_{\boldsymbol{a}}$ is translation along $\boldsymbol{a}$:

$$\Psi_{t,T_{\boldsymbol{a}}}(\boldsymbol{y}_0) = \boldsymbol{y}_0 + t\boldsymbol{a}.$$

The set of all translations can obviously be used to advance the numerical solution in any desired direction on $\mathbb{R}^N$. Note that translations *commute*:

$$\Psi_{t_1,T_{\boldsymbol{a}}} \circ \Psi_{t_2,T_{\boldsymbol{b}}}(\boldsymbol{y}_0) = \boldsymbol{y}_0 + t_1\boldsymbol{a} + t_2\boldsymbol{b} = \Psi_{t_2,T_{\boldsymbol{b}}} \circ \Psi_{t_1,T_{\boldsymbol{a}}}(\boldsymbol{y}_0).$$

Generally flows do not commute. We will later see that a major difference between traditional numerical integrators and Lie-group methods is that the former are based on a set of commuting flows whereas the latter allow more general flows as basic movements to advance the solution.

The degree to which the flows of two vector fields fail to commute is measured by the *commutator* of the two vector fields. Consider two general vector fields F and G and their flows $\Psi_{s,F}$ and $\Psi_{t,G}$. To investigate commutativity, we form the following composition of the flows:

$$\Phi_{s,t} = \Psi_{s,F} \circ \Psi_{t,G} \circ \Psi_{-s,F} \circ \Psi_{-t,G} \equiv \exp(sF) \circ \exp(tG) \circ \exp(-sF) \circ \exp(-tG) \tag{2.3}$$

(see Figure 2.2). Obviously the flows commute if $\Phi_{s,t}(\boldsymbol{y}) = \boldsymbol{y}$ for all $s, t \geq 0$ and all $\boldsymbol{y} \in \mathbb{R}^N$. However, this is a nonlinear condition that in many cases may be difficult to compute or verify. One of the fundamental ideas due to Sophus Lie was the observation that such nonlinear conditions can be turned

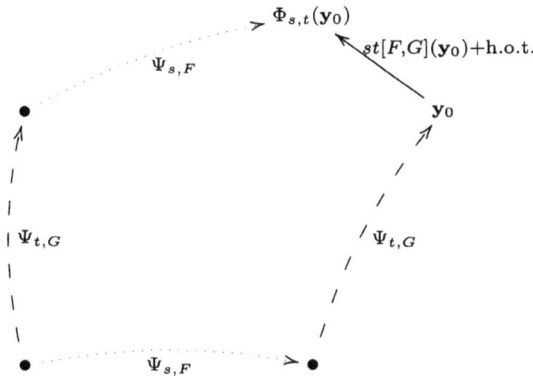

Fig. 2.2. A geometric interpretation of the commutator

into equivalent *linear infinitesimal* conditions. To accomplish this we want to linearize $\Phi_{s,t}$ for small s and t. Since $\Phi_{0,t}(\boldsymbol{y}) = \Phi_{s,0}(\boldsymbol{y}) = \boldsymbol{y}$ for all $\boldsymbol{y}$, we must have

$$\Phi_{s,t}(\boldsymbol{y}) = \boldsymbol{y} + stH(\boldsymbol{y}) + \mathcal{O}(s^2 t) + \mathcal{O}(st^2) \qquad (2.4)$$

for some vector field H. This vector field H is called the *commutator*, or the *Jacobi bracket* of F and G, and it is written as $H = [F, G]$.

Lemma 2.1. Given two vector fields F, G on $\mathbb{R}^N$, the commutator $H = [F, G]$ can be computed componentwise at a given point $\boldsymbol{y} \in \mathbb{R}^N$ as

$$H_i(\boldsymbol{y}) = \sum_{j=1}^{N} \left\{ G_j(\boldsymbol{y}) \frac{\partial F_i(\boldsymbol{y})}{\partial y_j} - F_j(\boldsymbol{y}) \frac{\partial G_i(\boldsymbol{y})}{\partial y_j} \right\}. \qquad (2.5)$$

(Note that many authors define the commutator with an opposite sign. The commutator as given here is often called *the (-)Jacobi bracket*.)

Proof. From (2.3) and (2.4) we get

$$H = \frac{\partial^2}{\partial s \partial t} \exp(sF) \circ \exp(tG) \circ \exp(-sF) \circ \exp(-tG) \Big|_{s=t=0}.$$

Since $\frac{\partial}{\partial s} \exp(sF) \circ \exp(tG) \circ \exp(-sF) \big|_{t=0} = 0$, this simplifies to

$$H = [F, G] = \frac{\partial^2}{\partial s \partial t} \exp(sF) \circ \exp(tG) \circ \exp(-sF) \Big|_{s=t=0}. \qquad (2.6)$$

Neglecting higher-order terms in s and t, Euler's integration scheme yields

$$\Psi_{s,F} \circ \Psi_{t,G} \circ \Psi_{-s,F}(\boldsymbol{y}) = \boldsymbol{y} - sF(\boldsymbol{y}) + tG(\boldsymbol{y}_1) + sF(\boldsymbol{y}_2) + \text{h.o.t.},$$

where

$$y_1 = y - sF(y) , \quad y_2 = y - sF(y) + tG(y_1).$$

Hence (2.6) implies that

$$H_i(y) = \left. \frac{\partial}{\partial s} G_i(y_1) + \frac{\partial}{\partial t} F_i(y_2) \right|_{s=t=0}$$

$$= \sum_{j=1}^{N} \left\{ -\frac{\partial G_i(y)}{\partial y_j} F_j(y) + \frac{\partial F_i(y)}{\partial y_j} G_j(y) \right\}.$$

□

We will now review the most salient properties of the commutator.

Lemma 2.2. If $F, G \in \mathfrak{X}(\mathcal{M})$ then $H = [F, G] \in \mathfrak{X}(\mathcal{M})$.

Proof. The function

$$\rho(t) = \exp(\sqrt{t}F) \circ \exp(\sqrt{t}G) \circ \exp(-\sqrt{t}F) \circ \exp(-\sqrt{t}G)(y_0)$$

is a curve that evolves on $\mathcal{M}$. The tangent defined by this curve is $[F, G](y_0)$.

□

Dividing the polygon of Figure 2.2 into infinitesimally small rectangles, we can verify that

Lemma 2.3. Two flows $\Psi_{s,F}$ and $\Psi_{t,G}$ commute if and only if $[F, G] = 0$.

From (2.5) one may prove the following important features of the commutator which should be familiar in the special case (which we will encounter again soon) of a commutator of two matrices.

Lemma 2.4. The commutator of vector fields satisfies the identities

$$
\begin{array}{llr}
[F, G] = -[G, F] & \text{(skew symmetry)}, & (2.7) \\
[\alpha F, G] = \alpha[F, G] \text{ for } \alpha \in \mathbb{R}, & & (2.8) \\
[F + G, H] = [F, H] + [G, H] & \text{(bilinearity)}, & (2.9) \\
0 = [F, [G, H]] + [G, [H, F]] & \text{(Jacobi's identity)}. & (2.10) \\
\quad + [H, [F, G]] & &
\end{array}
$$

Example 2.3. Let L_A denote the linear vector field on $\mathbb{R}^N$, given by some matrix A, that is $L_A(y) = Ay$. The solution of the linear equation $y' = Ay$ is given as

$$y(t) = \sum_{j=0}^{\infty} \frac{(tA)^j}{j!} y_0 = \text{expm}(tA)y_0,$$

where expm denotes the classical matrix exponential. Hence

$$\Psi_{t,L_A}(\boldsymbol{y}_0) \equiv \exp(tL_A)(\boldsymbol{y}_0) = \operatorname{expm}(tA)\boldsymbol{y}_0.$$

(This motivates the name 'exponentiation' for computing the flow.) Now let us compute the commutator of two linear vector fields from (2.5),

$$
\begin{aligned}
[L_A, L_B]_i(\boldsymbol{y}) &= \sum_{j,k,l} \left(B_{j,k} y_k \frac{\partial A_{i,l} y_l}{\partial y_j} - A_{j,k} y_k \frac{\partial B_{i,l} y_l}{\partial y_j} \right) \\
&= \sum_{j,k} (A_{i,j} B_{j,k} - B_{i,j} A_{j,k}) \, y_k.
\end{aligned}
$$

Thus

$$[L_A, L_B] = L_C \quad \text{where} \quad C = AB - BA, \tag{2.11}$$

the familiar definition of a commutator from linear algebra. Note that linear vector fields constitute a complete family of vector fields, closed under commutators and linear combinations, $L_A + L_B = L_{A+B}$ and $\alpha L_A = L_{\alpha A}$.

In applications of Lie-group integrators to partial differential equations (PDEs) it is often useful to consider a more general version of (2.5). Let y be a point in a (finite- or infinite-dimensional) linear space $\mathcal{M}$. By a vector field F on $\mathcal{M}$ we mean some operator (linear or nonlinear) such that the (ordinary or partial) differential equation

$$\frac{\partial y}{\partial t} = F(y)$$

is well defined. An infinite-dimensional example is a parabolic PDE, where y belongs to some function space on a domain and F is a spatial differentiation operator, for instance $F(y) = \nabla^2 y$. If F and G are two vector fields on $\mathcal{M}$ then

$$[F, G](y) = \frac{\partial}{\partial s}[F(y + sG(y)) - G(y + sF(y))]\Big|_{s=0}. \tag{2.12}$$

The proof is very similar to that of Lemma 2.1 and it is straightforward to verify that in the finite-dimensional case (2.12) reduces to (2.5). Note that if F and G are linear operators then (2.12) immediately yields $[F, G](y) = F(G(y)) - G(F(y))$, as we saw in Example 2.3.

Example 2.4. Let $\boldsymbol{y} \in \mathbb{R}^N$ and consider the set of all *affine linear vector fields*

$$F_{(A,\boldsymbol{a})}(\boldsymbol{y}) = A\boldsymbol{y} + \boldsymbol{a},$$

where A is an $N \times N$ matrix and $\boldsymbol{a} \in \mathbb{R}^N$. Let us compute the commutator of two such vector fields $F_{(A,a)}$ and $F_{(B,b)}$. By inserting these vector fields in (2.12) we obtain

$$\left[F_{(A,a)}, F_{(B,b)}\right](y)$$

$$= \frac{\partial}{\partial s}\{A(y + s(By + b)) + a - B(y + s(Ay + a)) - b\}\Big|_{s=0}$$

$$= ABy + Ab - BAy - Ba.$$

Thus

$$\left[F_{(A,a)}, F_{(B,b)}\right] = F_{(C,c)}, \quad \text{where} \quad (C, c) = (AB - BA, Ab - Ba). \quad (2.13)$$

We note that the set of all affine linear vector fields is yet another example of a collection of vector fields closed under linear combination and commutation.

2.2. Lie algebras, Lie groups and Lie-group actions

A problem, fundamental to the numerical analysis of differential equations on manifolds, is *to determine the set of all possible flows that can be obtained by composing a given set of basic flows.* If we restrict the discussion to 'sufficiently small t', important information is provided by the so-called *BCH formula.*

Theorem 2.5. (Baker–Campbell–Hausdorff) For sufficiently small $t \geq 0$ we have

$$\exp(tF) \circ \exp(tG) = \exp(tH),$$

where $H = \mathrm{bch}(F, G)$ can be constructed from iterated commutators of F and G. The first few terms are

$$H = F + G + \tfrac{1}{2}t[F, G] + \tfrac{1}{12}t^2\left([F, [F, G]] + [G, [G, F]]\right) + \mathcal{O}(t^3).$$

Higher-order terms can be obtained by recursion (Varadarajan 1984).

Definition 2.5. A *Lie algebra of vector fields* is a collection of vector fields which is closed under linear combination and commutation. In other words, letting $\mathfrak{g}$ denote the Lie algebra,

$$B \in \mathfrak{g} \quad \Rightarrow \quad \alpha B \in \mathfrak{g} \text{ for all } \alpha \in \mathbb{R},$$

$$B_1, B_2 \in \mathfrak{g} \quad \Rightarrow \quad B_1 + B_2, [B_1, B_2] \in \mathfrak{g}.$$

Given a collection of vector fields $B = \{B_1, B_2, \dots\}$, the smallest Lie algebra of vector fields containing B is called *the Lie algebra generated by B.*

We arrive at the following conclusion. Let $\mathfrak{g}$ be the Lie algebra generated by the set $B = \{B_1, B_2, \dots\}$ of vector fields. For small t, the combination of flows of vector fields in $\boldsymbol{B}$ yields the flow of a vector field in $\mathfrak{g}$. Furthermore, the flow of any vector field in $\mathfrak{g}$ can, provided $t \geq 0$ is small enough, be approximated arbitrarily well by composing flows of vector fields in $\boldsymbol{B}$. Thus *the Lie algebra contains (for small $t \geq 0$) all the information about composition of flows.*

Until now we have discussed vector fields and flows *on a manifold.* Yet, whether we wish to engage in mathematical analysis or produce software for solving ODEs, it is natural to ask, 'Are there important properties of vector fields and flows that can be specified (and possibly programmed) independently of which manifold they are acting upon?' This question will lead us to the abstract definition of Lie algebras and Lie groups. The 'glue' that connects an abstract Lie algebra to concrete vector fields on a manifold is called a *Lie-algebra homomorphism*, and abstract Lie groups are connected to flows on a manifold via a *Lie-group action*. To elucidate this state of affairs, let us commence with an important example.

Example 2.5. In Example 2.3 we have shown that there exists a natural correspondence between $N \times N$ matrices and linear vector fields defined on $\mathbb{R}^N$. We can illustrate this as

$$
\begin{aligned}
A &\mapsto L_A, \\
\alpha A &\mapsto \alpha L_A, \\
A + B &\mapsto L_A + L_B, \\
AB - BA &\mapsto [L_A, L_B].
\end{aligned}
$$

A linear subspace of matrices closed under *matrix commutation*, $[A, B] \equiv AB - BA$, is called a *matrix Lie algebra*. The arrow is an example of a Lie-algebra homomorphism, a linear map between two Lie algebras which preserves commutators.

Even the computation of flows of linear vector fields and compositions of such flows can be transformed into linear algebra operations. To achieve this, we must specify how a given $N \times N$ matrix P corresponds to a motion on our domain $\mathbb{R}^N$. The simplest possible choice is motions by matrix–vector products. Define thus the map

$$
\Lambda(P, \boldsymbol{y}) = P \boldsymbol{y}.
$$

Identifying a matrix P with the motion $\Lambda(P, \cdot)$ leads to the correspondence

$$
\begin{aligned}
P &\mapsto \Lambda(P, \cdot), \\
\operatorname{expm}(sA) &\mapsto \Lambda(\operatorname{expm}(sA), \cdot) = \Psi_{s,L_A}, \\
\operatorname{expm}(sA) \operatorname{expm}(tB) &\mapsto \Lambda(\operatorname{expm}(sA) \operatorname{expm}(tB), \cdot) = \Psi_{s,L_A} \circ \Psi_{t,L_B}.
\end{aligned}
$$

Note that the latter of these identifications relies on the associative property of the map Λ, that is,

$$\Lambda(P, \Lambda(R, \boldsymbol{y})) = \Lambda(PR, \boldsymbol{y}).$$

Motivated by this example, we now proceed to precise mathematical definition of the underlying concepts in a more abstract setting.

Definition 2.6. A *Lie algebra* is a linear space V equipped with a *Lie bracket*, a bilinear, skew-symmetric mapping

$$[\,\cdot\,,\,\cdot\,] : V \times V \to V$$

that obeys identities (2.7)–(2.11) from Lemma 2.4.

Definition 2.7. A *Lie-algebra homomorphism* is a linear map between two Lie algebras, $\varphi : \mathfrak{g} \to \mathfrak{h}$, satisfying the identity

$$\varphi([v, w]_{\mathfrak{g}}) = [\varphi(v), \varphi(w)]_{\mathfrak{h}}, \qquad v, w \in \mathfrak{g}.$$

An invertible homomorphism is called an *isomorphism*.

Definition 2.8. A *Lie group* is a differential manifold $\mathcal{G}$ equipped with a product $\cdot : \mathcal{G} \times \mathcal{G} \to \mathcal{G}$ satisfying

$p \cdot (q \cdot r) = (p \cdot q) \cdot r \ \ \forall \, p, q, r \in \mathcal{G}$	(associativity),
$\exists \boldsymbol{I} \in \mathcal{G}$ such that $\boldsymbol{I} \cdot p = p \cdot \boldsymbol{I} = p \ \ \forall \, p \in \mathcal{G}$	(identity element),
$\forall p \in \mathcal{G} \ \exists \, p^{-1} \in \mathcal{G}$ such that $p^{-1} \cdot p = \boldsymbol{I}$	(inverse),
the maps $(p, r) \mapsto p \cdot r$ and $p \mapsto p^{-1}$ are smooth functions	(smoothness).

Definition 2.9. An *action* of a Lie group $\mathcal{G}$ on a manifold $\mathcal{M}$ is a smooth map $\Lambda : \mathcal{G} \times \mathcal{M} \to \mathcal{M}$ satisfying

$$\begin{aligned}
\Lambda(\boldsymbol{I}, y) &= y \quad \forall y \in \mathcal{M}, \\
\Lambda(p, \Lambda(r, y)) &= \Lambda(pr, y) \quad \forall p, r \in \mathcal{G}, \ y \in \mathcal{M}.
\end{aligned} \tag{2.14}$$

If this relation does hold only in a local sense, for all elements p and r *sufficiently close to the identity* $\boldsymbol{I} \in \mathcal{G}$, we say that Λ is *local action*.

2.3. From finite to infinitesimal and back

Given a set of flows on a domain we can find their vector fields by differentiation. From the discussion above we know that if the flows are closed under composition then the vector fields are closed under Jacobi brackets and linear combinations, and hence form a Lie algebra. On the other hand, provided we know the vector fields, we may recover the corresponding flows by integrating differential equations. A similar correspondence between the finite and the infinitesimal is fundamental in abstract Lie theory. We will review here a number of basic results, interpreting them in the terminology of group actions on a manifold. Let $\mathcal{G}$ be a Lie group, acting on a manifold $\mathcal{M}$ through $\Lambda : \mathcal{G} \times \mathcal{M} \to \mathcal{M}$ and let $\rho(t) \in \mathcal{G}$ be a curve such that $\rho(0) = \boldsymbol{I}$, the identity of $\mathcal{G}$. This curve produces a flow $\Lambda(\rho(t), \cdot)$ on $\mathcal{M}$ and by differentiation we find a vector field

$$F(y) = \left. \frac{\mathrm{d}}{\mathrm{d}t} \Lambda(\rho(t), y) \right|_{t=0} .$$

The collection of all such vector fields forms a Lie algebra. Note that in order to produce F it is only necessary to know the tangent to $\rho(t)$ at $t = 0$. Thus the set of all tangents at identity can be endowed with a structure of a Lie algebra.

Definition 2.10. The Lie algebra $\mathfrak{g}$ of a Lie group $\mathcal{G}$ is defined as the linear space of all tangents to $\mathcal{G}$ at the identity $\boldsymbol{I}$. The Lie bracket in $\mathfrak{g}$ is defined as

$$[a, b] = \left. \frac{\partial^2}{\partial s \partial t} \rho(s) \sigma(t) \rho(-s) \right|_{s=t=0} \tag{2.15}$$

where $\rho(s)$ and $\sigma(t)$ are two smooth curves on $\mathcal{G}$ such that $\rho(0) = \sigma(0) = \boldsymbol{I}$, $\rho'(0) = a$ and $\sigma'(0) = b$.

Note that the bracket defined in (2.15) is essentially the same as (2.6). From this it is straightforward to verify that the correspondence between elements in $\mathfrak{g}$ and vector fields on $\mathcal{M}$ is an algebra homomorphism.

Lemma 2.6. Let $\lambda_* : \mathfrak{g} \to \mathfrak{X}(\mathcal{M})$ be defined as

$$\lambda_*(a)(y) = \left. \frac{\mathrm{d}}{\mathrm{d}s} \Lambda(\rho(s), y) \right|_{s=0}, \tag{2.16}$$

where $\rho(s)$ is a curve in $\mathcal{G}$ such that $\rho(0) = \boldsymbol{I}$ and $\rho'(0) = a$. Then λ_* is a linear map between Lie algebras such that

$$[a, b]_{\mathfrak{g}} = [\lambda_*(a), \lambda_*(b)]_{\mathfrak{X}(\mathcal{M})}.$$

Thus we can go from the finite to the infinitesimal (from groups and group actions to algebras and algebra homomorphisms) by differentiation. To do the opposite and move from the infinitesimal to the finite, we must somehow compute flows of vector fields. A discussion of this process leads to the general definition of the exponential mapping.

Suppose $\mathcal{G}$ is a Lie group with Lie algebra $\mathfrak{g}$ and let $\Lambda : \mathcal{G} \times \mathcal{M} \to \mathcal{M}$ be a group action. A given fixed element $a \in \mathfrak{g}$ corresponds to a vector field $\lambda_*(a) \in \mathfrak{X}(\mathcal{M})$. We want to compute the flow of this field. Let us first assume for simplicity that $\mathcal{G}$ is a matrix group.

Lemma 2.7. For a fixed $A \in \mathfrak{g}$ the flow of $\lambda_*(A)$, that is, the solution of

$$y'(t) = \lambda_*(A)(y(t)) \quad \text{for} \quad y(0) = y_0 \in \mathcal{M},$$

can be expressed in the form

$$y(t) = \Lambda(S(t), y_0)$$

where the curve $S(t) \in \mathcal{G}$ satisfies the matrix differential equation

$$S'(t) = AS, \quad t \geq 0, \qquad S(0) = I,$$

which has the explicit solution

$$S(t) = \text{expm}(tA), \qquad t \geq 0.$$

Proof. We assume that $y(t) = \Lambda(S(t), y_0)$. Differentiation results in $y'(t) = \partial_1 \Lambda(S'(t), y_0)$, where ∂_1 is the derivative with respect to the first argument. On the other hand, to compute $\lambda_*(A)$ we pick a curve $R(s) \in \mathcal{G}$ such that $R'(0) = A$ and $R(0) = I$. From (2.14) we get

$$y'(t) = \lambda_*(A)(y(t)) = \left. \frac{\partial}{\partial s} \Lambda(R(s), \Lambda(S(t), y_0)) \right|_{s=0}$$

$$= \left. \frac{\partial}{\partial s} \Lambda(R(s)S(t), y_0) \right|_{s=0} = \partial_1 \Lambda(R'(0)S(t), y_0) = \partial_1 \Lambda(AS(t), y_0).$$

Thus

$$S'(t) = AS(t).$$

Obviously $S(0) = I$. The explicit solution is easily verified. $\square$

Lemma 2.7 holds unaltered for a general group $\mathcal{G}$ if we define the product of an element of an algebra $a \in \mathfrak{g}$ with an element of a group $\sigma \in \mathcal{G}$ as

$$a\sigma \equiv \left. \frac{\mathrm{d}}{\mathrm{d}s} \rho(s)\sigma \right|_{s=0}, \tag{2.17}$$

where $\rho(s) \in \mathcal{G}$ is a smooth curve such that $\rho'(0) = a$ and $\rho(0) = I$. We also define the exponential mapping so that the flow of $\lambda_*(a)$ is of the form $\Lambda(\exp(ta), \cdot)$.

Definition 2.11. Let $\mathcal{G}$ be a Lie group and $\mathfrak{g}$ its Lie algebra. The exponential mapping $\exp : \mathfrak{g} \to \mathcal{G}$ is defined as $\exp(a) = \sigma(1)$ where $\sigma(t) \in \mathcal{G}$ satisfies the differential equation

$$\sigma'(t) = a\sigma(t), \qquad \sigma(0) = I.$$

These definitions lead to the following general form of Lemma 2.7.

Theorem 2.8. Let $\Lambda : \mathcal{G} \times \mathcal{M} \to \mathcal{M}$ be a group action and $\lambda_* : \mathfrak{g} \to \mathfrak{X}(\mathcal{M})$ the corresponding Lie-algebra homomorphism (2.16). For any $a \in \mathfrak{g}$ the flow of the vector field $F = \lambda_*(a)$, that is, the solution of the equation

$$y'(t) = F(y(t)) = \lambda_*(a)(y(t)), \quad t \geq 0, \qquad y(0) = y_0 \in \mathcal{M},$$

is given as

$$y(t) = \Lambda(\exp(ta), y_0).$$

Let us at this point make a small detour to introduce the *adjoint representation* which is fundamental in many contexts. By splitting (2.15) into two smaller steps, we obtain the following.

Definition 2.12. Let $p \in \mathcal{G}$ and let $\sigma(t)$ be a smooth curve on $\mathcal{G}$ such that $\sigma(0) = \boldsymbol{I}$ and $\sigma'(0) = b \in \mathfrak{g}$. The *adjoint representation* is defined as

$$\mathrm{Ad}_p(b) = \left. \frac{\mathrm{d}}{\mathrm{d}t} p\sigma(t)p^{-1} \right|_{t=0}. \tag{2.18}$$

The derivative of Ad with respect to the first argument is denoted ad. Let $\rho(s)$ be a smooth curve on $\mathcal{G}$ such that $\rho(0) = \boldsymbol{I}$ and $\rho'(0) = a$. Definition 2.10 now yields

$$\mathrm{ad}_a(b) \equiv \left. \frac{\mathrm{d}}{\mathrm{d}s} \mathrm{Ad}_{\rho(s)}(b) \right|_{s=0} = [a, b]. \tag{2.19}$$

The following formulae show that Ad is both a linear group action (of $\mathcal{G}$ on $\mathfrak{g}$) and also that for a fixed argument p it is a Lie-algebra isomorphism of $\mathfrak{g}$ onto itself:

$$\mathrm{Ad}_p(a) \in \mathfrak{g}, \qquad \text{for all } p \in \mathcal{G}, \, a \in \mathfrak{g}, \tag{2.20}$$
$$\mathrm{Ad}_p \circ \mathrm{Ad}_q = \mathrm{Ad}_{pq}, \tag{2.21}$$
$$\mathrm{Ad}_p(a + b) = \mathrm{Ad}_p(a) + \mathrm{Ad}_p(b), \tag{2.22}$$
$$\mathrm{Ad}_p([a, b]) = [\mathrm{Ad}_p(a), \mathrm{Ad}_p(b)]. \tag{2.23}$$

Note that, according to (2.22), both Ad_p and ad_a are linear in their second argument, hence they may be regarded as matrices acting on the linear space $\mathfrak{g}$. This gives meaning to the following important formula relating Ad, ad and the exponential mapping:

$$\mathrm{Ad}_{\exp(a)} = \mathrm{expm}(\mathrm{ad}_a). \tag{2.24}$$

2.4. Differential equations on manifolds

We wish to return to general differential equations on manifolds, as given in (2.1). In order to construct and implement numerical solvers for this equation, we require a concrete way of representing the vector field $F(y)$.

Herewith we describe a very general approach presented in Munthe-Kaas and Zanna (1997).

Assumption 2.1. Given a differential equation $y'(t) = F(t, y)$ on a manifold $\mathcal{M}$, we assume the existence of a Lie algebra $\mathfrak{g}$, a Lie-algebra homomorphism $\lambda_* : \mathfrak{g} \to \mathfrak{X}(\mathcal{M})$ and a function $a : t \times \mathcal{M} \to \mathfrak{g}$ such that the equation can be written in the form

$$y'(t) = \lambda_*(a(t, y))(y). \tag{2.25}$$

If λ_* is known from the context and no confusion is likely, we will usually write equation (2.25) in the shorthand form

$$y'(t) = a(t, y)y.$$

In many important examples the function a depends only on t and not on y. These equations, $y'(t) = a(t)y$, are called equations of Lie type, or *linear-type* Lie-group equations. Some of the algorithms to be presented later are aimed at the general equation (2.25), while others are aimed at exploiting the special structure of linear equations.

Given an equation to be solved, an important challenge is to find a 'good' homomorphism λ_*. It is not difficult to see that *any* differential equation can be written in the form (2.25). We might for example let $\mathfrak{g} = \mathfrak{X}(\mathcal{M})$ and choose λ_* as the identity map, which would trivially render any equation in this form. However, in order to construct practical solution algorithms we need to make some additional assumptions about $\mathfrak{g}$. We will usually assume that either *all the elements* of $\mathfrak{g}$, or at least *a particular basis* of $\mathfrak{g}$ can be exponentiated efficiently. To achieve this, one might embed $\mathcal{M}$ in a linear space $\mathbb{R}^N$, and let $\mathfrak{g}$ be the set of all translations on $\mathbb{R}^N$, since translations are trivial to exponentiate. This choice will, however, fail to capture much of the structure of the equations to be solved. In fact we will see that for this choice most of our numerical solution techniques will reduce to classical Runge–Kutta methods. The task of finding a 'good action' is in many respects similar to the task of finding a good preconditioner in the theory of iterative methods for solving linear algebraic equations $A\boldsymbol{x} = \boldsymbol{b}$. In both cases we want to find some approximation to our original equation which is both simple to solve and which captures some important structural feature of the equation. The two extreme choices, on the one hand $\mathfrak{g} = \mathfrak{X}(\mathcal{M})$ and on the other hand $\mathfrak{g}$ as a set of all translations on $\mathbb{R}^N$, are similar to preconditioning a linear system with the matrix A itself or on the other hand choosing the identity matrix as preconditioner.

Let us now examine briefly a number of examples of equations presented in the form of Assumption 2.1. A useful approach to finding a good action is the following.

(1) Given a differential equation written in a familiar form, look at the terms and see if it is possible to find related equations that are simpler to integrate.

(2) Check that the family of simpler equations forms a Lie algebra. Find a suitable representation $\mathfrak{g}$ for the algebra and the corresponding homomorphism λ_*.

(3) Check that the original equation can be written in the form (2.25). This is not possible only when there exist some points on $\mathcal{M}$ where the vector fields in the Lie algebra do not generate the direction of the original equation. In this case one must search for a larger Lie algebra.

Example 2.6. (Orthogonal matrix flows) Matrix differential equations of the form

$$Y' = A(t, Y)Y, \quad t \geq 0, \qquad Y(0) = Y_0 \in \mathrm{O}(N), \qquad (2.26)$$

where $A : \mathbb{R}^+ \times \mathrm{O}(N) \to \mathfrak{so}(N)$, are called *orthogonal flows* – we have already encountered such a flow in (1.3). (See Section 2.5 for unfamiliar notation.) It is well known that the exact solution $Y(t)$ is an orthogonal matrix for all $t \geq 0$ (Dieci, Russell and van Vleck 1994). A simpler family of equations is given by

$$Y' = CY, \quad t \geq 0, \qquad Y(0) = Y_0 \in \mathrm{O}(N), \qquad (2.27)$$

where C is any constant matrix in $\mathfrak{so}(N)$. The solution of (2.27) is given as

$$Y(t) = \mathrm{expm}(tC)Y_0, \quad t \geq 0.$$

From (2.11) we find that if $F(Y) = CY$ and $G(Y) = DY$ then $[F, G](Y) = (CD - DC)Y$. Thus, the family of simple vector fields is a Lie algebra isomorphic to $\mathfrak{so}(N)$. We have $\lambda_*(C)(Y) = CY$, and hence (2.26) is in the form (2.25) if $a(t, Y) = A(t, Y)$. Note also that the flow of (2.27) is orthogonal, hence any numerical method based on the composition of such flows will yield a solution that retains orthogonality.

Example 2.7. (Isospectral matrix flows) We have already encountered such flows in Section 1, in (1.1). With slightly greater generality, we write them in the form

$$Y' = B(t, Y)Y - YB(t, Y), \quad t \geq 0, \qquad Y(0) = Y_0 \in \mathrm{S}_N, \qquad (2.28)$$

where $B : \mathbb{R}^+ \times \mathrm{S}_N \to \mathfrak{so}(N)$. The analytical solution $Y(t)$ is a family of matrices with eigenvalues invariant under the flow. A simpler family of equations is given by

$$Y'(t) = CY - YC, \quad t \geq 0, \qquad Y(0) = Y_0 \in \mathrm{S}_N, \qquad (2.29)$$

where again $C \in \mathfrak{so}(N)$ is constant. The solution of this equation is given explicitly in the form

$$Y(t) = \text{expm}(tC) Y_0 \, \text{expm}(-tC), \quad t \geq 0. \tag{2.30}$$

We may now proceed, exactly like in the previous example, to find the brackets of such vector fields. An alternative route, already anticipated in (1.2), is to note that the basic flows in (2.30) are given by the orthogonal matrix group $\text{O}(N)$ acting on $Y \in S_N$. We have the action $\Lambda : \text{O}(N) \times S_N \to S_N$ given by

$$\Lambda(Q, X) = QXQ^T \quad \text{for any} \quad Q \in \text{O}(N).$$

By differentiation, as in Lemma 2.6, we obtain $\lambda_*(C)(Y) = CY - YC$. Hence (2.28) is in the form (2.25) if $a(t, Y) = B(t, Y)$. The basic flow (2.30) is isospectral since it is a similarity transformation, hence any numerical method based on this flow is also automatically isospectral.

Note that, essentially, these two examples are identical, except for the action involved. Hence in an algorithm we might re-use the implementation of the Lie algebra and the Lie group, while simply changing the action: there is no need to develop a separate computational approach to orthogonal and isospectral flows! This illustrates the importance of working with abstractly defined groups and algebras rather than tying these concepts to flows and vector fields on particular manifolds.

Example 2.8. (ODEs on $\mathbb{R}^N$)   Consider an ordinary differential equation in the familiar form required by all classical numerical integrators,

$$\boldsymbol{y}'(t) = \boldsymbol{g}(t, \boldsymbol{y}), \quad t \geq 0, \qquad \boldsymbol{y}(0) = \boldsymbol{y}_0 \in \mathbb{R}^N, \tag{2.31}$$

where $\boldsymbol{g} : \mathbb{R} \times \mathbb{R}^N \to \mathbb{R}^N$. A simple family of basic flows is given by translations, for instance

$$\boldsymbol{z}' = \boldsymbol{a},$$

where $\boldsymbol{a} \in \mathbb{R}^N$ is constant. The algebra of these flows can be identified with $\mathbb{R}^N$ and, since translations commute, we obtain the trivial bracket $[\boldsymbol{a}, \boldsymbol{b}] = 0$ for all $\boldsymbol{a}, \boldsymbol{b} \in \mathbb{R}^N$. The algebra homomorphism is given here simply as the identity map; hence the equation is in the form (2.25) if $a(t, \boldsymbol{y}) = \boldsymbol{g}(t, \boldsymbol{y})$. This choice yields nothing new compared to classical integration schemes.

A more interesting choice is choosing linear vector fields as basic flows,

$$\boldsymbol{z}' = A\boldsymbol{z}.$$

We already know from Example 2.3 that the Lie algebra of these is $\mathfrak{gl}(N)$, the set of all $N \times N$ matrices with the usual matrix bracket. However, these flows cannot produce everywhere all possible tangent directions. Indeed, at $\boldsymbol{y} = \boldsymbol{0}$ these flows cannot produce *any* nonzero tangent. Therefore, using this action, it is generally not possible to write (2.31) in the form (2.25).

We might finally consider a combination of translations and linear maps, *i.e.*, the set of all *affine* linear maps given by equations of the form

$$\boldsymbol{z}' = A\boldsymbol{z} + \boldsymbol{b} \qquad \text{for} \qquad A \in \mathfrak{gl}(N), \quad \boldsymbol{b} \in \mathbb{R}^N.$$

The Lie algebra can be identified with all pairs $(A, \boldsymbol{b})$ where A is a matrix and $\boldsymbol{b}$ a vector and we have already seen in Example 2.4 that the bracket is given as $[(A, \boldsymbol{b}), (C, \boldsymbol{d})] = (AC - CA, A\boldsymbol{d} - C\boldsymbol{b})$. The algebra homomorphism is

$$\lambda_*(A, \boldsymbol{b})(\boldsymbol{y}) = A\boldsymbol{y} + \boldsymbol{b}.$$

In this situation there are many possible choices of a function $a(t, \boldsymbol{y})$ such that (2.31) acquires the form (2.25). No matter what we pick as the matrix part, it is always possible to adjust the vector so that

$$\lambda_*(a(t, \boldsymbol{y}))(\boldsymbol{y}) = \boldsymbol{g}(t, \boldsymbol{y}).$$

A natural choice is *local linearization*, namely, letting the matrix part be the Jacobian of $\boldsymbol{g}$ at $\boldsymbol{y}$,

$$J\boldsymbol{g}(t, \boldsymbol{y})_{i,j} = \frac{\partial \boldsymbol{g}_i(t, \boldsymbol{y})}{\partial y_j}.$$

The resulting $a : \mathbb{R} \times \mathcal{M} \to \mathfrak{g}$ is

$$a(t, \boldsymbol{y}) = (J\boldsymbol{g}(t, \boldsymbol{y}), \boldsymbol{g}(t, \boldsymbol{y}) - J\boldsymbol{g}(t, \boldsymbol{y})\boldsymbol{y}).$$

Possible advantages of using affine motions to advance the solution of stiff ODEs are a subject of ongoing research.

Example 2.9. (ODEs on a sphere) Many mechanical problems involve rotations in a 3-space. In Appendix B we list useful formulae for fast computations in SO(3). As a simple example we will consider a motion on the surface of a sphere,

$$\boldsymbol{y}'(t) = \boldsymbol{a}(\boldsymbol{y}(t)) \times \boldsymbol{y}(t), \tag{2.32}$$

where $\boldsymbol{a}, \boldsymbol{y} \in \mathbb{R}^3$, and $\times$ is the standard vector product in $\mathbb{R}^3$. If $\|\boldsymbol{y}(0)\|_2 = 1$ then $\boldsymbol{y}(t)$ evolves on the unit sphere $\mathbb{S}^2 \subset \mathbb{R}^3$. A simpler system that also evolves on $\mathbb{S}^2$ is given by

$$\boldsymbol{y}'(t) = \boldsymbol{c} \times \boldsymbol{y}(t), \tag{2.33}$$

for any fixed vector $\boldsymbol{c}$. To compute the commutator of two equations of the form (2.33), we employ the *hat map* (B.1) taking a 3-vector c to a 3×3 matrix $\widehat{c}$, such that $c \times y = \widehat{c} y$. Using (B.2) we see that the commutator of the vector fields $c \times y$ and $d \times y$ is given by $(c \times d) \times y$. Thus the Lie algebra in this example may be identified with $(\mathbb{R}^3, \times)$, the real 3-space with the Lie bracket

$$[c, d] = c \times d$$

given by a vector product. The simplified equation (2.33) is a matrix equation $y' = \widehat{c} y$ with solution

$$y(t) = \text{expm}(t\widehat{c})y_0,$$

where $\text{expm}(t\widehat{c})$ can be computed rapidly using the Rodrigues formula (B.10).

Example 2.10. (Parabolic PDEs)   The final example in this section is chosen to illustrate the diversity of problems that may be tackled with the machinery of Lie-group methods. This example involves infinite-dimensional Lie algebras, a topic that is technically more demanding than the finite-dimensional case. There are several ways to circumvent the mathematical problems: we may either discuss the problem after it has been discretized in space (and has become finite-dimensional), or we may plunge straight ahead using the available techniques, disregarding possible mathematical difficulties. We will henceforth follow the latter approach, being aware that the resulting algorithms must be verified by other means.

Suppose that we wish to integrate a parabolic PDE with coefficients varying in space and time, for instance the *heat equation*

$$\frac{\partial u(t, x)}{\partial t} = \nabla \cdot (\mu(x) \nabla u(t, x)),$$

where u is the temperature and μ the heat conductivity of the material. With greater generality, consider equations of the form

$$\frac{\partial u(t, x)}{\partial t} = \mathcal{L}(u), \tag{2.34}$$

where $\mathcal{L}$ is an elliptic operator. To simplify the discussion, we wish to trivialize boundary conditions, so suppose that u is defined on the unit square with periodic boundary conditions. Thus, u should be thought of as a 'point' on the infinite-dimensional manifold $\mathcal{M} = C^\infty(T)$, the collection of all smooth functions on a torus. It is well known that (classical) explicit integrators for parabolic PDEs are typically *stiff* and stability analysis leads to severe step-size restriction: $\Delta t < c(\Delta x)^2$ for explicit finite-difference methods. A family of simpler equations, which can be solved exactly, explicitly and very efficiently with the *Fast Fourier Transform* (FFT), is the set of all parabolic equations with constant coefficients of the form

$$\frac{\partial u(t, \boldsymbol{x})}{\partial t} = \bar{\mu}\nabla^2 u(t, \boldsymbol{x}),$$

where $\bar{\mu}$ is constant. However, just as in Example 2.8, these equations cannot move an arbitrary point $u \in \mathcal{M}$ in an arbitrary direction. (In other words, they do not define a *transitive action* on $\mathcal{M}$.) Hence, we enlarge the family of simplified equations by adding an inhomogeneous term,

$$\frac{\partial u(t, \boldsymbol{x})}{\partial t} = \bar{\mu}\nabla^2 u(t, \boldsymbol{x}) + b(\boldsymbol{x}), \qquad (2.35)$$

where $b \in C^\infty(T)$. This equation is also easy to solve using FFTs. Letting the flows of (2.35) define our group action on $\mathcal{M}$, we see that the corresponding Lie algebra $\mathfrak{g}$ can be identified with pairs $(\bar{\mu}, b)$. The Lie-algebra action is given by

$$\lambda_*((\bar{\mu}, b))(u) = \bar{\mu}\nabla^2 u + b.$$

Using this action we see that any equation of the form (2.34) can be cast into the form (2.25) by choosing the function $a : \mathbb{R} \times \mathcal{M} \to \mathfrak{g}$ to be

$$a(u) = (\bar{\mu}, \mathcal{L}(u) - \bar{\mu}\nabla^2 u) \qquad (2.36)$$

for some choice of $\bar{\mu}$. For example, if $\mathcal{L}(u) = \nabla \cdot (\mu(\boldsymbol{x})\nabla u(t, \boldsymbol{x}))$, we would let $\bar{\mu}$ be some averaged value of $\mu(\boldsymbol{x})$.

In order to define the entire structure of $\mathfrak{g}$, we need to determine the Lie bracket. Let F and G be two vector fields on $\mathcal{M}$ defined at a point u by

$$F(u) = \bar{\mu}\nabla^2 u + f(\boldsymbol{x}),$$
$$G(u) = \bar{\nu}\nabla^2 u + g(\boldsymbol{x}).$$

Using (2.12), we obtain

$$[F, G](u)$$
$$= \frac{\partial}{\partial s}\left[\bar{\mu}\nabla^2(u + s(\bar{\nu}\nabla^2 u + g)) + f - \bar{\nu}\nabla^2(u + s(\bar{\mu}\nabla^2 u + f)) - g\right]\Big|_{s=0}$$
$$= 0 \cdot \nabla^2 u + \bar{\mu}\nabla^2 g - \bar{\nu}\nabla^2 f.$$

Thus, the bracket on $\mathfrak{g} = \mathbb{R} \times C^\infty(T)$ is

$$[(\bar{\mu}, f), (\bar{\nu}, g)] = (0, \bar{\mu}\nabla^2 g - \bar{\nu}\nabla^2 f).$$

It is interesting to note that the bracket can also be computed efficiently using FFTs.

This type of equation is considered in greater detail and various numerical examples are given in Munthe-Kaas and Lodden (2000). It turns out that, at least in some cases, it is possible to construct explicit integrators based on this action that are not subject to any step-size restriction involving the spatial discretization Δx. Thus, the methods are stable regardless of how spatial discretization is chosen. This is a topic of ongoing research, and many aspects of these integrators are as yet incompletely understood.

We will return to more examples and to numerical experiments in Section 11. Before discussing numerical algorithms, we need to study some important properties of the exponential map.

2.5. Much Ado about something

In this section we have emphasized a general view of differential equations on manifolds, based on Lie groups *acting* on manifolds. This outlook is important not just for the sake of mathematical beauty or abstraction but, as we hope to have persuaded the reader, also from the point of view of applications and computation. However, in so far as clarity of exposition is concerned, it is often better to restrict ourselves to the far simpler, familiar and more intuitive matrix theory.

In fact, it turns out that for all the algorithms that we present in this paper it is quite straightforward to translate results derived in matrix setting to the more general setting of local Lie-group actions on some domain. The following theorem of Ado underscores the importance of studying the matrix case (Olver 1995, Varadarajan 1984).

Theorem 2.9. (Ado's theorem)   Every finite-dimensional Lie algebra is isomorphic to a subalgebra of the matrix algebra $\mathfrak{gl}(N)$ for some $N \geq 1$.

Although a similar result does not hold for all finite-dimensional Lie groups, it is true that, whenever we are given a finite-dimensional *local* Lie-group action, we can always find an equivalent local action by a matrix Lie group. More information on these topics can be found in Olver (1995) and Varadarajan (1984).

Aware of the danger of rules of thumb being mathematically imprecise, it is nonetheless worthwhile to summarize these results as follows.

For practically any concept in general Lie theory there exists a corresponding concept within matrix Lie theory. Vice versa, practically any result that holds in the matrix case remains valid within the general Lie theory.

The above remark is even more important in a numerical context, since computation always takes place in a finite-dimensional setting. Even if the original equation evolves on an infinite-dimensional manifold, its practical computation must ultimately involve a discretization to a finite-dimensional formulation.

At this point it is time to wake the readers who have surfed through the general theory in a relaxed manner. We will restate the main definitions in the concrete form in which they appear within the matrix theory. It is worthwhile to compare these definitions to the corresponding general definitions above.

Definition 2.13. A real *matrix Lie group* is a smooth subset $\mathcal{G} \subseteq \mathbb{R}^{N \times N}$, closed under matrix products and matrix inversion. We let $\boldsymbol{I} \in \mathcal{G}$ denote the identity matrix.

Definition 2.14. The Lie algebra $\mathfrak{g}$ of a matrix Lie group $\mathcal{G}$ is the linear subspace $\mathfrak{g} \subseteq \mathbb{R}^{N \times N}$ consisting of all matrices of the form

$$\mathfrak{g} = \left\{ A \in \mathbb{R}^{N \times N} : A = \left. \frac{\mathrm{d}\rho(s)}{\mathrm{d}s} \right|_{s=0} \right\},$$

where $\rho(s) \in \mathcal{G}$ is a smooth curve such that $\rho(0) = \boldsymbol{I}$. The space $\mathfrak{g}$ is closed under matrix additions, scalar multiplication and the matrix commutator

$$[A, B] = AB - BA. \tag{2.37}$$

Complex matrix Lie groups and algebras are defined similarly.

The time has come to introduce some of the main *dramatis personae* of our survey: concrete examples of Lie groups and algebras. In each case it is easy to verify that all the axioms of a group or an algebra, as the case might be, are fulfilled, and we leave this as an exercise to the reader.

- The set of all real $N \times N$ nonsingular matrices is a (multiplicative) Lie group, the *general linear group* $\mathrm{GL}(N)$. The corresponding Lie algebra is the set $\mathbb{R}^{N \times N}$ of all $N \times N$ real matrices which, in keeping with our terminology, we denote by $\mathfrak{gl}(N)$.

  The general linear group and algebra can be defined over other fields than $\mathbb{R}$, in which case we communicate this in the second argument. For example, $\mathrm{GL}(N; \mathbb{C})$ consists of all nonsingular $N \times N$ complex matrices.

- All members of $\mathrm{GL}(N)$ with unit determinant form the *special linear group* $\mathrm{SL}(N)$. Its Lie algebra, $\mathfrak{sl}(N)$, consists of all matrices in $\mathfrak{gl}(N)$ with zero trace.

- $N \times N$ real orthogonal matrices form the *orthogonal group* $\mathrm{O}(N)$, whose Lie algebra $\mathfrak{so}(N)$ consists of $N \times N$ skew-symmetric matrices.

  The set $\mathrm{SO}(N) = \mathrm{SL}(N) \cap \mathrm{O}(N)$, consisting of $N \times N$ real orthogonal matrices with unit determinant, is the *special orthogonal group*. Its Lie algebra is $\mathfrak{so}(N)$, which we have just encountered. This is not contradictory: we never claimed that two different Lie groups must have different Lie algebras! As a matter of fact, more is true: if $\mathcal{G}$ is a Lie group and $\mathcal{G}_{Id}$ is the connected component of $\mathcal{G}$ containing the identity I – a Lie subgroup – then they produce the same Lie algebra. (This is precisely the situation with O(N) and SO(N).)

- The set of all $(2N) \times (2N)$ real matrices X such that $XJX^{\mathrm{T}} = J$, where

$$J = \begin{bmatrix} O_N & I_N \\ -I_N & O_N \end{bmatrix},$$

is the *symplectic group* and denoted by Sp(N). (The Jacobian of the flow of a Hamiltonian ODE system evolves in Sp(N).) The corresponding Lie algebra, $\mathfrak{sp}(N)$, consists of $F \in \mathfrak{gl}(2N)$ such that $FJ + JF^{\mathrm{T}} = O$.

- All the matrices $X \in$ SL(4) such that $XJX^{\mathrm{T}} = J$, where

$$J = \begin{bmatrix} 1 & 0 & 0 & 0 \\ 0 & 1 & 0 & 0 \\ 0 & 0 & 1 & 0 \\ 0 & 0 & 0 & -1 \end{bmatrix}$$

form the *Lorenz group* SO(3, 1). Its Lie algebra $\mathfrak{so}(3, 1)$ is made out of all $F \in \mathfrak{gl}(4)$ such that $FJ + JF^{\mathrm{T}} = O$.

- As an example of complex Lie groups, we mention the *unitary group* U(N; $\mathbb{C}$) of all $N \times N$ complex unitary matrices: $X \in$ U(N; $\mathbb{C}$) if and only if $XX^{\mathrm{H}} = I$. The Lie algebra corresponding to U(N; $\mathbb{C}$) is the set $\mathfrak{u}(N; \mathbb{C})$ of all skew-Hermitian matrices in $\mathfrak{gl}(N; \mathbb{C})$.

  The unitary group should not be confused with O(N; $\mathbb{C}$), the group of all $N \times N$ complex orthogonal matrices, whose Lie algebra is $\mathfrak{so}(N; \mathbb{C})$.

- As for O(N) and SO(N), we obtain the *special unitary group* intersecting U(N; $\mathbb{C}$) with SL(N; $\mathbb{C}$). Its Lie algebra, $\mathfrak{su}(N; \mathbb{C})$, consists of $N \times N$ complex skew-Hermitian matrices with zero trace.

Definition 2.15. A differential equation on a matrix Lie group is an equation of the form

$$Y' = A(t, Y)Y, \quad t \geq 0, \qquad Y(0) \in \mathcal{G}, \tag{2.38}$$

where $A : \mathbb{R} \times \mathcal{G} \rightarrow \mathfrak{g}$ and AY is the usual matrix product between $A \in g$ and $Y \in \mathcal{G}$.

The reader may verify that this is the special case of the general form of a differential equation on a manifold, given in Assumption 2.1, where $\mathcal{M} = \mathcal{G}$ is a matrix Lie group and the action Λ is taken to be the left (matrix) multiplication in $\mathcal{G}$,

$$\Lambda(R, Y) = RY.$$

Table 2.1. *Correspondence between the matrix case and general Lie theory*

Matrix case	General case
AY, $A \in \mathfrak{g}$, $Y \in \mathcal{G}$	$\lambda_*(a)(y)$, $a \in \mathfrak{g}$, $y \in \mathcal{M}$
$Y' = A(Y,t)Y$	$y' = \lambda_*(a(y,t))(y)$
RY, $R, Y \in \mathcal{G}$	$\Lambda(r,y)$, $r \in \mathcal{G}$, $y \in \mathcal{M}$
$\mathrm{expm}(A) = \sum_{j=0}^{\infty} A^j/j!$	Definition 2.11
$[A,B] = AB - BA$	Definition 2.10
PAP^{-1}, $P \in \mathcal{G}$, $A \in \mathfrak{g}$	Definition 2.12

Using (2.16) we find

$$\lambda_*(A)(Y) = AY.$$

Since $\mathfrak{g}$ is defined as the collection of *all* tangent directions at $I \in \mathcal{G}$ and matrix multiplication by Y is an invertible mapping, we see that any tangent at Y can be written in the form AY and all differential equations on $\mathcal{G}$ can be written in the form (2.38).

Definition 2.16. The exponential mapping $\mathrm{expm} : \mathfrak{g} \to \mathcal{G}$ is defined as

$$\mathrm{expm}(A) = \sum_{j=0}^{\infty} \frac{A^j}{j!}. \tag{2.39}$$

Note that $\mathrm{expm}(O) = I$, and that for A sufficiently near $O \in \mathfrak{g}$ the exponential has a smooth inverse given by the matrix logarithm $\mathrm{logm} : \mathcal{G} \to \mathfrak{g}$.

Definition 2.17. The *adjoint representation*, Ad, and its derivative, ad, are given by the formulae

$$\mathrm{Ad}_P(A) = PAP^{-1}, \tag{2.40}$$

$$\mathrm{ad}_A(B) = AB - BA = [A,B]. \tag{2.41}$$

It is easy to verify that (2.20)–(2.23) hold in the matrix case, while (2.24) is far from being an obvious identity even in that case.

Table 2.1 summarizes the correspondence between the matrix case and the general case.

2.6. *The differential of the exponential map*

We have introduced the exponential mapping $\exp : \mathfrak{g} \to \mathcal{G}$ as the fundamental solution of the equation $y' = ay$, or more explicitly in the matrix case as $\exp A = \sum_{j=0}^{\infty} A^j/j!$. For development of numerical algorithms it is essential to discuss the derivative of the exponential map. This will first be used to deduce an infinitesimal version of the BCH formula and in the following chapters to derive a variety of different numerical algorithms for differential equations on Lie groups.

To simplify the exposition we will restrict the proofs to matrix theory. The results are, however, valid in an abstract setting. The reader is referred to Varadarajan (1984) for proofs in a general context.

Given a scalar function $a(t) \in \mathbb{R}$, the derivative of the exponential is given by $\mathrm{d}\exp(a(t))/\mathrm{d}t = a'(t)\exp(a(t))$. One might have hoped for a similar result when $A(t)$ is a matrix. However, since in general $[A, A'] \neq O$, this is not the case and we must correct this formula. Note that $\mathrm{d}\exp(A(t))/\mathrm{d}t$ must be tangent to $\mathcal{G}$ in the point $P(t) = \exp(A(t))$. We have seen in Section 2.5 that *any* such tangent can be written as $C(t)P(t)$, where $C(t) \in \mathfrak{g}$. Furthermore, general properties of $\mathrm{d}/\mathrm{d}t$ imply that $C(t)$ must depend only on $A(t)$ and $A'(t)$, and that the dependence on A' is linear. This function is denoted by dexp.

Definition 2.18. The differential of the exponential mapping is defined as the 'right trivialized' tangent of the exponential map, that is, as a function $\mathrm{dexp} : \mathfrak{g} \times \mathfrak{g} \to \mathfrak{g}$ such that

$$\frac{\mathrm{d}}{\mathrm{d}t}\exp(A(t)) = \mathrm{dexp}_{A(t)}(A'(t))\exp(A(t)). \tag{2.42}$$

Just like the functions Ad_A and ad_A defined in (2.40) and (2.41) respectively, dexp_A is also linear in its second argument for a fixed A. Hence we may regard all these as being matrices acting on $\mathfrak{g}$. In fact dexp_A is an analytic function of the matrix transformation ad_A:

$$\mathrm{dexp}_A = \frac{\mathrm{expm}(\mathrm{ad}_A) - I}{\mathrm{ad}_A}. \tag{2.43}$$

This formula should be read as a power series in the following manner. Since

$$\frac{\mathrm{e}^x - 1}{x} = 1 + \tfrac{1}{2!}x + \tfrac{1}{3!}x^2 + \tfrac{1}{4!}x^3 \cdots + \tfrac{1}{(j+1)!}x^j + \cdots,$$

we obtain

$$\mathrm{dexp}_A(C) = C + \tfrac{1}{2!}[A, C] + \tfrac{1}{3!}[A, [A, C]] + \tfrac{1}{4!}[A, [A, [A, C]]] + \cdots$$
$$= \sum_{j=0}^{\infty} \frac{1}{(j+1)!}\mathrm{ad}_A^j C. \tag{2.44}$$

The fact that dexp_A is an analytic function in ad_A makes it easy to invert the matrix dexp_A simply by inverting the analytic function,

$$\text{dexp}_A^{-1} = \frac{\text{ad}_A}{\text{expm}(\text{ad}_A) - I}. \tag{2.45}$$

Recall that

$$\frac{x}{e^x - 1} = 1 - \tfrac{1}{2}x + \tfrac{1}{12}x^2 - \tfrac{1}{720}x^4 + \cdots = \sum_{j=0}^{\infty} \frac{B_j}{j!}x^j,$$

where B_j are *Bernoulli numbers* (Abramowitz and Stegun 1970). Thus,

$$\text{dexp}_A^{-1}(C) = C - \tfrac{1}{2}[A, C] + \tfrac{1}{12}[A, [A, C]] + \cdots = \sum_{j=0}^{\infty} \frac{B_j}{j!}\text{ad}_A^j(C). \tag{2.46}$$

Note that, except for B_1, all odd-indexed Bernoulli numbers vanish. Hence $\text{dexp}_A^{-1} + \tfrac{1}{2}\text{ad}_A$ is an even function of ad_A. We have based the formulae here on *right trivializations*, that is, tangents at a point $P \in G$ being written as CP, $C \in \mathfrak{g}$. It is equally possible to derive formulae based on left trivializations, tangents written in the form $P\tilde{C}$. If $P\tilde{C} = CP$, we observe that $\tilde{C} = P^{-1}CP = \text{Ad}_{P^{-1}}(C)$. Using (2.24), we compute the left-trivialized formulae as

$$\text{Ad}_{\exp(-A)} \text{dexp}_A = \exp(\text{ad}_{-A})\frac{\exp(\text{ad}_A) - I}{\text{ad}_A} = \frac{I - \exp(\text{ad}_{-A})}{\text{ad}_A} = \text{dexp}_{-A}.$$

Hence

$$\frac{\text{d}}{\text{dt}} \exp(A(t)) = \text{dexp}_{A(t)}(A'(t)) \exp(A(t)) \tag{2.47}$$
$$= \exp(A(t)) \text{dexp}_{-A(t)}(A'(t)).$$

Thus we can arrive at the left versions by changing the sign of every commutator. Note that $\text{dexp}_A^{-1}(C)$ and $\text{dexp}_{-A}^{-1}(C)$ differ only in the sign of the term $\pm\tfrac{1}{2}[A, C]$.

The definition of dexp in Definition 2.18 can be generalized to any smooth function $\psi : \mathfrak{g} \to \mathcal{G}$:

Definition 2.19. Given a smooth function $\psi : \mathfrak{g} \to \mathcal{G}$, we define *the right trivialized tangent* of ψ as the function $\text{d}\psi : \mathfrak{g} \times \mathfrak{g} \to \mathfrak{g}$ defined such that

$$\frac{\text{d}}{\text{dt}}\psi(A(t)) = \text{d}\psi_{A(t)}(A'(t))\psi(A(t)). \tag{2.48}$$

The function $\text{d}\psi$ is always linear in the second argument, A'.

Let us now apply dexp and dexp^{-1} to obtain a differential equation for the BCH formula. Going back to Theorem 2.5, we define a function $\text{bch} : \mathfrak{g} \times \mathfrak{g} \to \mathfrak{g}$ such that

$$\text{expm}(\text{bch } AB) = \text{expm}(A) \text{expm}(B).$$

We may compute $C = \text{bch}(A, B)$ by integrating a differential equation. Let $C(t) = \text{bch}(tA, B)$. Clearly $C(0) = B$, and we seek $C(1)$. Writing

$$\text{expm}(C(t)) = \text{expm}(tA)\,\text{expm}(B),$$

we find by differentiation that

$$\text{dexp}_{Ct}(C'(t))\exp(C(t)) = \text{dexp}_{tA}(A)\exp(tA)\exp(B) = A\exp(C(t)),$$

whence $C'(t) = \text{dexp}^{-1}_{C(t)}(A)$.

We have proved the following result.

Lemma 2.10. The function $C = \text{bch}(A, B)$ can be computed by integrating the differential equation

$$C'(t) = \text{dexp}^{-1}_{C(t)}(A), \quad 0 \le t \le 1, \qquad C(0) = B, \tag{2.49}$$

from $t = 0$ to $t = 1$.

In this light the dexp^{-1} function may be regarded as a kind of 'infinitesimal BCH generator'. A more concise presentation of this idea is given in Engø (2000). A symbolic algorithm to compute the BCH from this formula can be found in Munthe-Kaas and Owren (1999).

2.7. Crouch–Grossman methods

The discipline of Lie-group methods owes a great deal to the pioneering work of Peter Crouch and his co-workers, who were the first to introduce in a systematic, mathematically sophisticated manner ODE methods that evolve on manifolds. It is interesting to note that their work was primarily motivated by problems in robotics and control theory.

The main algorithm originating in this circle of ideas is the method of *rigid frames* of Crouch and Grossman (1993). It was originally stated in a more general formalism of differential equations evolving on manifolds. To fit the method into our narrative and to simplify its exposition we restrict our discussion to Lie-group equations.

In essence, the Crouch–Grossman approach is an attempt to apply a Runge–Kutta method to (2.1) (or, in a Lie-group context, to (2.38)) by repeatedly freezing and thawing coefficients and keeping the flow in the correct configuration space. The solution of a 'frozen' Lie-group equation,

$$Y' = A(\tilde{t}, \tilde{Y})Y \quad t \ge t_*, \qquad Y(t_*) = Y_*,$$

is simply $\text{expm}((t - t_*)A(\tilde{t}, \tilde{Y}))Y_*$. Freezing (2.38) at t_n and letting $Y_{n+1} = \text{dexp}(hA(t_n, Y_n))Y_n$, where $Y_m \approx Y(t_m)$, $t_m = mh$, results in a first-order method of very little merit. This can be remedied in the following procedure.

We choose constants c_k, b_l, $a_{k,l}$, $1 \leq l < k \leq \nu$ and let

$$\left. \begin{array}{l} X_k = e^{ha_{k,k-1}F_{k-1}} e^{ha_{k,k-2}F_{k-2}} \cdots e^{ha_{k,1}F_1} Y_n, \\[2mm] F_k = A(t_n + c_k h, X_k), \\[2mm] Y_{n+1} = e^{hb_\nu F_\nu} e^{hb_{\nu-1}F_{\nu-1}} \cdots e^{hb_1 F_1} Y_n. \end{array} \right\} \quad k = 1, 2, \ldots, \nu, \quad (2.50)$$

In other words, we model the solution as a product of ν 'frozen' steps. Note that $X_k \in \mathcal{G}$, $F_k \in \mathfrak{g}$; hence, as required, $Y_{n+1} \in \mathcal{G}$.

To the initiated, method (2.50) might appear to be a 'Lie-group version' of a Runge–Kutta scheme, an analogy that we have deliberately reinforced by employing notation that will be reserved later to RK schemes. Yet, order conditions are considerably more challenging than in the classical RK case: they are nonlinear in the weights $b_1, b_2, \ldots, b_\nu$ and there are more of them!

Moderate headway can be made by elementary means and a great deal of algebra. Thus, Crouch and Grossman (1993) have derived three-stage methods of order three, for example

$$\begin{array}{ll} X_1 = Y_n, & F_1 = A(t_n, X_1), \\[2mm] X_2 = e^{\frac{3}{4}hF_1} Y_n, & F_2 = A(t_n + \frac{3}{4}h, X_2), \\[2mm] X_3 = e^{\frac{17}{108}hF_2} e^{\frac{119}{216}hF_1} Y_n, & F_3 = A(t_n + \frac{17}{24}h, X_3), \\[2mm] Y_{n+1} = e^{\frac{13}{51}hF_3} e^{-\frac{2}{3}hF_2} e^{\frac{24}{17}hF_1} Y_n. \end{array}$$

Inquiry into higher-order methods, though, requires more than algebra and elbow grease. The situation is further complicated by the fact that (2.50) is typically formulated in a considerably more abstract manner, in a manifold setting: this does not make order analysis any simpler!

The order of classical RK methods is nowadays determined by a method due to Butcher (1963), which identifies expansion terms of both the exact and the approximate solution with rooted trees. Remarkably, a similar approach can be generalized to Crouch–Grossman methods and this has been accomplished by Owren and Marthinsen (1999b). Details of their work are outside the scope of this survey and we refer the readers to the primary source. Let us just mention that, unlike the classical RK case, there are no fourth-order methods of this kind with four stages and $\nu = 5$ is required. Moreover, Owren and Marthinsen (1999b) extended (2.50) to *implicit* methods, whereby $a_{k,l}$ is given for all $k, l = 1, 2, \ldots, \nu$ and

$$X_k = e^{ha_{k,\nu}F_\nu} e^{ha_{k,\nu-1}F_{\nu-1}} \cdots e^{ha_{k,1}F_1} Y_n, \qquad k = 1, 2, \ldots, \nu.$$

This allows for better order/stages ratio but the downside is the need to solve nonlinear equations in Lie groups. Classical methods, for instance the Newton–Raphson technique, are of little use here since, unless iterated to convergence, they are unlikely to deliver a solution that resides in the

Lie group $\mathcal{G}$. Recently, however, Owren and Welfert (1996) developed two variants of Newton's method that always evolve in $\mathcal{G}$. This brings implicit versions of (2.50) within the realm of computation, although they are expensive.

The mainstream of research into Lie-group methods has moved in the last few years away from the Crouch–Grossman approach. The main reason is that the RK-MK methods, the theme of the next section, provide a considerably more convenient, intuitive and easy-to-analyse means of translating Runge–Kutta formalism to a Lie-group setting. Yet, it would be unfair to pronounce Crouch–Grossman methods as inviable or of purely historical interest. Firstly, in their more general setting, Crouch–Grossman methods can be made (with some effort!) to evolve on arbitrary smooth manifolds, while the scope of RK-MK is restricted to group actions. Secondly, at the present stage in the lifetime of geometric integration theory, we are denied the comfort of discarding lines of inquiry simply because of our current, incomplete understanding.

3. Runge–Kutta on manifolds and RK-MK

In this section we describe a class of numerical integration schemes for computing (2.38), or more generally (2.25). The methods have become known under the name of *RK-MK-type* schemes. We will later see that they might just as well be called *integration schemes based on canonical coordinates of the first kind*. These methods were originally developed in a series of four papers, Munthe-Kaas (1995), Munthe-Kaas and Zanna (1997), Munthe-Kaas (1998) and Munthe-Kaas (1999).

A major motivation behind the first of these papers was an attempt to understand and specify the basic operations underlying classical Runge–Kutta methods for integration of differential equations. Abstract specifications of mathematical structures are fundamental in theoretical computer science as a tool for structuring software. An *object-oriented* program consists of a collection of program modules which interact in a well-specified manner. A module could, for instance, represent some mathematical structure, like a linear space, a Lie algebra or a Lie group. The basic idea of object-oriented programming is that particular *representations* of the mathematical structure to be modelled should be *hidden* within the program module, and that interactions between different program modules should be independent of particular representations. Although this approach to programming has been very successful for discrete problems, considerably less has been done within areas of continuous mathematics, such as integration of differential equations. Much insight about the important underlying structures can be gained by studying 'pure' mathematical definitions, since these focus more on *what* the essential mathematical structures are, rather than on *how* they can be represented. Thus it is natural, for example, to specify that domains of differential equations should be differential manifolds and that the right-hand side of the equation should be a vector field on the manifold.

Seen from this perspective, classical integration schemes such as Runge–Kutta methods contain 'type errors' in their formulation. As an example, consider the trapezoidal rule

$$\boldsymbol{y}_{n+1} = \boldsymbol{y}_n + \frac{h}{2}[\boldsymbol{f}(\boldsymbol{y}_n) + \boldsymbol{f}(\boldsymbol{y}_{n+1})].$$

All the operations are valid if $\boldsymbol{y}$ and $\boldsymbol{f}$ are vectors. However, if $\boldsymbol{y}$ is a point on a manifold and $\boldsymbol{f}$ a vector field, then this expression involves addition of tangent vectors at different base points, and also the addition of a point on a manifold and a tangent to the manifold, both being invalid operations in the context of general manifolds.

In Munthe-Kaas (1995) classical Runge–Kutta methods are reformulated using coordinate-independent operations on a Lie group. It is shown there that the Butcher theory for order conditions of Runge–Kutta methods (see Butcher (1963), Hairer, Nørsett and Wanner (1993)) can be reformulated in a geometrical language, replacing the 'Butcher trees' with commutators in a Lie algebra. The outcome is a so-called *Lie–Butcher theory*. Although the resulting algorithms respect Lie-group structure, they can, in the simplest version, attain at most order two for a general non-commutative Lie group. In the sequel, Munthe-Kaas (1998), the Lie–Butcher order theory is improved, order conditions derived to arbitrary order in general Lie groups and explicit methods of Runge–Kutta type presented up to order four. The paper Munthe-Kaas and Zanna (1997) generalizes the theory from equations on Lie groups to equations evolving on more general manifolds acted upon by a Lie group. In the last of these papers, Munthe-Kaas (1999), it is shown that similar methods of arbitrarily high order can be constructed and analysed in a relatively simple manner, without employing the Lie–Butcher theory.

In this section our goal is to arrive at the main ideas of the algorithms while employing a minimal amount of formal theory. We have therefore decided not to discuss the general Lie–Butcher theory since this would require a significant amount of Lie theory, beyond what we have already introduced in Section 2. The interested reader is referred to Munthe-Kaas (1998) for details on Lie–Butcher series. We will continue along the lines of Munthe-Kaas (1999), but instead of proofs for the general equation (2.25) we restrict the discussion mainly to the simple matrix case (2.38).

We have seen that classical Runge–Kutta methods are valid and 'type-correct' only for differential equations evolving in a linear space V, since then the configuration space and the space of vector fields coincide. If the analytical solution of the differential equation evolves on some linear subspace $W \subset V$, then it is easy to show that a consistent numerical integrator will also evolve on W (up to a departure due to round-off errors). On the other hand, if the differential equation evolves on some nonlinear manifold embedded in V, it is much more difficult to devise numerical algorithms that stay on the right submanifold. It is well known that traditional ν-stage Runge–Kutta methods preserve quadratic submanifolds if the coefficients satisfy the condition

$$b_k b_l = b_k a_{k,l} + b_l a_{l,k}, \qquad k, l = 1, 2, \ldots, \nu$$

(Cooper 1987), while Calvo et al. (1997) show that this condition is also necessary. In the same paper it is also shown that it is essentially impossible to devise classical Runge–Kutta methods that preserve arbitrary cubic submanifolds. Linear multistep methods or truncated Taylor expansions cannot in general preserve even quadratic submanifolds (Iserles 1997).

In the case of equations on a Lie group $\mathcal{G}$, recall that the local structure in a neighbourhood of any point can be described by the Lie algebra $\mathfrak{g}$, which is a *linear space*. Even better, if $\mathcal{H}$ is a Lie subgroup of $\mathcal{G}$, then there exists a (linear!) subalgebra $\mathfrak{h}$ of $\mathfrak{g}$ describing the local structure of $\mathcal{H}$. Given a differential equation evolving on $\mathcal{H}$, it is in general impossible to devise classical integration scheme that will evolve on $\mathcal{H}$. On the other hand, if an equation is evolving on $\mathfrak{h}$, so will almost *any* reasonable numerical integrator. It thus seems a good idea to try to solve a differential equation in the Lie algebra rather than in the Lie group!

Given the equation $Y' = A(t, Y)Y$, $Y(0) = Y_0$, we call the map

$$\mathfrak{g} \ni A \mapsto \mathrm{expm}(A)Y_0 \in \mathcal{G} \tag{3.1}$$

canonical coordinates of the first kind (Varadarajan 1984). This defines a smooth invertible map between a neighbourhood of $O \in \mathfrak{g}$ and a neighbourhood of $Y_0 \in G$. We say that these coordinates are *centred* about $Y_0 \in \mathcal{G}$. A crucial step is 'pulling back' the equation from $\mathcal{G}$ to $\mathfrak{g}$ using this map.

Lemma 3.1. For small $t \geq 0$ the solution of (2.38) is given by

$$Y(t) = \mathrm{expm}(\Theta(t))Y_0,$$

where $\Theta \in \mathfrak{g}$ satisfies the differential equation

$$\Theta'(t) = \mathrm{dexp}^{-1}_{\Theta(t)}(A(t, Y)) \qquad \Theta(0) = O \tag{3.2}$$

and the dexp^{-1} operator has been defined in (2.46).

Proof. Differentiation of $Y(t) = \exp(\Theta(t))Y_0$ yields

$$Y'(t) = \mathrm{dexp}_{\Theta(t)}(\Theta'(t)) \exp(\Theta(t))Y_0 = \mathrm{dexp}_{\Theta(t)}(\Theta'(t))Y(t).$$

The lemma follows from $Y'(t) = A(t, Y)Y(t)$. □

Equation (3.2) is absolutely crucial to the entire business of Lie-group methods. It was originally stated by Felix Hausdorff (1906), although some attribute it to John Edward Campbell, who might have published it a few years earlier. (The names of both, together with Henry Frederick Baker, have been immortalized in the BCH formula; *cf.* Theorem 2.5.) The corresponding result for the general case (2.25) is given in Munthe-Kaas (1999).

Note that the proof uses no other property of the exponential mapping than the definition of dexp. Hence the argument can easily be generalized, replacing exp with a general coordinate map ψ and dexp with dψ as discussed in Definition 2.19. This will be used in Section 6.

The simplest and most natural solution strategy is to apply a classically formulated Runge–Kutta method to (3.2), *rather than to the original equation* (2.38). At each step we choose a coordinate system of the form (3.1), centred at the last known point Y_n. Let us consider briefly the details of this algorithm. Recall that a ν-stage Runge–Kutta method is defined by constants $\{a_{k,l}\}_{k,l=1}^{\nu}$, $\{b_l\}_{l=1}^{\nu}$, $\{c_k\}_{k=1}^{\nu}$, usually written as a *Butcher tableau*:

$$
\begin{array}{c|cccc}
c_1 & a_{1,1} & a_{1,2} & \cdots & a_{1,\nu} \\
c_2 & a_{2,1} & a_{2,2} & \cdots & a_{2,\nu} \\
\vdots & \vdots & \vdots & & \vdots \\
c_\nu & a_{\nu,1} & a_{\nu,2} & \cdots & a_{\nu,\nu} \\
\hline
& b_1 & b_2 & \cdots & b_\nu
\end{array}
$$

(Hairer et al. 1993). Applied to a standard vector equation $\boldsymbol{y}' = \boldsymbol{f}(t, \boldsymbol{y})$, a single step of length h from $\boldsymbol{y}_n$ to $\boldsymbol{y}_{n+1}$ is given by

$$
\left.
\begin{aligned}
\boldsymbol{\theta}_k &= \boldsymbol{y}_n + \sum_{l=1}^{\nu} a_{k,l} \boldsymbol{f}_l, \\
\boldsymbol{f}_k &= h \boldsymbol{f}(t_n + c_k h, \boldsymbol{\theta}_k), \\
\boldsymbol{y}_{n+1} &= \boldsymbol{y}_n + \sum_{l=1}^{\nu} b_l \boldsymbol{f}_l.
\end{aligned}
\right\} \quad k = 1, \dots, \nu,
\tag{3.3}
$$

Applying this scheme to (3.2), we obtain the *RK-MK algorithm*. The following equations describe a single RK-MK step from $Y_n \in \mathcal{G}$ to $Y_{n+1} \in \mathcal{G}$:

$$
\left.
\begin{aligned}
\Theta_k &= \sum_{l=1}^{\nu} a_{k,l} F_l, \\
A_k &= h A(t_n + c_k h, \operatorname{expm}(\Theta_k) Y_n), \\
F_k &= \operatorname{dexp}_{\Theta_k}^{-1}(A_k), \\
\Theta &= \sum_{l=1}^{\nu} b_l F_l, \\
Y_{n+1} &= \operatorname{expm}(\Theta) Y_n.
\end{aligned}
\right\} \quad k = 1, \dots, \nu,
\tag{3.4}
$$

The same algorithm also integrates the general equation (2.28), provided we replace $\text{expm}(\Theta)Y_n$ with its general form $\Lambda(\exp(\Theta), Y_n)$.

In order to complete this algorithm, we need to provide practical means for computing $\text{dexp}_{\Theta_k}^{-1}(A_k)$. In some cases there exists fast direct algorithms for this: see Appendix B. Note that even if dexp^{-1} is approximated, the resulting algorithm will evolve on the correct manifold. In general one may use the expansion (2.46), truncated to the order of the underlying Runge–Kutta scheme, and the resulting algorithm will obtain the same order as the underlying Runge–Kutta scheme, while staying on the correct manifold. For high-order methods, a significant number of commutators must be computed if dexp^{-1} is computed using (2.46). In Section 5.3 the structure of so-called *free Lie algebras* is used to dramatically reduce the number of commutators.

Examples of specific methods based on (3.4) feature in Appendix A, where, in a more general setting, it is redesignated as (A.1). Here we just stress again the main difference between (3.3) and (3.4): the latter acts in the Lie algebra $\mathfrak{g}$, which is a linear space, thereby respecting Lie-group structure.

4. Magnus and Fer expansions

There are several possible points of departure for our description of Magnus and Fer expansions. Perhaps the simplest and the most intuitive is the scalar linear differential equation

$$y' = a(t)y, \quad t \geq 0, \qquad y(0) = y_0.$$

Its solution, $y(t) = \exp\left(\int_0^t a(\xi)\,d\xi\right)y_0$, is familiar to all well-trained mathematics undergraduates. Bearing in mind that the definition of the exponential can easily be extended from scalars to matrices, one might have perhaps hoped that its obvious generalization,

$$\text{expm}\left(\int_0^t A(\xi)\,d\xi\right)Y_0, \tag{4.1}$$

is the solution of the matrix linear system

$$Y' = A(t)Y, \quad t \geq 0, \qquad Y(0) = Y_0. \tag{4.2}$$

Unless $A(t_1)$ and $A(t_2)$ commute with each other for all $t_1, t_2 \geq 0$, this is, unfortunately, misplaced hope. Before offering possible remedies to this state of affairs, we mention that if $Y_0 \in \mathcal{G}$, a Lie group, and A lies in its Lie algebra $\mathfrak{g}$ (whence (4.2) is a Lie-group equation) then (4.1) evolves in $\mathcal{G}$. Bearing in mind the advisability of respecting Lie-group structure, we need to 'correct' (4.1) without losing this feature.

Two possible remedies suggest themselves. Firstly, we may seek a correction $\Delta(t)$ evolving *in the Lie algebra* $\mathfrak{g}$ so that

$$Y(t) = \mathrm{expm}\left(\int_0^t A(\xi)\,\mathrm{d}\xi + \Delta(t)\right) Y_0.$$

Alternatively, we may correct with $V(t)$ *in the Lie group* $\mathcal{G}$,

$$Y(t) = \mathrm{expm}\left(\int_0^t A(\xi)\,\mathrm{d}\xi\right) V(t).$$

This gives rise to *Magnus* and *Fer* expansions, respectively.

Both Magnus (1954) and Fer (1958) expansions originated within a non-numerical context and they have found extensive use, for instance in mathematical physics, quantum chemistry, control theory and stochastic differential equations as a perturbative tool in the investigation of linear systems (4.2). Fashioning them into a numerical weapon is nontrivial and will occupy us in this and the following sections.

4.1. Magnus expansions and rooted trees

Our point of departure is the *dexpinv equation* (2.46) which we recall for completeness of exposition: the solution of (4.2) can be written in the form $Y(t) = \mathrm{expm}(\Theta(t))Y_0$, $t \geq 0$, where

$$\Theta' = \mathrm{dexp}_\Theta^{-1} A = \sum_{k=0}^{\infty} \frac{\mathrm{B}_k}{k!} \mathrm{ad}_\Theta^k A, \quad t \geq 0, \qquad \Theta(0) = \boldsymbol{O}, \qquad (4.3)$$

$\{\mathrm{B}_k\}_{k \in \mathbb{Z}^+}$ being Bernoulli numbers. As a first step, we attempt to solve (4.3) by Picard iteration,

$$\Theta^{[0]}(t) \equiv \boldsymbol{O},$$

$$\Theta^{[m+1]}(t) = \int_0^t \mathrm{dexp}_{\Theta^{[m]}(\xi)}^{-1} A(\xi)\,\mathrm{d}\xi$$

$$= \sum_{k=0}^{\infty} \frac{\mathrm{B}_k}{k!} \int_0^t \mathrm{ad}_{\Theta^{[m]}(\xi)}^k A(\xi)\,\mathrm{d}\xi, \quad m = 0, 1, \ldots.$$

Rearranging terms for simplicity, we obtain

$$\Theta^{[1]}(t) = \int_0^t A(\xi_1)\,\mathrm{d}\xi_1,$$

$$\Theta^{[2]}(t) = \int_0^t A(\xi_1)\,\mathrm{d}\xi_1 - \tfrac{1}{2}\int_0^t \left[\int_0^{\xi_1} A(\xi_2)\,\mathrm{d}\xi_2, A(\xi_1)\right]\mathrm{d}\xi_1$$

$$+ \tfrac{1}{12}\int_0^t \left[\int_0^{\xi_1} A(\xi_2)\,\mathrm{d}\xi_2, \left[\int_0^{\xi_1} A(\xi_2)\,\mathrm{d}\xi_2, A(\xi_1)\right]\right]\mathrm{d}\xi_1 + \cdots,$$

$$\Theta^{[3]}(t) = \int_0^t A(\xi_1)\,\mathrm{d}\xi_1 - \tfrac{1}{2}\int_0^t \left[\int_0^{\xi_1} A(\xi_2)\,\mathrm{d}\xi_2, A(\xi_1)\right]\mathrm{d}\xi_1$$

$$+ \tfrac{1}{12}\int_0^t \left[\int_0^{\xi_1} A(\xi_2)\,\mathrm{d}\xi_2, \left[\int_0^{\xi_1} A(\xi_2)\,\mathrm{d}\xi_2, A(\xi_1)\right]\right]\mathrm{d}\xi_1$$

$$+ \tfrac{1}{4}\int_0^t \left[\int_0^{\xi_1}\left[\int_0^{\xi_2} A(\xi_3)\xi_3, A(\xi_2)\right]\mathrm{d}\xi_2, A(\xi_1)\right]\mathrm{d}\xi_1$$

$$- \tfrac{1}{24}\int_0^t \left[\int_0^{\xi_1}\left[\int_0^{\xi_2} A(\xi_3)\,\mathrm{d}\xi_3, \left[\int_0^{\xi_2} A(\xi_3)\,\mathrm{d}\xi_3, A(\xi_2)\right]\right]\mathrm{d}\xi_2, A(\xi_1)\right]\mathrm{d}\xi_1$$

$$- \tfrac{1}{24}\int_0^t \left[\int_0^{\xi_1}\left[\int_0^{\xi_2} A(\xi_3)\,\mathrm{d}\xi_3, A(\xi_2)\right]\mathrm{d}\xi_2, \left[\int_0^{\xi_1} A(\xi_2)\,\mathrm{d}\xi_2, A(\xi_1)\right]\right]\mathrm{d}\xi_1$$

$$- \tfrac{1}{24}\int_0^t \left[\int_0^{\xi_1} A(\xi_2)\,\mathrm{d}\xi_2, \left[\int_0^{\xi_1}\left[\int_0^{\xi_2} A(\xi_3)\,\mathrm{d}\xi_3, A(\xi_2)\right]\mathrm{d}\xi_2, A(\xi_1)\right]\right]\mathrm{d}\xi_1$$

$$+ \cdots$$

and so on. The Picard theorem implies that $\Theta(t) = \lim_{m\to\infty}\Theta^{[m]}(t)$ exists in a suitably small neighbourhood of the origin and the above first few iterations indicate that it can be expanded as a linear combination of terms that are composed from *integrals* and *commutators* acting recursively on the matrix A. This is the *Magnus expansion*

$$\Theta(t) = \sum_{k=0}^{\infty} H_k(t), \qquad (4.4)$$

where each H_k is a linear combination of terms that include exactly $k+1$ integrals (or – and later we will see that it boils down to the same thing – k commutators). Thus,

$$H_0(t) = \int_0^t A(\xi_1)\,\mathrm{d}\xi_1,$$

$$H_1(t) = -\tfrac{1}{2}\int_0^t \left[\int_0^{\xi_1} A(\xi_2)\,\mathrm{d}\xi_2, A(\xi_1)\right]\mathrm{d}\xi_1,$$

$$H_2(t) = \tfrac{1}{12}\int_0^t \left[\int_0^{\xi_1} A(\xi_2)\,\mathrm{d}\xi_2, \left[\int_0^{\xi_1} A(\xi_2)\,\mathrm{d}\xi_2, A(\xi_1)\right]\right]\mathrm{d}\xi_1$$

$$+ \tfrac{1}{4}\int_0^t \left[\int_0^{\xi_1}\left[\int_0^{\xi_2} A(\xi_3)\,\mathrm{d}\xi_3, A(\xi_2)\right]\mathrm{d}\xi_2, A(\xi_1)\right]\mathrm{d}\xi_1,$$

$$H_3(t) = -\tfrac{1}{24}\int_0^t \left[\int_0^{\xi_1}\left[\int_0^{\xi_2} A(\xi_3)\,\mathrm{d}\xi_3, \left[\int_0^{\xi_2} A(\xi_3)\,\mathrm{d}\xi_3, A(\xi_2)\right]\right]\mathrm{d}\xi_2, A(\xi_1)\right]\mathrm{d}\xi_1$$

$$- \tfrac{1}{24}\int_0^t \left[\int_0^{\xi_1}\left[\int_0^{\xi_2} A(\xi_3)\,\mathrm{d}\xi_3, A(\xi_2)\right]\mathrm{d}\xi_2, \left[\int_0^{\xi_1} A(\xi_2)\,\mathrm{d}\xi_2, A(\xi_1)\right]\right]\mathrm{d}\xi_1$$

$$-\frac{1}{24}\int_0^t\left[\left[\int_0^{\xi_1}A(\xi_2)\,\mathrm{d}\xi_2,\left[\int_0^{\xi_1}\left[\int_0^{\xi_2}A(\xi_3)\,\mathrm{d}\xi_3,A(\xi_2)\right]\mathrm{d}\xi_2,A(\xi_1)\right]\right]\right]\mathrm{d}\xi_1$$

$$-\frac{1}{8}\int_0^t\left[\left[\int_0^{\xi_1}A(\xi_2)\,\mathrm{d}\xi_2,\left[\int_0^{\xi_1}A(\xi_2)\,\mathrm{d}\xi_2,\left[\int_0^{\xi_1}A(\xi_2)\,\mathrm{d}\xi_2,A(\xi_1)\right]\right]\right]\right]\mathrm{d}\xi_1.$$

It is possible to derive the next few sets H_k with enough perseverance and perhaps a little help from a computer algebra package. Yet, it is evident that the terms are becoming increasingly complex. A considerably more transparent form of the Magnus expansion, amenable for easy recursive derivation and easier discussion of computational issues, can be obtained by associating each term in the expansion with a *rooted binary tree*, an approach that has been pioneered by Iserles and Nørsett (1999).

Let us briefly recall relevant terminology of graph theory (Harary 1969).

- Let $V = \{v_1, v_2, \dots, v_r\}$ be a finite set of distinct *vertices* and $E \in V \times V$ a set of *edges*. (The edges (v_i, v_j) and (v_j, v_i) are identified with each other.) We say that $G = \langle V, E \rangle$ is a *graph*.

- The ordered set $\{(v_{s_l}, v_{t_l}) : l = 1, 2, \dots, r\}$ of edges is a *path* from $v_i \in V$ to $v_j \in V$, $i \neq j$, if $s_1 = i$, $t_l = s_{l+1}$, $l = 1, 2, \dots, r - 1$ and $t_r = j$.

- The graph is said to be *connected* if there is a path between any two distinct vertices. It is a *tree* if exactly one path links every two vertices.

- A *rooted tree* is the pair $T = (G, w)$, where G is a tree and $w \in V$ is its *root*. There exists natural partial order on T: we say that $v_i \prec v_j$ if v_i precedes v_j in the unique path extending from the root w to v_j. In that case v_i is the *ancestor* of v_j, while v_j is the *successor* of v_i. (Thus, the root is the ancestor of all vertices in $V \setminus \{w\}$.)

- If $v_i \prec v_j$ and there is no $v_k \in V$ such that $v_i \prec v_k \prec v_j$, we say that v_i is the *parent* of v_j (most graph-theory texts adopt a more sexist definition) and v_j the *child* of v_i. Childless vertices are called *leaves*.

- If each vertex in a rooted tree has at most two children, T is called a *binary tree*. If each vertex has either exactly two children or is a leaf, T is said to be a *strictly binary tree*.

It is always worthwhile to draw a pictorial representation of a graph, whereby edges are merely undirected lines extending between vertices, which are denoted by black discs. The graph-theoretical convention is that the root is always placed at the bottom, although computer scientists occasionally defy gravity by locating it at the top. We will follow the mathematical convention.

We commence our investigation of the Magnus expansion by rewriting (4.4) in the form

$$\Theta(t) = \sum_{k=0}^{\infty} H_k(t) = \sum_{k=0}^{\infty} \sum_{\tau \in \mathbb{T}_k} \alpha(\tau) \int_0^t \mathcal{C}_\tau(\xi)\,\mathrm{d}\xi, \quad t \geq 0, \qquad (4.5)$$

Thus, each $\mathcal{C}_\tau$ for $\tau \in \mathbb{T}_k$ is made out of exactly k integrals and k commutators, while each $\alpha(\tau)$ is a scalar constant. Recalling how the expansion has been obtained from Picard's iteration, we observe that the terms $\mathcal{C}_\tau$ can be obtained by just two *composition rules*.

(1) The index set $\mathbb{T}_0$ is a singleton, $\mathbb{T}_0 = \{\tau_\circ\}$, say, and $\mathcal{C}_{\tau_\circ}(t) = A(t)$.

(2) If $\tau_1 \in \mathbb{T}_{m_1}$ and $\tau_2 \in \mathbb{T}_{m_2}$ then there exists $\tau \in \mathbb{T}_{m_1+m_2+1}$ such that

$$\mathcal{C}_\tau(t) = \left[\int_0^t \mathcal{C}_{\tau_1}(\xi)\,\mathrm{d}\xi, \mathcal{C}_{\tau_2}(t) \right]. \qquad (4.6)$$

Although this observation makes the expansion somewhat more transparent, much greater transparency is obtained by identifying the index sets $\mathbb{T}_k$ with subsets of binary rooted trees, in a manner that makes the above composition rules stand out pictorially. We express the relationship between the index τ and the term $\mathcal{C}_\tau(t)$ by the map $\tau \rightsquigarrow \mathcal{C}_\tau(t)$.

(1) We let $\mathbb{T}_0$ consist of the single rooted tree with one vertex, $\bullet$, and

$$\bullet \rightsquigarrow A(t).$$

(2) Suppose that $\mathbb{T}_{m_1} \ni \tau_1 \rightsquigarrow \mathcal{C}_{\tau_1}(t)$ and $\mathbb{T}_{m_2} \ni \tau_2 \rightsquigarrow \mathcal{C}_{\tau_2}(t)$. Then

$$\mathbb{T}_{m_1+m_2+1} \ni \overset{\tau_1}{\underset{}{\diagdown}} \overset{\tau_2}{\diagup} \rightsquigarrow \left[\int_0^t \mathcal{C}_{\tau_1}(\xi)\,\mathrm{d}\xi, \mathcal{C}_{\tau_2}(t) \right]. \qquad (4.7)$$

Thus, (4.6) is 'coded' in graph terminology by denoting integration by adding a root to a tree, while commutation is denoted by joining two trees with a common root. It is possible to show that all the terms in the Magnus expansion can be obtained in this manner (Iserles and Nørsett 1999).

To derive $\mathbb{T}_1$ we have just one option, $m_1 = m_2 = 0$, and the outcome is a single tree,

$$\tau_1 = \bullet, \qquad\qquad \tau_2 = \bullet \qquad \Rightarrow \qquad \tau = \diagdown\!\!\diagup.$$

There are two possibilities in $\mathbb{T}_2$, namely $m_1 = 0$, $m_2 = 1$ and $m_1 = 1$,

$m_2 = 0$. They yield

$$\tau_1 = \bullet, \qquad \tau_2 = \bigvee \qquad \Rightarrow \qquad \tau = \bigvee,$$

$$\tau_1 = \bigvee, \qquad \tau_2 = \bullet \qquad \Rightarrow \qquad \tau = \bigvee.$$

Next, we construct $\mathbb{T}_3$:

$$\tau_1 = \bullet, \qquad \tau_2 = \bigvee \qquad \Rightarrow \qquad \tau = \bigvee,$$

$$\tau_1 = \bullet, \qquad \tau_2 = \bigvee \qquad \Rightarrow \qquad \tau = \bigvee,$$

$$\tau_1 = \bigvee, \qquad \tau_2 = \bigvee \qquad \Rightarrow \qquad \tau = \bigvee,$$

$$\tau_1 = \bigvee, \qquad \tau_2 = \bullet \qquad \Rightarrow \qquad \tau = \bigvee,$$

$$\tau_1 = \bigvee, \qquad \tau_2 = \bullet \qquad \Rightarrow \qquad \tau = \bigvee.$$

The principle should be quite clear by now, as should the correspondence between trees and expansion terms which, indeed, can be 'read' directly from the graph. Thus, the last tree can be 'recited' as 'the integral of A, commuted with A, integrated, commuted with A, integrated and commuted with A'.

We now have a recursive algorithm to derive the $\mathcal{C}_\tau$s but recall that, to complete our description of (4.5), we also require the constants $\alpha(\tau)$. Fortunately, this too can be deduced from the tree formalism. We thus commence by letting $\alpha(\bullet) = 1$ and continue by observing that each tree in $\cup_{k\in\mathbb{N}}\mathbb{T}_k$ can be written in a unique form as

$$\tau = \hspace{6cm} (4.8)$$

for some $s \geq 1$. We can assume by induction that $\alpha(\tau_i)$ are already known for $i = 1, 2, \ldots, s$. In that case it has been proved in Iserles and Nørsett (1999) that

$$\alpha(\tau) = \frac{\mathrm{B}_s}{s!} \prod_{i=1}^{s} \alpha(\tau_i). \qquad (4.9)$$

Note that $\mathrm{B}_{2m+1} = 0$ for $m \geq 1$. This implies that many (although, unfortunately, not too many ...) terms in (4.5) vanish.

We can now write the first terms of the Magnus expansion in a tree formalism, using the convention that linear combination of trees corresponds to a linear combination of expansion terms,

$$\Theta(t) \;=\; \bullet \;-\; \tfrac{1}{2}\; \mathsf{Y} \;+\; \tfrac{1}{12}\; \mathsf{Y} \;+\; \tfrac{1}{4}\; \mathsf{Y} \;-\; \tfrac{1}{8}\; \mathsf{Y}$$

$$(4.10)$$

$$-\; \tfrac{1}{24}\; \mathsf{Y} \;-\; \tfrac{1}{24}\; \mathsf{Y} \;-\; \tfrac{1}{24}\; \mathsf{Y} \;+\; \cdots .$$

Note that the coefficient corresponding to the last tree in $\mathbb{T}_3$ vanishes, this being a consequence of (4.9) and of $\mathrm{B}_3 = 0$.

It is relatively easy to continue the expansion to higher-order terms. Moreover, it is easy to prove that

$$\tau \in \mathbb{T}_k \quad \Rightarrow \quad \int_0^t \mathcal{C}_\tau(\xi)\,\mathrm{d}\xi = \mathcal{O}\left(t^{k+1}\right), \qquad k \in \mathbb{Z}^+ \qquad (4.11)$$

for every sufficiently smooth matrix function A. On the face of it, this gives a handy device to truncate the Magnus expansion to obtain an approximant of given order – we will bother later about the calculation of multivariate integrals. This, however, is grossly misleading, since the naive estimate (4.11) can be improved a very great deal: far fewer terms are required!

4.2. Convergence of the Magnus expansion

Before we proceed to improve the estimate (4.11) and even consider the question of designing a realistic numerical algorithms based on the Magnus expansion, we need to examine the issue of convergence.

It has been proved in Iserles and Nørsett (1999) that convergence takes place for sufficiently small $t \geq 0$, but the result was unduly pessimistic. A considerably better (and in a well-defined sense optimal) estimate has been obtained by Blanes, Casas, Oteo and Ros (1998). Herewith we present briefly a short and elegant proof due to Moan (1998).

Theorem 4.1. Suppose that the Lie algebra $\mathfrak{g}$ is equipped with the norm $\| \cdot \|$. The Magnus expansion (4.4) absolutely converges in this norm for every $t \geq 0$ such that

$$\int_0^t \|A(\xi)\|\,\mathrm{d}\xi \leq \int_0^{2\pi} \frac{\mathrm{d}\xi}{4 + \xi[1 - \cot(\xi/2)]} \approx 1.086868702. \qquad (4.12)$$

Proof. Integrating (4.3) and taking norms, we have by the triangle inequality and the trivial bound $\|\mathrm{ad}_B^k A\| \leq (2\|B\|)^k\|A\|$ that

$$\|\Theta(t)\| = \left\| \int_0^t \mathrm{dexp}_{\Theta(\xi)}^{-1} A(\xi)\,\mathrm{d}\xi \right\| \leq \int_0^t \|\mathrm{dexp}_{\Theta(\xi)}^{-1} A(\xi)\|\,\mathrm{d}\xi$$

$$\leq \int_0^t \sum_{k=0}^\infty \frac{|B_k|}{k!}(2\|\Theta(\xi)\|)^k\|A(\xi)\|\,\mathrm{d}\xi = \int_0^t g(2\|\Theta(\xi)\|)\|A(\xi)\|\,\mathrm{d}\xi,$$

where

$$g(x) = 2 + \frac{x}{2}\left(1 - \cot\frac{x}{2}\right).$$

We now use a Bihari-type inequality from Moan (1998): suppose that $h, g, v \in C(0, t^*)$ are positive and that g is nondecreasing. Then

$$h(t) \leq \int_0^t g(h(\xi))v(\xi)\,\mathrm{d}\xi, \qquad t \in (0, t^*)$$

implies that

$$h(t) \leq \tilde{g}^{-1}\left(\int_0^t v(\xi)\,\mathrm{d}\xi\right), \quad t \in (0,t^{**}), \qquad \text{where} \qquad \tilde{g}(x) = \int_0^x \frac{\mathrm{d}\xi}{g(\xi)}$$

and $t^{**} \in (0,t^*]$ is such that $\tilde{g}\left(\int_0^t v(\xi)\,\mathrm{d}\xi\right)$ is bounded in $(0,t^{**})$. In our case $h(t) = 2\|\Theta(t)\|$, $v(t) = \|A(t)\|$ and $g(t)$ are all positive and the latter is nondecreasing for $t \in (0,2\pi)$. Therefore,

$$\|\Theta(t)\| \leq \tfrac{1}{2}\tilde{g}^{-1}\left(\int_0^t \|A(\xi)\|\,\mathrm{d}\xi\right)$$

and $\|\Theta(t)\|$ is bounded, provided that $\tilde{g}\left(\int_0^t \|A(\xi)\|\,\mathrm{d}\xi\right)$ is bounded. The latter holds as long as condition (4.12) is satisfied. $\qquad\square$

The condition of Theorem 4.1 has recently been improved for a more relaxed convergence framework. Moan (2000) proved that the Magnus expansion *converges in norm* for all $t \in (0,t^*)$ with regard to the Euclidean norm,

$$\lim_{m\to\infty}\left\|\Theta(t) - \sum_{k=0}^m H_m(t)\right\|_2 = 0,$$

provided that

$$\int_0^{t^*}\|A(\xi)\|_2\,\mathrm{d}\xi < \pi. \tag{4.13}$$

Magnus expansions cannot be expected to converge *always*. In its numerical implementation, this means that the expansion (like any other numerical method for ODEs) needs to be advanced in a time-stepping fashion, rather than being applied globally. Yet, the condition of Theorem 4.1 is not unduly restrictive and (4.13) is even less so. It might be problematic for stiff systems, a subject that has not received much attention in the study of Lie-group methods. An important exception is the use of Magnus expansions in the calculation of Sturm–Liouville spectra, where an elegant device allows us to integrate the equations way beyond the formal upper bound (4.12) (*cf.* Section 11.2 and Moan (1998)).

4.3. Power of a tree and time symmetry

Following Iserles, Nørsett and Rasmussen (1998), we say that a tree $\tau \in \cup_{k\in\mathbb{Z}^+}\mathbb{T}_k$ is of *power* m if $m \geq 0$ is the smallest integer such that

$$C_\tau(t) = \mathcal{O}(t^m)$$

for all sufficiently smooth matrix functions A. Letting $\mathbb{F}_m$ be the set of all trees of power m and truncating the Magnus expansion,

$$\Theta_p(t) = \sum_{m=0}^{p-1} \sum_{\tau \in \mathbb{F}_m} \alpha(\tau) \int_0^t \mathcal{C}_\tau(\xi) \, \mathrm{d}\xi, \qquad (4.14)$$

it is trivial to verify that $\Theta_p(t) = \Theta(t) + \mathcal{O}(t^{p+1})$ and we have an order-p approximant.

We already know from (4.11) that $\tau \in \mathbb{T}_k$ implies $\tau \in \mathbb{F}_m$ for some $m \geq k$. How large can m get, though? The main mechanism that increases order is commutation. Thus, suppose that

$$\mathcal{C}_{\tau_i}(t) = D_i t^{m_i} + E_i t^{m_i+1} + F_i t^{m_i+2} + \cdots, \qquad i = 1, 2.$$

Then

$$
\begin{aligned}
\left[\int_0^t \mathcal{C}_{\tau_1}(\xi) \, \mathrm{d}\xi, \mathcal{C}_{\tau_2}(t) \right] &= \frac{1}{m_1+1}[D_1, D_2] t^{m_1+m_2+1} \\
&+ \left(\frac{1}{m_1+1}[D_1, E_2] + \frac{1}{m_1+2}[E_1, D_2] \right) t^{m_1+m_2+2} \\
&+ \left(\frac{1}{m_1+1}[D_1, F_2] + \frac{1}{m_1+2}[E_1, E_2] + \frac{1}{m_1+3}[F_1, D_2] \right) t^{m_1+m_2+3} \\
&+ \cdots.
\end{aligned}
$$

In general, $[D_1, D_2] \neq O$, hence the new term resides in $\mathbb{F}_{m_1+m_2+1}$. However, cancellation takes place in the important special case $\tau_1 = \tau_2$, whence

$$
\begin{aligned}
\left[\int_0^t \mathcal{C}_{\tau_1}(\xi) \, \mathrm{d}\xi, \mathcal{C}_{\tau_1}(t) \right] &= \frac{1}{(m_1+1)(m_1+2)}[D_1, E_1] t^{2m_1+2} \\
&+ \frac{2}{(m_1+1)(m_1+3)}[D_1, F_1] t^{2m_1+3} + \cdots
\end{aligned}
$$

and the new term is in $\mathbb{F}_{2m_1+2}$, a gain of one unit.

While $\mathbb{F}_0 = \{\bullet\}$, we observe that $\mathbb{F}_1 = \emptyset$ and

$$\mathbb{F}_4 \;=\; \left\{ \;\cdot\cdot\cdot\cdot\; , \;\cdot\cdot\cdot\cdot\; , \;\cdot\cdot\cdot\cdot\; , \;\cdot\cdot\cdot\cdot\; \right\}.$$

In general, the number of terms counted according to power is significantly smaller than similar enumeration by the number of commutators. It is possible to prove that

$$\limsup_{k\to\infty}(\#\mathbb{T}_k)^{1/k} = 4, \qquad \limsup_{m\to\infty}(\#\mathbb{F}_m)^{1/m} \approx 3.11674 \qquad (4.15)$$

(Iserles and Nørsett 1999, Iserles et al. 1998). As an example of the reduction in cardinality, compare $\#\mathbb{T}_6 = 132$ with $\#\mathbb{F}_6 = 21$.

Using the 'truncation by power' (4.14) is thus aptly justified. However, before we rush to pronounce this as the 'correct' truncated Magnus expansion, we need to pay attention to yet another device that reduces the number of terms in the expansion.

Let Φ_t be the *flow* corresponding to the differential equation (4.2), $\Phi_t(Y_0) = Y(t)$. It is obvious that the flow is *time-symmetric*, $\Phi_{-t} \circ \Phi_t = \mathrm{Id}$, since integrating from 0 to t and back to 0 returns us to the original initial value. What is far less obvious, yet has been proved in Iserles et al. (1998), is that the truncation by power (4.14) respects time symmetry. In other words, let

$$\tilde{\Phi}_t(Y_0) = \mathrm{e}^{\Theta_p(t)}Y_0, \qquad t \geq 0.$$

Then $\tilde{\Phi}_{-t} \circ \tilde{\Phi}_t = \mathrm{Id}$. This is remarkable, since any analytic time-symmetric map S_t can be represented in the form $S_t = \mathrm{e}^{F_t}$ where the map F_t is expandable in *odd powers of t only* (Hairer et al. 1993). This fits our framework perfectly.

Theorem 4.2. The function Θ_p can be expanded in odd powers of t and

$$\Theta_{2q-1}(t) = \Theta(t) + \mathcal{O}\!\left(t^{2q+1}\right), \qquad q \in \mathbb{N}.$$

Therefore, truncating by power with odd p leads to a gain of an extra unit of order! This is a critical observation which leads to substantial savings in high-order Magnus expansions.

It is important to realize that it is not true that individual elements in H_{2q-1}, $q \geq 2$, are $\mathcal{O}(t^{2q+1})$: it is their linear combination that knocks out the $\mathcal{O}(t^{2q})$ term!

We may now re-examine the expansion (4.10), truncating by power and identifying the order. Reverting from trees to standard notation, we have

$$\Theta(t) \;=\; \int_0^t A(\xi)\,\mathrm{d}\xi \quad \dots\dots\dots\dots\dots\dots\dots\dots\dots\dots\text{order 2}\;(4.16)$$

$$-\tfrac{1}{2}\int_0^t\!\!\int_0^{\xi_1}[A(\xi_2),A(\xi_1)]\,\mathrm{d}\boldsymbol{\xi}\quad\dots\dots\dots\dots\dots\dots\text{order 4}\;(4.17)$$

$$+\tfrac{1}{12}\int_0^t\!\!\int_0^{\xi_1}\!\!\int_0^{\xi_1}[A(\xi_2),[A(\xi_3),A(\xi_1)]]\,\mathrm{d}\boldsymbol{\xi}$$

$$+\tfrac{1}{4}\int_0^t\!\!\int_0^{\xi_1}\!\!\int_0^{\xi_2}[[A(\xi_3),A(\xi_2)],A(\xi_1)]\,\mathrm{d}\boldsymbol{\xi}$$

$$-\tfrac{1}{24}\int_0^t\!\!\int_0^{\xi_1}\!\!\int_0^{\xi_1}\!\!\int_0^{\xi_3}[A(\xi_2),[[A(\xi_4),A(\xi_3)],A(\xi_1)]]\,\mathrm{d}\boldsymbol{\xi}$$

$$-\tfrac{1}{24}\int_0^t\!\!\int_0^{\xi_1}\!\!\int_0^{\xi_2}\!\!\int_0^{\xi_2}[[A(\xi_3),[A(\xi_4),A(\xi_2)]],A(\xi_1)]\,\mathrm{d}\boldsymbol{\xi}$$

$$-\tfrac{1}{8}\int_0^t\!\!\int_0^{\xi_1}\!\!\int_0^{\xi_2}\!\!\int_0^{\xi_3}[[[A(\xi_4),A(\xi_3)],A(\xi_2)],A(\xi_1)]\,\mathrm{d}\boldsymbol{\xi}\quad\text{order 6}\;(4.18)$$

$$+\cdots.$$

Very often entries of A and its commutators can be integrated explicitly, for instance when they are polynomials or trigonometric functions. In that case the truncated Magnus expansions (4.16)–(4.18), say, can be computed explicitly. However, a more comprehensive numerical approach to Magnus expansions requires the computation of multivariate integrals. Although at first glance this may appear to be a very formidable task, it turns out that the special structure of 'Magnus integrals' renders them amenable to very effective and affordable numerical treatment. We defer the discussion of this issue to Section 5.

4.4. Fer expansions

At the first instance, we wish to represent the solution of (4.2) in the form

$$Y(t) = \mathrm{e}^{\int_0^t A(\xi)\,\mathrm{d}\xi}V(t), \qquad t \geq 0. \tag{4.19}$$

Direct differentiation yields

$$V' = \left[\frac{\mathrm{d}}{\mathrm{d}t}\mathrm{e}^{-\int_0^t A(\xi)\,\mathrm{d}\xi}\right]Y + \mathrm{e}^{-\int_0^t A(\xi)\,\mathrm{d}\xi}Y'$$

$$= \left[\frac{\mathrm{d}}{\mathrm{d}t}\mathrm{e}^{-\int_0^t A(\xi)\,\mathrm{d}\xi}\right]\mathrm{e}^{\int_0^t A(\xi)\,\mathrm{d}\xi}V + \mathrm{e}^{-\int_0^t A(\xi)\,\mathrm{d}\xi}A\mathrm{e}^{\int_0^t A(\xi)\,\mathrm{d}\xi}V$$

$$= \left[\mathrm{Ad}_{\mathrm{expm}(-\int_0^t A(\xi)\,\mathrm{d}\xi)}A(t) - \mathrm{dexp}_{\mathrm{expm}(-\int_0^t A(\xi)\,\mathrm{d}\xi)}A(t)\right]V.$$

Recalling from (2.24) and (2.43) that

$$\text{Ad}_{\text{expm}(E)} D = e^{\text{ad}_E} D,$$

$$\text{dexp}_E D = \frac{e^{\text{ad}_E} - I}{\text{ad}_E} D,$$

we deduce that the correction term V itself obeys a linear differential equation,

$$V' = \left[\frac{(I + \text{ad}_E) \, e^{-\text{ad}_E} - I}{\text{ad}_E} D \right] V, \quad t \geq 0, \qquad V(0) = Y_0,$$

where

$$D = A(t), \qquad E = \int_0^t A(\xi) \, d\xi.$$

This is indeed the main step in constructing the *Fer expansion*: the correction V in (4.19) satisfies the equation

$$V' = \left[\sum_{k=1}^{\infty} (-1)^k \frac{k}{(k+1)!} \text{ad}^k_{\int_0^t A(\xi) \, d\xi} A(t) \right] V, \quad t \geq 0, \qquad V(0) = Y_0. \quad (4.20)$$

The idea is now to iterate (4.20). Thus, we let $B_0 = A$ and generate the sequence $\{B_m\}_{m \in \mathbb{Z}^+}$ recursively, where

$$B_m(t) = \sum_{k=1}^{\infty} (-1)^k \frac{k}{(k+1)!} \text{ad}^k_{\int_0^t B_{m-1}(\xi) \, d\xi} B_{m-1}(t), \qquad t \geq 0, \, m \in \mathbb{N}.$$

$$(4.21)$$

The *Fer expansion* of the solution of (4.2) is

$$Y(t) = e^{\int_0^t B_0(\xi) \, d\xi} e^{\int_0^t B_1(\xi) \, d\xi} e^{\int_0^t B_2(\xi) \, d\xi} \cdots Y_0, \qquad t \geq 0. \quad (4.22)$$

This expansion was introduced by Fer (1958) who, remarkably, did not recognize that it respects Lie-group structure. It was rediscovered by Iserles (1984) in a numerical context but, again, Lie groups were not mentioned. Finally, Zanna (1996) recognized (4.22) as a Lie-group solver. In Zanna and Munthe-Kaas (1997) it is shown that this expansion can be understood as a version of so-called Lie reduction. From this it follows that if $\mathfrak{g}$ is *solvable* (*cf.* Section 6.5) then the expansion in (4.2) is exact for a *finite* product of exponentials.

The first step in a numerical implementation of the Fer expansion (4.22) is truncation of the infinite product. To this end it is vital to recognize the rate of decay of the matrices B_m. Assuming that $B_m(t) = \mathcal{O}(t^{p_m})$, $m \in \mathbb{Z}^+$, it is easy to verify from (4.21) that $p_m = 2p_{m-1} + 2$. This, in tandem with

$p_0 = 0$, yields $p_m = 2^{m+1} - 2$ and we deduce that

$$e^{\int_0^t B_0(\xi)\,d\xi} e^{\int_0^t B_1(\xi)\,d\xi} \ldots e^{\int_0^t B_{s-1}(\xi)\,d\xi} Y_0 = Y(t) + \mathcal{O}\big(t^{2^{s+1}-1}\big). \qquad (4.23)$$

Thus, we obtain an approximant of order $2^{s+1} - 2$: the order grows exponentially with s (Iserles 1984)!

Of course, if we are interested in an order-p Fer approximant to the solution of (4.2), there is no need to carry out the summation in (4.21) *ad infinitum*. Systematic analysis of order conditions necessary for the formation of Fer approximants of various orders has been carried out by Zanna (1998) using the same binary rooted trees that we have already encountered in our analysis of the Magnus expansion. Identifying A with the single-vertex tree, we have the following explicit form of a Fer approximant of any given order.

Order 2: $s = 1$,

$\quad B_0(t):\quad \bullet$.

Order 3: $s = 2$,

$\quad B_0(t):\quad \bullet$,

$\quad B_1(t):\quad \frac{1}{2}\;$ [tree diagram] .

Order 4: $s = 2$,

$\quad B_0(t):\quad \bullet$,

$\quad B_1(t):\quad \frac{1}{2}\;$ [tree diagram] $+\,\frac{1}{3}\;$ [tree diagram] .

Order 5: $s = 2$,

$\quad B_0(t):\quad \bullet$,

$\quad B_1(t):\quad \frac{1}{2}\;$ [tree diagram] $+\,\frac{1}{3}\;$ [tree diagram] $+\,\frac{1}{8}\;$ [tree diagram] .

Order 6: $s = 2$,

$\quad B_0(t):\quad \bullet$,

$\quad B_1(t):\quad \frac{1}{2}\;$ [tree diagram] $+\,\frac{1}{3}\;$ [tree diagram] $+\,\frac{1}{8}\;$ [tree diagram] $+\,\frac{1}{30}\;$ [tree diagram] .

Order 7: $s = 3$,

$B_0(t)$:   •,

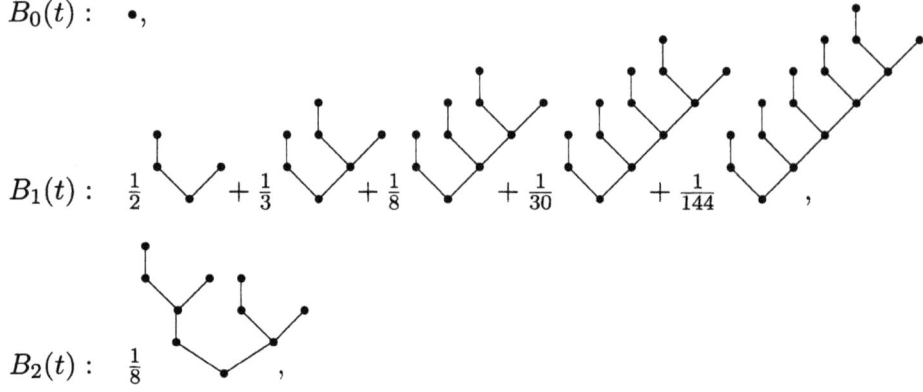

$B_1(t)$: $\frac{1}{2}$ $\bigvee$ $+ \frac{1}{3}$ $\bigvee$ $+ \frac{1}{8}$ $\bigvee$ $+ \frac{1}{30}$ $\bigvee$ $+ \frac{1}{144}$ $\bigvee$,

$B_2(t)$: $\frac{1}{8}$,

and so on. Note that only a subset of 'Magnus trees' occurs in the expansion. This representation of expansion terms as linear combination of trees is central to the application of the multivariate quadrature algorithms of the next section and derivation of practical numerical methods for the Fer expansion.

5. Quadrature and graded algebras

5.1. *Multivariate quadrature over polytopes*

Casting our eyes again over the Magnus method (4.16)–(4.18) and considering its numerical implementation, let us discuss the computation of the first four integrals, noting that each needs to be carried out over a different polytope, as follows.

$$I_1(t) = \int_0^t A(\xi)\, d\xi \qquad\qquad \text{over the line segment}$$

$$I_2(t) = \int_0^t \int_0^{\xi_1} [A(\xi_2), A(\xi_1)]\, d\boldsymbol{\xi} \qquad \text{over the triangle}$$

$$I_3(t) = \int_0^t \int_0^{\xi_1} \int_0^{\xi_1} [A(\xi_2), [A(\xi_3), A(\xi_1)]]\, d\boldsymbol{\xi} \quad \text{over the prism}$$

$$I_4(t) = \int_0^t \int_0^{\xi_1} \int_0^{\xi_2} [[A(\xi_3), A(\xi_2)], A(\xi_1)]\, d\boldsymbol{\xi} \quad \text{over the pyramid}$$

Unless we can replace all these integrals by affordable and accurate quadrature, the Magnus method (and by the same token the Fer method) of non-trivial order is of little but theoretical value. Yet, multivariate quadrature is notoriously expensive in terms of function evaluations (Cools 1997). Fortunately, the special nature of integrals occurring within the context of Magnus and Fer expansions renders them particularly suitable for numerical quadrature with a remarkably small number of function evaluations (Iserles and Nørsett 1999).

We commence by observing that, time-stepping with step $h > 0$, each Magnus or Fer expansion term is of the form

$$I(h) = \int_{\mathcal{S}} \boldsymbol{L}(A(\xi_1), A(\xi_2), \dots, A(\xi_s)) \, \mathrm{d}\boldsymbol{\xi}, \tag{5.1}$$

where $\boldsymbol{L}$ is a *multilinear form*, while $\mathcal{S}$ is a polytope of a special form,

$$\mathcal{S} = \{\boldsymbol{\xi} \in \mathbb{R}^s : \xi_1 \in [0, h], \ \xi_l \in [0, \xi_{m_l}], \ l = 2, 3, \dots, s\},$$

where $m_l \in \{1, 2, \dots, l-1\}$, $l = 2, 3, \dots, s$. Thus, for example,

$$\int_0^t \int_0^{\xi_1} \int_0^{\xi_1} [A(\xi_2), [A(\xi_3), A(\xi_1)]] \, \mathrm{d}\boldsymbol{\xi}$$

$$\text{and} \quad \int_0^t \int_0^{\xi_1} \int_0^{\xi_2} [[A(\xi_3), A(\xi_2)], A(\xi_1)] \, \mathrm{d}\boldsymbol{\xi}$$

yield $s = 3$ and

$$\boldsymbol{L}(E_1, E_2, E_3) = [E_2, [E_3, E_1]], \qquad m_2 = 1, \ m_3 = 1$$

and

$$\boldsymbol{L}(E_1, E_2, E_3) = [[E_3, E_2], E_1], \qquad m_2 = 1, \ m_3 = 2,$$

respectively.

Following Iserles and Nørsett (1999), we propose to discretize $I(h)$ as follows. Choose ν distinct quadrature points, $c_1, c_2, \dots, c_\nu \in [0, 1]$, evaluate $A_k = hA(c_k h)$, $k = 1, 2, \dots, \nu$ and form the quadrature

$$K(h) = \sum_{\boldsymbol{k} \in C_s^\nu} b_{\boldsymbol{k}} \boldsymbol{L}(A_{k_1}, A_{k_2}, \dots, A_{k_s}), \tag{5.2}$$

where C_s^ν is the set of all combinations of length s from the set $\{1, 2, \dots, \nu\}$. The weights are

$$b_{\boldsymbol{k}} = \int_{\tilde{\mathcal{S}}} \prod_{i=1}^s \ell_{k_i}(\xi_i) \, \mathrm{d}\boldsymbol{\xi}, \tag{5.3}$$

where

$$\tilde{\mathcal{S}} = \{\boldsymbol{\xi} \in \mathbb{R}^s : \xi_1 \in [0, 1], \ \xi_l \in [0, \xi_{m_l}], \ l = 2, 3, \dots, s\}$$

is the polytope $\mathcal{S}$ scaled to the unit cube and

$$\ell_j(x) = \prod_{\substack{i=1 \\ i \neq j}}^{\nu} \frac{x - c_j}{c_i - c_j}, \qquad j = 1, 2, \ldots, \nu,$$

are the familiar *cardinal polynomials of Lagrange's interpolation*. Note that (5.3) are *interpolatory weights*: they follow naturally by substituting the interpolation polynomial at the quadrature points,

$$\tilde{A}(t) = h^{-1} \sum_{k=1}^{\nu} \ell_k \left(\frac{t}{h} \right) A_k \tag{5.4}$$

in place of $A(t)$ in (5.1) and carrying out the integration explicitly. Conceptually, (5.2) recycles ν function values at all possible combinations at the s 'slots' of the multilinear function $\boldsymbol{L}$.

How well does the quadrature (5.2) approximate the integral (5.1)? The answer is straightforward in the case $s = 1$, $\boldsymbol{L}(E) = E$, since we recover standard univariate interpolatory quadrature, which is of order $\nu + m$, where $m \geq 0$ is the largest integer so that

$$\int_0^1 \xi^{i-1} c(\xi) \, d\xi = 0, \quad i = 1, 2, \ldots, m \qquad \text{where} \quad c(t) = \prod_{k=1}^{\nu} (t - c_k) \tag{5.5}$$

is the *collocation polynomial*.

Theorem 5.1. The orthogonality condition (5.5) implies that the quadrature rule (5.2) is of order $\nu + m$ for all polytopes $\mathcal{S}$ and all multilinear forms $\boldsymbol{L}$. In particular, if $c_1, c_2, \ldots, c_\nu$ are the roots of the *Legendre polynomial* P_ν, shifted to the interval $[0, 1]$ (*Gauss–Legendre points*), then the quadrature is of order 2ν.

We do not propose to present the proof from Iserles and Nørsett (1999), which is long, technical and not particularly illuminating. Later, in the context of Magnus methods for nonlinear Lie-group equations, we describe a much clearer argument due to Zanna (1999) that explains why a suitable linear combination of quadratures (5.2) approximates the truncated Magnus expansion to the order reported in Theorem 5.1. Instead, we present an example, the quadrature of the four integrals that have opened this subsection using (5.2) at three Gauss–Legendre points.

Letting $\nu = 3$, hence order six, we have

$$c_1 = \tfrac{1}{2} - \tfrac{\sqrt{15}}{10}, \qquad c_2 = \tfrac{1}{2}, \qquad c_3 = \tfrac{1}{2} + \tfrac{\sqrt{15}}{10},$$

whence

$$
\ell_1(x) = \frac{5 + \sqrt{15}}{6} - \frac{10 + \sqrt{15}}{3}x + \frac{10}{3}x^2,
$$

$$
\ell_2(x) = -\frac{2}{3} + \frac{20}{3}x - \frac{20}{3}x^2,
$$

$$
\ell_3(x) = \frac{5 - \sqrt{15}}{6} - \frac{10 - \sqrt{15}}{3}x + \frac{10}{3}x^2.
$$

In so far as the univariate integral I_1 is concerned, we have the familiar Gauss–Legendre quadrature,

$$
I_1(h) \approx \tfrac{1}{18}(5A_1 + 8A_2 + 5A_3),
$$

while, after taking into account skew symmetry of commutators, the quadrature of the planar integral is

$$
I_2(h) \approx \tfrac{\sqrt{15}}{54}(2[A_1, A_2] + [A_1, A_3] + 2[A_2, A_3]).
$$

For the two cubic integrals, we are interested (*cf.* (4.18)) in their linear combination which, after a great deal of simplification, yields

$$
\begin{aligned}
\tfrac{1}{12}I_3(h) + \tfrac{1}{4}I_4(h) \approx\ & \tfrac{1}{27216}(94[A_1, [A_1, A_2]] + 45[A_1, [A_1, A_3]] \\
& + 194[A_1, [A_2, A_3]] - 152[A_2, [A_1, A_2]] + 152[A_2, [A_2, A_3]] \quad (5.6) \\
& - 194[A_3, [A_1, A_2]] - 45[A_3, [A_1, A_3]] - 94[A_3, [A_2, A_3]]).
\end{aligned}
$$

Before any attempts are made either to anoint (5.2) as a new wonder-quadrature or to expand efforts to optimize its calculation, we hasten to say that it is suboptimal! Indeed, it is an immediate consequence of Theorem 4.2 that $\tfrac{1}{12}I_3(h) + \tfrac{1}{4}I_4(h) = \mathcal{O}(h^4)$, rather than $\mathcal{O}(h^3)$. Moreover, some commutators can be expressed in terms of other commutators. Although all this can be done on an *ad hoc* basis, it is significantly better and more efficient to understand this phenomenon mathematically and apply the fruits of our understanding not just to truncated Magnus method but also to other Lie-group solvers.

Quadrature (5.2) scores exceedingly well in terms of function evaluations: *the number of function evaluations required to compute all the integrals in a Magnus expansion to requisite order is the same as the cost of the corresponding univariate Gauss–Legendre quadrature!* The trade-off, though, is that this approach requires a very large volume of linear-algebra calculations, since the number of all combinations is inordinately large. Fortunately, the cost of linear algebra can be reduced a very great deal by exploiting the theory of free Lie algebras. This salutary example of a mathematical theory purer than a driven snow finding a very practical application in the design of numerical algorithms is told in the remainder of this section.

5.2. The self-adjoint basis

The first step in the effort to improve the multivariate quadrature formula (5.2) is a change of basis. As suggested first by Munthe-Kaas and Owren (1999), we choose $c_1 < c_2 < \cdots < c_\nu$ *symmetric with respect to* $\frac{1}{2}$ and replace the function values $A_1, A_2, \ldots, A_\nu$ with the solution of the *Vandermonde system*

$$\sum_{l=1}^{\nu} (c_k - \tfrac{1}{2})^{l-1} B_l = A_k, \qquad k = 1, 2, \ldots, \nu. \tag{5.7}$$

We say that $\{B_1, B_2, \ldots, B_\nu\}$ is a *self-adjoint basis* and note for future reference that Gauss–Legendre points in $[0, 1]$, being symmetric with respect to $\frac{1}{2}$, lead to such a basis.

Proposition 5.2.   Given a sufficiently smooth matrix function A, it is true that $B_l = c_l h^l A^{(l-1)}(\frac{1}{2}h) + \mathcal{O}(h^{l+1})$, where $c_l \neq 0$ is a scalar constant, $l = 1, 2, \ldots, \nu$. Moreover, each $h^{-l}B_l$ can be expanded in even powers of h.

It follows from the proposition that the interpolating polynomial can be written in the form

$$\tilde{A}(t) = h^{-1} \sum_{l=0}^{\nu-1} B_{l+1}(\tfrac{t}{h} - \tfrac{1}{2})^l. \tag{5.8}$$

Substituting this into (5.1) allows us to rephrase the quadrature formula (5.2) in a considerably more convenient form,

$$\bar{K}(h) = \sum_{\boldsymbol{l} \in \mathbf{C}_s^\nu} \bar{b}_{\boldsymbol{l}} \boldsymbol{L}(B_{l_1}, B_{l_2}, \ldots, B_{l_s}), \tag{5.9}$$

where

$$\bar{b}_{\boldsymbol{l}} = \int_{\bar{S}} \prod_{i=1}^{s} (\xi_i - \tfrac{1}{2})^{l_i - 1} \, \mathrm{d}\boldsymbol{\xi}. \tag{5.10}$$

Proposition 5.3. (Munthe-Kaas and Owren 1999)   Suppose that a linear combination of integrals $I(h)$ can be expanded in odd powers of h. Then so can the linear combination of quadratures (5.9).

Recall from Section 4.3 that Magnus series, truncated by power, conform with the assumptions of Proposition 5.3: at a stroke, roughly half the commutators go away. Yet, this is but the first of three important savings that are a consequence of the change of a basis.

An element $F \in \mathfrak{g}$ constructed from the basis terms $B_1, B_2, \ldots, B_\nu$ by the standard Lie-algebra operations of linear combination and commutation is said to be of *grade* m if $F = \mathcal{O}(h^m)$ for all sufficiently smooth matrix functions A. This is denoted by $\omega(F) = m$. We note from Proposition 5.2 that $\omega(B_l) = l$, $l = 1, 2, \ldots, \nu$. Moreover, the grade is inherited under commutation, for instance, $\omega([B_k, B_l]) = \omega(B_k) + \omega(B_l)$ and, with greater generality,

$$\omega(\boldsymbol{L}(B_{l_1}, B_{l_2}, \ldots, B_{l_s})) = |\boldsymbol{l}| = \sum_{i=1}^{\nu} l_i.$$

By the definition of the grade, this is equivalent to

$$\boldsymbol{L}(B_{l_1}, B_{l_2}, \ldots, B_{l_s}) = \mathcal{O}(h^{|\boldsymbol{l}|}).$$

Thus, as long as we are interested in an order-p quadrature, we can discard higher-order terms in (5.10). The outcome is

$$\hat{K}(h) = \sum_{\boldsymbol{l} \in \hat{C}_s^{\nu,p}} \bar{b}_l \boldsymbol{L}(B_{l_1}, B_{l_2}, \ldots, B_{l_s}), \qquad (5.11)$$

where $\hat{C}_s^{\nu,p} \subseteq C_s^{\nu}$ such that

$$\boldsymbol{l} \in \hat{C}_s^{\nu,p} \qquad \Leftrightarrow \qquad |\boldsymbol{l}| \leq p.$$

Let us recall the four integrals from Section 5.1. We presently obtain using (5.11) the following order-6 quadrature formulae using Gauss–Legendre points with $\nu = 3$. In line with (5.7), we let $B_1 = A_2$, $B_2 = \frac{\sqrt{15}}{3}(A_3 - A_1)$ and $B_3 = \frac{10}{3}(A_3 - 2A_2 + A_1)$:

$$I_1(h) \approx B_1 + \tfrac{1}{12}B_3,$$
$$I_2(h) \approx \tfrac{1}{6}[B_2, B_1] - \tfrac{1}{120}[B_3, B_2],$$
$$I_3(h) \approx -\tfrac{1}{8}[B_1, [B_2, B_1]] + \tfrac{1}{80}[B_2, [B_2, B_1]] - \tfrac{1}{120}[B_1, [B_3, B_1]]$$
$$\qquad + \tfrac{1}{480}[B_1, [B_3, B_2]] + \tfrac{1}{240}[B_2, [B_3, B_1]] - \tfrac{7}{480}[B_3, [B_2, B_1]]$$
$$\qquad - \tfrac{1}{160}[B_1, [B_3, B_1]],$$
$$I_4(h) \approx \tfrac{1}{24}[B_1, [B_2, B_1]] + \tfrac{1}{80}[B_2, [B_2, B_1]] - \tfrac{1}{120}[B_1, [B_3, B_1]]$$
$$\qquad - \tfrac{1}{1440}[B_1, [B_3, B_2]] - \tfrac{1}{720}[B_2, [B_3, B_1]] + \tfrac{7}{1440}[B_3, [B_2, B_1]]$$
$$\qquad + \tfrac{1}{480}[B_1, [B_3, B_1]].$$

Moreover, consistently with Proposition 5.3, we have

$$\tfrac{1}{12}I_3(h) + \tfrac{1}{4}I_4(h) \approx \tfrac{1}{240}[B_2, [B_2, B_1]] - \tfrac{1}{360}[B_1, [B_3, B_1]].$$

Just two terms survive!

Taken together, throwing away terms of high enough grade and the removal of terms with even $|\boldsymbol{l}|$ after summation removes a high proportion of commutators. Having said this, the most important feature of the self-adjoint basis that allows us to reduce the computational cost has not been mentioned yet!

We have already exploited skew symmetry of the commutator in the derivation and 'beautification' of our integration formulae. This is a fairly transparent procedure. However, let us recall that the commutator is also subject to the *Jacobi identity* (2.11). This allows for a very powerful mechanism to express commutators as linear combinations of other commutators and leads to results that are of importance not just to Magnus expansions, but also to RK-MK methods and the evaluation of the BCH formula. This is the theme of the next subsection.

5.3. Free Lie algebras

In the previous subsections we have seen how to construct an approximation to the solution Y of the Lie-group differential equation $Y' = A(t)Y$ by means of linear combinations of matrices $G_1, G_2, \ldots, G_\nu \in \mathfrak{g}$, whereby the G_i are either 'samples' of the matrix function $A(t)$, in which case $G_i = hA(c_ih)$, or terms of the self-adjoint basis, in which case the G_is coincide with the matrices B_i of Section 5.2.

It is clear that, if we want to make the best out of the properties of the commutator (skew symmetry and Jacobi identity) it is useful to depart from specific representations (A_is and B_is) and treat the G_is as abstract objects in an abstract algebra $\mathfrak{g}$ that embodies the structure that is common to all Lie algebras but nothing more. This is the main idea behind *free Lie algebras*, a formalization of a Lie algebra whose terms can be generated by means of brackets of pairwise elements and such that there are no reducing mechanisms other than skew symmetry and the Jacobi identity of the commutator.

More precisely, the following definition formalizes the concept of free Lie algebras presented above (Munthe-Kaas and Owren 1999).

Definition 5.1.   Let I be a set of indices, either finite or countable. A Lie algebra $\mathfrak{g}$ is *free* over the set I if:

   (i) for every index $i \in I$ there exists $G_i \in \mathfrak{g}$;
   (ii) for any Lie algebra $\mathfrak{h}$ and any function $i \in I \mapsto H_i \in \mathfrak{h}$ there exists a unique Lie-algebra homomorphism $\pi : \mathfrak{g} \to \mathfrak{h}$ such that $\pi(G_i) = H_i$ for all $i \in I$.

Moreover, $\mathcal{S} = \{G_i\}_{i \in I}$ is called the *set of generators* of the free Lie algebra $\mathfrak{g}$.

In our exposition it is useful to think of a free Lie algebra (FLA) as a linear space and to describe it in terms of a basis. One of the most popular is a *Hall basis* $\mathcal{H}$ that contains the generators, $\mathcal{S} \subseteq \mathcal{H}$ (Bourbaki 1975). All elements of $\mathcal{H}$ are produced by recursive commutation of generators. We can associate a *length* function to each element in the following fashion: $\ell(G_i) = 1$ for $G_i \in \mathcal{S}$ and $\ell(H) = \ell(H_1) + \ell(H_2)$ for all $H \in \mathcal{H} \setminus \mathcal{S}$, where $H = [H_1, H_2]$. Intuitively, we may say that the length of H corresponds to how many commutators of generators are needed to construct H. In other words, the length function merely counts commutators.

The Hall basis $\mathcal{H}$ can be endowed with a total ordering defined recursively on the length ℓ of its elements. In general, we say that $G \prec H$ if $\ell(G) < \ell(H)$. If $\ell(G) = \ell(H)$ then $G \prec H$ if G precedes H in lexicographic order. Moreover, to take into account skew symmetry and the Jacobi identity, we require that

- elements of length two $[G_i, G_j]$ are included in $\mathcal{H}$ if $G_i \prec G_j$;
- elements of length greater or equal to three are included in $\mathcal{H}$ if they are of the form $[H_i, [H_j, H_k]]$, with $H_i, H_j, H_k, [H_j, H_k] \in \mathcal{H}$ and moreover $H_j \preceq H_i \prec [H_j, H_k]$.

An example of the first terms of a Hall basis generated by three elements G_1, G_2, G_3 is given by

$$G_1, \quad G_2, \quad G_3, \quad [G_1, G_2], \quad [G_1, G_3],$$
$$[G_2, G_3], \quad [G_1, [G_1, G_2]], \quad [G_1, [G_1, G_3]], \quad [G_2, [G_1, G_2]], \quad [G_2, [G_1, G_3]],$$
$$[G_2, [G_2, G_3]], \quad [G_3, [G_1, G_2]], \quad [G_3, [G_1, G_3]], \quad [G_3, [G_2, G_3]], \quad \ldots .$$

We shall not go into details of algorithmic construction of the Hall basis, which can be found in Bourbaki (1975), and Munthe-Kaas and Owren (1999). It is interesting, however, to mention how fast the number of elements of the Hall basis grows: assuming that I is finite and consists of ν indices (equivalently, $\mathcal{S}$ consists of ν generators), the linear subspace of terms of length exactly equal to m has dimension

$$\rho_m = \frac{1}{m} \sum_{d|m} \mu(d) \nu^{m/d},$$

the sum being carried over all integers d dividing m, a result known as *Witt's formula*. The function μ is the *Möbius function*, defined as follows: assume that d can be factorized as $d = d_1^{n_1} d_2^{n_2} \cdots d_q^{n_q}$, with each d_i a prime number and $n_i \geq 1$. Then

$$\mu(d) = \begin{cases} 1, & d = 1, \\ (-1)^q, & n_i = 1 \text{ for all } i = 1, 2, \ldots, q, \\ 0, & \text{otherwise.} \end{cases}$$

The number ρ_m grows quite fast, as illustrated in Table 5.1 for $\nu = 3$.

Why is all this relevant to our discussion? Let us represent $G_1 = A_1, G_2 = A_2, \ldots, G_\nu = A_\nu$, where $A_i = hA(c_i h)$, for $i = 1, \ldots, \nu$. Since $A_i = \mathcal{O}(h)$, a term containing exactly m commutators corresponds to a combination that

Table 5.1. *Dimension of linear spaces of words of length equal to m in the Hall basis generated by G_1, G_2 and G_3*

m	1	2	3	4	5	6	7	8	9	10
ρ_m	3	3	8	18	48	116	312	810	2184	5580

is at least of order $\mathcal{O}(h^m)$. Thus, in order to have a numerical approximation of order p, the number of linearly independent terms that we need to take into account is bounded by $\sum_{m=1}^{p} \rho_m$ or, because of Theorem 4.2, by $\sum_{m=1}^{p-1} \rho_m$ if the c_i are symmetrically distributed with respect to $\frac{1}{2}$. For instance, for a sixth-order Gauss–Legendre scheme ($\nu = 3$), this leads to 80 linearly independent terms of the Hall basis, not counting the number of commutators involved!

The growth of ρ_m can be reduced introducing a *grading* ω of the FLA $\mathfrak{g}$, as suggested by Munthe-Kaas and Owren (1999). Assume that

$$\omega(G_i) = \omega_i \in \mathbb{N}, \qquad G_i \in \mathcal{S}, \quad i = 1, 2, \dots, \nu.$$

The grading propagates in a natural manner in the Hall basis $\mathcal{H}$: for all $H \in \mathcal{H}$ of the form $H = [H_1, H_2]$ we let

$$\omega(H) = \omega(H_1) + \omega(H_2).$$

A consequence of the grading is that the Hall basis $\mathcal{H}$ splits into a disjoint union of sets of grade m,

$$\mathcal{H} = \bigcup_{m=1}^{\infty} \mathcal{H}_m, \qquad \mathcal{H}_m = \{H \in \mathcal{H} : \omega(H) = m\},$$

consequently $\mathfrak{g}$ becomes a direct sum of subspaces,

$$\mathfrak{g} = \bigoplus_{m=1}^{\infty} \mathfrak{g}_m, \qquad \mathfrak{g}_m = \operatorname{span} \mathcal{H}_m,$$

and we say that $\mathfrak{g}$ is a *graded FLA algebra*.

The fundamental result on the dimension of $\mathfrak{g}_m$ is due to Munthe-Kaas and Owren (1999).

Theorem 5.4. Let $\mathfrak{g}$ be the graded FLA generated by $\mathcal{S} = \{G_1, \dots, G_\nu\}$, with grades $\omega_1, \dots, \omega_\nu$ respectively. Denote by $\lambda_1, \dots, \lambda_r$ the roots of the rth degree polynomial

$$p(z) = 1 - \sum_{i=1}^{\nu} z^{\omega_i}, \qquad r = \max_{1 \le i \le \nu} \omega_i.$$

Table 5.2. *Dimension of linear spaces of terms of weight exactly equal to m (top) and weight at most m (bottom) in the Hall basis generated by G_1, G_2 and G_3 with weights $1, 2$ and 3, respectively*

m	1	2	3	4	5	6	7	8	9	10
$\bar{\rho}_m$	1	1	2	2	4	5	10	15	26	42
$\sum_{i=1}^{m} \bar{\rho}_i$	1	2	4	6	10	15	25	40	66	108

Then

$$\dim \mathfrak{g}_m = \bar{\rho}_m = \frac{1}{m} \sum_{d|m} \left(\sum_{i=1}^{r} \lambda_i^{m/d} \right) \mu(d). \qquad (5.12)$$

To illustrate the benefits of the grading, Table 5.2 displays $\dim \mathfrak{g}_m$, for $m = 1, 2, \ldots, 10$ for a graded algebra generated by G_1, G_2, G_3 with weights $\omega_1 = 1, \omega_2 = 2, \omega_3 = 3$. Comparison with Table 5.1 reveals a significant reduction in the number of linearly independent terms. The results have been obtained using the MATLAB package *DiffMan*, which will be further discussed in Section 10.2 (Engø et al. 1999).

Linking again with the theory of Sections 5.1–5.2, the case of a graded algebra with generators $G_1, \ldots, G_\nu$ and weights $1, 2, \ldots, \nu$, corresponds to the realization

$$G_i = \frac{h^i}{(i-1)!} A^{(i-1)}(\xi_i), \qquad i = 1, 2, \ldots, \nu,$$

where $A^{(i-1)}(\xi_i)$ is the $(i-1)$th derivative of the function A for some $\xi_i \in (0, h)$. The weight ω_i merely indicates that G_i is an $\mathcal{O}(h^i)$ term. Moreover,

$$\tilde{A}(t) = h^{-1} \sum_{i=1}^{\nu} G_i (\tfrac{t}{h})^i$$

is exactly the collocation polynomial (5.4), the information about the nodes c_i being hidden in the G_is. The equivalence is revealed by means of the Vandermonde transformation

$$A_i = \sum_{j=1}^{\nu} c_j^{j-1} G_j, \qquad i = 1, \ldots, \nu.$$

This procedure applies to all kinds of collocation, whether with symmetric nodes or otherwise. In particular we deduce from Table 5.2 that with three collocation points it is possible to obtain order six with at most 15 terms! A substantial saving, compared with 80 using an ungraded algebra . . . .

To reduce further the dimension of the graded FLA, we exploit the argument of Theorem 5.2, assuming that the collocation points are symmetric in [0,1] with respect to $\frac{1}{2}$. We construct

$$\tilde{A}(t) = h^{-1} \sum_{i=1}^{\nu} G_i(\tfrac{t}{h} - \tfrac{1}{2})^i,$$

where $G_i \equiv B_i$ are the *self-adjoint bases* introduced in Section 5.2. Theorem 5.2 implies that *only terms with odd grades need be considered*! As an example, if $\mathcal{S} = \{G_1, G_2, G_3\}$, with weights $1, 2$ and 3 respectively, then the growth of the dimension of the graded FLA is given by Table 5.3. Comparison with Table 5.2 is remarkable. In particular, we deduce that for methods based on three symmetric collocation points (for instance a sixth-order Gauss–Legendre), we need at most seven terms of the graded Hall basis $\mathcal{H}$. Letting $G_i = B_i$, the seven terms are

$$B_1, \; B_3, \; [B_1, B_2], \; [B_2, B_3], \; [B_1, [B_1, B_3]], \; [B_2, [B_1, B_2]], \; [B_1, [B_1, [B_1, B_2]]].$$

Bounds on the number of independent terms for different orders are given in Table 5.4 for methods based on Gauss–Legendre quadrature. The sharpest bound corresponds to the case of a graded FLA including only odd terms. It is important to note that

$$\sum_{i=1}^{p-1} \bar{\rho}_i \text{ odd} \qquad \left(\sum_{i=1}^{p} \rho_i \text{ respectively} \right)$$

in the case of the self-adjoint (non-self-adjoint respectively) basis is an *upper bound* on the number of linearly independent commutators required for a method of order p.

We have seen the advantage of changing the basis in the case of linear equations $Y' = A(t)Y$. Similar savings can also be achieved in the case of explicit RK-MK methods for the general equation $Y' = A(t, Y)Y$. By combining the stage values A_i computed by the algorithm, we seek linear combinations B_i of highest-possible grade. However, in order to obtain an explicit method, we must require that B_i are related to A_i by a triangular

Table 5.3.

m	1	2	3	4	5	6	7	8	9	10
$\sum_{i=1,\ i\ \text{odd}}^{m} \bar{\rho}_i$	1	1	3	3	7	7	11	11	37	37

matrix. Optimal combinations can be found using linear algebra and the theory of B-series. The details can be found in Munthe-Kaas and Owren (1999). It turns out that in this case it is in general not possible to change basis in such a way that $\omega(B_i) = i$: the weights grow more slowly. The method (A.7) originates in the classical four-stage RK4 scheme. The resulting basis B_i has the grading $1, 2, 3, 3$, and the scheme has just two commutators. Similar savings have been realized for the seven-stage DOPRI5(4) method in Munthe-Kaas and Owren (1999).

5.4. Reducing further the number of commutators

The theory of free Lie algebras allows us to derive an upper bound on the number of linearly independent terms required to obtain numerical methods of given order, taking into account skew symmetry and the Jacobi identity of the commutator. It also provides algorithms to compute the requisite pattern of dependency, for instance in terms of the Hall basis.

Although the theory of free Lie algebras estimates the numbers of commutators for a method of order p, this by no means indicates the *least* number of commutators required for a method of a given order.

For general $N \times N$ matrices, the computation of the commutator is an $\mathcal{O}(N^3)$ operation, a cost that quickly adds up when we consider methods of order three and higher.

At present there is no theory that systematically reduces the number of commutators to minimum. However, we shall present a technique, due to Blanes, Casas and Ros (1999) that gives, true for today, the least number of commutators for methods based on Gauss–Legendre and Newton–Cotes quadratures up to order ten. Before proceeding further, it is important to remark that the content of this section applies to linear Lie-group problems $Y' = A(t)Y$ solved with Magnus-type/RK-MK methods based on symmetric nodes in $[0, 1]$. It is convenient to illustrate the procedure with an example.

Table 5.4.

ν (stages)	1	2	3	4	5
2ν (order of the method)	2	4	6	8	10
$\sum_{i=1}^{2\nu-1} \rho_i$	1	5	80	3304	> 10000
$\sum_{i=1}^{2\nu-1} \bar{\rho}_i$	1	3	10	33	111
$\sum_{i=1,\ i\ \mathrm{odd}}^{2\nu-1} \bar{\rho}_i$	1	2	7	22	73

Let us thus construct a sixth-order Gauss–Legendre method based on the collocation nodes

$$c_1 = \tfrac{1}{2} - \tfrac{\sqrt{15}}{10}, \qquad c_2 = \tfrac{1}{2}, \qquad c_3 = \tfrac{1}{2} + \tfrac{\sqrt{15}}{10}.$$

Using the toolbox of graded algebras, we obtain

$$\Theta = B_1 + \tfrac{1}{12}B_3 - \tfrac{1}{12}[B_1, B_2] + \tfrac{1}{240}[B_2, B_3] + \tfrac{1}{360}[B_1, [B_1, B_3]]$$
$$- \tfrac{1}{240}[B_2, [B_1, B_2]] + \tfrac{1}{720}[B_1, [B_1, [B_1, B_2]]].$$

Since in the self-adjoint basis $B_i = \mathcal{O}(h^i)$, note that Θ includes terms with odd powers of h only and, in this form, can be evaluated by computing just seven commutators.

Let us focus on the portion of Θ consisting of single commutators,

$$C_2 = -\tfrac{1}{12}[B_1, B_2] + \tfrac{1}{240}[B_2, B_3].$$

The first fundamental observation is that terms of the form $[B_i, B_j]$ can only be obtained if one of the indices is even and the other is odd. Therefore, we look for a linear combination

$$[b_1 B_1 + b_3 B_3, b_2 B_2] \tag{5.13}$$

for some real coefficients b_i, that equals C_2. This can be achieved by choosing for instance $b_1 = -\tfrac{1}{12}, b_3 = -\tfrac{1}{240}$ and $b_2 = 1$. Note that computing (5.13) requires one commutator only instead of two. In general, given $B_1, B_2, \dots, B_\nu$, terms of the form

$$\sum_{i=1}^{\nu-1} \sum_{j=i+1}^{\nu} k_{i,j}[B_i, B_j]$$

may be replaced by a *single* commutator,

$$\left[\sum_{i=1}^{\nu/2} b_{2i-1} B_{2i-1}, \sum_{j=1}^{\nu/2} b_{2j} B_{2j} \right],$$

provided that the coefficients $\{b_i\}_{1 \leq i \leq \nu}$ and $\{k_{i,j}\}_{1 \leq i < j \leq \nu}$ are compatible up to the order of the method.

Next, let us consider terms with double commutators. Let us focus on

$$-\tfrac{1}{240}[B_2, [B_1, B_2]].$$

Since $C_2 = -\tfrac{1}{12}[B_1, B_2] + \mathcal{O}(h^5)$, evaluating the term $-\tfrac{1}{20}[B_2, C_2]$ in place of $-\tfrac{1}{240}[B_2, [B_1, B_2]]$ amounts to the calculation of just one commutator. Note that $-\tfrac{1}{20}[B_2, C_2] = -\tfrac{1}{240}[B_2, [B_1, B_2]] + \mathcal{O}(h^7)$, hence the approximation retains the odd power of h expansion and the error introduced is subsumed in the local truncation error of the method.

Finally, let us consider the combination

$$\tfrac{1}{360}[B_1,[B_1,B_3]] + \tfrac{1}{720}[B_1,[B_1,[B_1,B_2]]]. \qquad (5.14)$$

Clearly, $[B_1, B_3]$ is not obtained from (5.13) using the linearity of the bracket, and needs to be taken into account. However, $[B_1, B_2]$ conforms with (5.13) and we can replace it with C_2, introducing only odd-powered error in h, which is also subsumed in the local truncation error. In summary, (5.14) can be replaced by the term

$$C_3 = [B_1,[B_1, \tfrac{1}{360}B_3 - \tfrac{1}{60}C_2]],$$

which requires the computation of two commutators. Therefore, Θ can be computed in the form

$$\Theta = B_1 + \tfrac{1}{12}B_3 + C_2 + C_3,$$

requiring four commutators only.

It is difficult to formalize the last two steps, involving two or more commutators. The reduction in the number of commutators has, in the present state of knowledge, to be done on a case-by-case analysis.

Some examples of methods for linear problems obtained by means of graded algebras, made more economic with the technique of Blanes et al. (1999), are described in Appendix A.2.

5.5. Nonlinear problems: collocation methods

Magnus and Fer expansions presented in Section 4 have been designed for linear problems, when the matrix function A depends on time only. When $A = A(t, Y)$, though, multivariate integrals appearing in (5.1) depend on $A_k = hA(c_k h, Y(c_k h))$, namely also on the value of the unknown variable Y at quadrature points. Assume that the quadrature points $c_1, c_2, \ldots c_\nu$ obey the orthogonality conditions (5.5) for some $m \le \nu$, hence the corresponding univariate interpolatory quadrature has order $p = \nu + m$. Since each A_k is a multiple of h, it is clear that the values A_k can be replaced by $hA(c_k h, X_k)$, where X_k is an approximation to $Y(c_k h)$ of order at least $p - 1$. Following this point of view, it is possible to derive $X_k \approx Y(c_k h)$ with a numerical method of order $p - 1$. For instance, using an RK-MK method of order $p - 1$ and a Magnus (or Fer) expansion of order p is a Lie-group equivalent of a *predictor–corrector* method in classical numerical ODE theory.

A more elegant approach, due to Zanna (1999), is to construct suitable approximants to X_ks using directly the underlying principle of collocation methods. Proceeding as in Section 5.1, we replace the function $A(t, Y)$ by its Lagrangian interpolating polynomial

$$\tilde{A}(t, Y) = h^{-1} \sum_{k=1}^{\nu} \ell_k \left(\frac{t}{h} \right) A_k$$

where $A_k = hA(c_k h, X_k)$, $k = 1, 2, \ldots, \nu$. The corresponding *dexpinv equation* is integrated by means of Picard iterations *à la* Section 4, obtaining the usual order trees for the Magnus and the Fer expansion. The main difference with the linear case is that now the same order trees are also employed to evaluate the integration coefficients for the internal stages X_k. At each internal stage we need to calculate quadratures of the form (5.2),

$$K_l(h) = \sum_{\boldsymbol{j} \in C_s^\nu} a_{l;\boldsymbol{j}} \boldsymbol{L}(A_{j_1}, A_{j_2}, \ldots, A_{j_s}),$$

where C_s^ν is the set of all combinations of length s from the set $\{1, 2, \ldots, \nu\}$. The integration weights are different from those in (5.2) and are defined as

$$a_{k;\boldsymbol{j}} = \int_{\tilde{S}_k} \prod_{i=1}^s \ell_{j_i}(\xi_i)\, \mathrm{d}\boldsymbol{\xi},$$

where

$$\tilde{S}_k = \{\boldsymbol{\xi} \in \mathbb{R}^s : \xi_1 \in [0, c_k], \ \xi_l \in [0, \xi_{m_l}], \ l = 2, 3, \ldots, s\}$$

is the polytope $\tilde{S}$ scaled to the $[0, c_k]$-cube instead of the unit cube. The weights $b_{\boldsymbol{j}}$ are recovered by substituting $c_k = 1$.

Theorem 5.5. Let $c_1, c_2, \ldots, c_\nu$ be ν collocation nodes and let $p = \nu + m$, where m is the largest index such that (5.5) is satisfied. Assume that the Magnus or Fer expansion is truncated to include all trees of power $q \leq p$ for the evaluation of Y_{n+1}, and of power $q \leq p - 1$ for the intermediate stages. Then the resulting scheme has order p.

We sketch the main idea and refer the reader to Zanna (1999) for details. The starting point is the *Alekseev–Gröbner lemma*, a nonlinear version of the variation of constants formula whose proof can be found in Nørsett and Wanner (1981), stating that, if $\boldsymbol{y}$ is the solution of the differential equation $\boldsymbol{y}' = \boldsymbol{f}(t, \boldsymbol{y})$ with initial condition $\boldsymbol{y}_0$, and if $\boldsymbol{w}(t)$ is a C^1 approximation to $\boldsymbol{y}$ such that $\boldsymbol{w}(t_0) = \boldsymbol{y}_0$, then

$$\boldsymbol{y}(t) - \boldsymbol{w}(t) = \int_{t_0}^t \Phi(t, \xi, \boldsymbol{w}(\xi))[\boldsymbol{f}(\xi, \boldsymbol{w}(\xi)) - \boldsymbol{w}'(\xi)]\, \mathrm{d}\xi,$$

where $\Phi(t, \xi, \boldsymbol{w}(\xi))$ is the partial derivative of the solution passing through $(\xi, \boldsymbol{w}(\xi))$ with respect to the initial condition $\boldsymbol{w}(\xi)$. In the usual classical collocation setting for RK methods, $\boldsymbol{w}$ corresponds to the case when the function $\boldsymbol{f}$ is replaced by a collocation polynomial: at the nodes it is true that $\boldsymbol{f}(t_n + c_k h, \boldsymbol{w}(t_n + c_k h)) - \boldsymbol{w}'(t_n + c_k h)) = \boldsymbol{0}$, hence the error reduces solely to quadrature error.

In our Lie-group setting the main difference consists in the fact that only the function $A(t, Y)$ is collocated, and not the whole right-hand side

$A(t,Y)Y$. However, this can be viewed as a collocation method in the algebra $\mathfrak{g}$, where classical analysis remains valid.

A typical example of such collocation methods is the fourth-order scheme

$$
\begin{aligned}
X_1 &= Y_n, \\
A_1 &= hA(t_n, X_1), \\
X_2 &= \operatorname{expm}\{\tfrac{5}{24}A_1 + \tfrac{1}{3}A_2 - \tfrac{1}{24}A_3 - \tfrac{1}{2}(\tfrac{11}{240}[A_1, A_2] + \tfrac{5}{576}[A_1, A_3] \\
&\qquad + \tfrac{1}{72}[A_2, A_3])\}Y_n, \\
A_2 &= hA(t_n + \tfrac{1}{2}h, X_2), \\
X_3 &= \operatorname{expm}\{\tfrac{1}{6}A_1 + \tfrac{2}{3}A_2 + \tfrac{1}{6}A_3 - \tfrac{1}{2}(\tfrac{2}{15}[A_1, A_2] + \tfrac{1}{30}[A_1, A_3] \\
&\qquad + \tfrac{2}{15}[A_2, A_3])\}Y_n, \\
A_3 &= hA(t_n + h, X_3), \\
Y_{n+1} &= \operatorname{expm}\{\tfrac{1}{6}A_1 + \tfrac{2}{3}A_2 + \tfrac{1}{6}A_3 - \tfrac{1}{2}(\tfrac{2}{15}[A_1, A_2] + \tfrac{1}{30}[A_1, A_3] \\
&\qquad + \tfrac{2}{15}[A_2, A_3])\}Y_n
\end{aligned}
$$

for $n \in \mathbb{Z}^+$, with the *Gauss–Lobatto* quadrature points $c_1 = 0$, $c_2 = \tfrac{1}{2}$ and $c_3 = 1$ (Zanna 1998). Note that the coefficients of the A_is are the classical Runge–Kutta coefficients of Lobatto collocation scheme with the same quadrature points (Hairer et al. 1993), while the coefficients of the commutator terms $[A_i, A_j]$ are evaluated by integrating

$$
a_{k;i,j} - a_{k;j,i} = \int_0^{c_k}\int_0^{\xi_1} \ell_i(\xi_1)\ell_j(\xi_2)\,\mathrm{d}\xi_2\,\mathrm{d}\xi_1 - \int_0^{c_k}\int_0^{\xi_1} \ell_j(\xi_1)\ell_i(\xi_2)\,\mathrm{d}\xi_2\,\mathrm{d}\xi_1,
$$

a term corresponding to the power-three tree in the Magnus expansion (*cf.* Section 4).

A useful formula for evaluating the $a_{k;i,j}$s corresponding to the power-three tree is given by

$$
\{a_{k;i,j}\}_{i,j=1}^{\nu} = a(c_k), \qquad k = 1, 2, \dots, \nu, \tag{5.15}
$$

where $a(\theta)$ is the $\nu \times \nu$ matrix function of the scalar argument θ defined as

$$
a(\theta) = V^{-\mathrm{T}}T(\theta)HJT(\theta)V^{-1},
$$

where $J = \operatorname{diag}(1, \tfrac{1}{2}, \tfrac{1}{3}, \dots \tfrac{1}{\nu})$, $T(\theta) = \operatorname{diag}(\theta, \theta^2, \dots, \theta^\nu)$, V is the Vandermonde matrix

$$
V = \begin{bmatrix}
1 & c_1 & \cdots & c_1^{\nu-1} \\
1 & c_2 & \cdots & c_2^{\nu-1} \\
\vdots & \vdots & & \vdots \\
1 & c_\nu & \cdots & c_\nu^{\nu-1}
\end{bmatrix},
$$

and finally H is the Hilbert matrix with entries $H_{i,j} = \frac{1}{i+j}$, $i, j = 1, 2, \ldots, \nu$. This formula is reminiscent of the matrix representation of the standard Runge–Kutta matrix of a collocation method (Nørsett and Wanner 1981). This is not a coincidence: the methods of Zanna (1998) generalize the concept of collocation to the special multivariate integrals that occur in Magnus or Fer expansions.

6. Alternative coordinates

All the methods so far, whether applied to Lie groups or in a homogeneous-space setting, have been based on the exponential map. In other words, we represented the solution as an exponential (RK-MK and Magnus methods) or as a product of exponentials (Crouch–Grossman and Fer methods). It is entirely legitimate to query to which extent this renders such methods unduly expensive and non-competitive.

Sometimes the exact computation of the exponential is easy: a case in point is the application of Magnus expansions to the computation of Sturm–Liouville spectra in Section 11.2, since the exponential of an element in $\mathfrak{sl}(2)$ can be evaluated exactly with great ease. In other cases the exponential can be replaced by a suitable approximation $\phi : \mathfrak{g} \to \mathcal{G}$, $\phi(z) \approx e^z$. This is the case with quadratic Lie groups: a Lie group G is *quadratic* if

$$\mathcal{G} = \{X \in \mathrm{GL}(N) : XPX^{\mathrm{T}} = P\}, \tag{6.1}$$

where $P \in \mathrm{GL}(N)$ is a given matrix. Many Lie groups that appear in applications are of this kind, for example $\mathrm{O}(N)$, $\mathrm{Sp}(N)$ and $\mathrm{O}(N, M)$. Moreover, some complex groups can be brought into this framework by replacing the transpose ^{T} by the Hermitian (*i.e.*, conjugate) transpose ^{H}.

The Lie algebra of the quadratic Lie group (6.1) is

$$\mathfrak{g} = \{B \in \mathfrak{gl}(N) : BP + PB^{\mathrm{T}} = O\} \tag{6.2}$$

and it is easy to prove that ϕ maps $\mathfrak{g}$ into $\mathcal{G}$ whenever $\phi(z) = e^{\gamma(z)}$ and γ is an odd function that is analytic in a neighbourhood of the origin (Celledoni and Iserles 1998). An important case occurs when $\gamma(z) = \log(q(z)/q(-z))$, where q is a polynomial, and it leads to rational functions $\phi(z) = q(z)/q(-z)$. In particular, *diagonal Padé approximants* to the exponential are of this form (Baker 1975) and they can be applied very effectively in place of the exponential.

Yet, for some algebras there exists no analytic nonconstant function $\phi : \mathfrak{g} \to \mathcal{G}$ except for the (scaled) exponential. This, in particular, is the case with $\mathrm{SL}(N)$ (Kang and Zai-jiu 1995). In yet other cases, although we may replace the exponential with, say, a Padé approximant with impunity, the sheer size of the system renders this impractical when the number of variables is large. In that case there exist two possibilities. Firstly, we may endeavour to approximate the exponential of a matrix by some nonstandard means while keeping the outcome in a Lie group. This is the theme of Section 8. In the present section we consider another approach, which disposes of the dexpinv equation (2.46) altogether.

The research into 'alternative coordinates' is in a fairly preliminary stage and just two surrogates to the dexpinv equation have been identified so far, the Cayley transform and canonical coordinates of the second kind. We consider them in detail in the remainder of this section.

6.1. The Cayley transform and RK–Cayley methods

Let $\mathcal{G}$ be a *quadratic* Lie group (Diele, Lopez and Peluso 1998). The main idea is to replace the exponential with the *Cayley transform*. Thus, given the Lie-group equation $Y' = A(t, Y)Y$, where $A : \mathbb{R} \times \mathcal{G} \to \mathfrak{g}$, we seek a solution in the form

$$Y(t) = \operatorname{cay}[\Delta(t)]Y_0 = [I - \tfrac{1}{2}\Delta(t)]^{-1}[I + \tfrac{1}{2}\Delta(t)]Y_0, \qquad t \geq 0, \qquad (6.3)$$

and at the first instance seek a differential equation for Δ. More generally, we may solve the homogeneous-space equation (2.26) replacing the exponential with the Cayley action but, for the sake of simplicity, the discussion is restricted to the 'straight' Lie-group case.

It is important to realize that our approach has no connection whatsoever with approximating the exponential. True, $\phi(z) = (1 + \tfrac{1}{2}z)/(1 - \tfrac{1}{2}z)$ is a special case of a Padé approximant to the exponential, $\phi(z) = \operatorname{expm}(z) + \mathcal{O}(z^3)$, but this is entirely coincidental: as a matter of fact, we could have replaced, at the cost of slightly more complicated coefficients, the number $\tfrac{1}{2}$ with an arbitrary nonzero constant. So far, everything is exact and no approximation has taken place.

It is an easy exercise, left to the reader, to ascertain that the function Δ in (6.3) obeys the differential equation

$$\Delta' = \operatorname{dcay}_\Delta^{-1} A(\operatorname{cay}(\Delta)Y_0, t) \qquad (6.4)$$
$$= A - \tfrac{1}{2}[\Delta, A] - \tfrac{1}{4}\Delta A\Delta, \qquad t \geq 0, \qquad \Delta(0) = \boldsymbol{O}.$$

Note the presence of the term $\Delta A\Delta$ in the above equation. In general, we cannot expect such a term to reside in $\mathfrak{g}$, but quadratic Lie algebras (6.2) are an exception. For future reference we note that the more general *symmetric triple product* $[\![D, E, F]\!]_3 = DEF + FED$ resides in $\mathfrak{g}$ for all $D, E, F \in \mathfrak{g}$ and quadratic Lie groups $\mathfrak{g}$. (The reason for this notation will be clear in Section 6.3.) Applying (6.2) thrice,

$$P(DEF)^{\mathrm{T}} = PF^{\mathrm{T}}E^{\mathrm{T}}D^{\mathrm{T}} = -FPE^{\mathrm{T}}D^{\mathrm{T}} = FEPD^{\mathrm{T}} = -FEDP.$$

Therefore

$$(DEF + FED)P + P(DEF + FED)^{\mathrm{T}}$$
$$= (DEF + FED)P - (FED + DEF)P = O,$$

and indeed $[\![D, E, F]\!]_3 \in \mathfrak{g}$. This confirms that no illicit terms have crept into (6.4) and Δ' evolves in $\mathfrak{g}$. (To be more precise, it evolves in $T\mathfrak{g}$, except that the latter can be identified with $\mathfrak{g}$.)

The simplest implementation of (6.3) to quadratic Lie-group solvers is by employing Runge–Kutta methods in the Lie algebra, *à la* RK-MK, except that expm and dexp^{-1} in (3.4) need to be replaced by cay and dcay^{-1} respectively. This has been accomplished systematically by Engø (2000).

6.2. Cayley expansions

Proceeding as in Section 4.1 and subjecting (6.4) to Picard iteration, we observe that the solution Δ can be expanded in a similar way to the Magnus expansion of the dexpinv equation (4.3),

$$
\begin{aligned}
\Delta(t) \;=\; & \int_0^t A(\xi)\,\mathrm{d}\xi \\
& - \tfrac{1}{2} \int_0^t \int_0^{\xi_1} [A(\xi_2), A(\xi_1)]\,\mathrm{d}\xi \\
& + \tfrac{1}{4} \int_0^t \int_0^{\xi_1} \int_0^{\xi_2} [[A(\xi_3), A(\xi_2)], A(\xi_1)]\,\mathrm{d}\xi_3\,\mathrm{d}\xi_2\,\mathrm{d}\xi_1 \\
& - \tfrac{1}{4} \int_0^t \int_0^{\xi_1} \int_0^{\xi_1} A(\xi_2)A(\xi_1)A(\xi_3)\,\mathrm{d}\xi_3\,\mathrm{d}\xi_2\,\mathrm{d}\xi_1 + \cdots .
\end{aligned}
$$

We seek to expand Δ, in greater generality, in a manner similar to (4.5),

$$\Delta(t) = \sum_{k=0}^{\infty} \sum_{\tau \in \mathbb{S}_k} \alpha(\tau) D_\tau(t), \qquad t \geq 0, \tag{6.5}$$

where each D_τ for $\tau \in \mathbb{S}_k$ is made out of exactly $k+1$ integrals. (Note that, unlike (4.5), the 'exterior' integral is already included in the expansion term – intuitively speaking, $D_\tau(t) = \int_0^t C_\tau(\xi)\,\mathrm{d}\xi$. This makes the notation somewhat simpler and more transparent.) Following the construction of Iserles (1999b), we identify three composition rules that are needed to assemble the terms in (6.5).

(1) $\mathbb{S}_0 = \{\tau_\circ\}$, and

$$D_{\tau_\circ}(t) = \int_0^t A(\xi)\,\mathrm{d}\xi.$$

(2) If $\tau_1 \in \mathbb{S}_{k-1}$, $k \geq 1$, then there exists $\tau \in \mathbb{S}_k$ such that

$$D_\tau(t) = \int_0^t \left[\int_0^t D_{\tau_1}(\xi), A(\xi) \right] d\xi. \qquad (6.6)$$

(3) If $k \geq 2$ and $\tau_1 \in \mathbb{S}_{k-j}$, $\tau_2 \in \mathbb{S}_{j-1}$ for some $1 \leq j \leq k$ then there exists $\tau \in \mathbb{S}_k$ such that

$$D_\tau(t) = \int_0^t D_{\tau_1}(\xi) A(\xi) D_{\tau_2}(\xi) \, d\xi. \qquad (6.7)$$

Note that the outcome resides in the Lie algebra as long as $\alpha(\tilde{\tau}) = \alpha(\tau)$ where τ has been given in (6.7) and

$$D_{\tilde{\tau}}(t) = \int_0^t D_{\tau_2}(\xi) A(\xi) D_{\tau_1}(\xi) \, d\xi$$

is the *conjugate term* of $D_{\tilde{\tau}}(t)$. (The existence of such a term is assured by the third composition rule.)

As for the association between rooted binary trees and terms in the Magnus expansion, we wish to render the structure of the above composition rules clearer by using graph theory. The presence of three, rather then two, composition rules makes this goal different and 'plain' rooted binary trees are no longer adequate for the task in hand. Instead, following Iserles (1999b), we employ rooted *bicolour* binary trees: each vertex can be one of two colours, black or white. The composition rules are interpreted in the following manner, borrowing as much as possible from the construction in Section 4.

(1) $\mathbb{S}_0 = \{ \mathbf{I} \}$ and

$$\mathbf{I} \rightsquigarrow \int_0^t A(\xi) \, d\xi.$$

(2) If $\mathbb{S}_{k-1} \ni \tau_1 \rightsquigarrow D_{\tau_1}(t)$ then (6.6) corresponds to

$$\mathbb{S}_k \ni \rightsquigarrow \int_0^t [D_{\tau_1}(\xi), A(\xi)] \, d\xi.$$

(3) Letting $\mathbb{S}_{k-j} \ni \tau_1 \rightsquigarrow D_{\tau_1}(t)$ and $\mathbb{S}_j \ni \tau_1 \rightsquigarrow D_{\tau_2}(t)$, (6.7) corresponds to

$$\mathbb{S}_k \ni \rightsquigarrow D_\tau(t) = \int_0^t D_{\tau_1}(\xi) A(\xi) D_{\tau_2}(\xi) \, d\xi.$$

Unlike the case of the Magnus expansion, the derivation of the coefficients $\alpha(\tau)$ is straightforward and does not require any recursion. Given $\tau \in \cup \mathbb{S}_k$, we denote by $\gamma(\tau)$ the number of *white* nodes therein. It is possible to prove that $\alpha(\tau) = (-1)^{k+\gamma(\tau)} 2^{-k}$ and the outcome is the *Cayley expansion*

$$\Delta(t) = \sum_{k=0}^{\infty} \frac{(-1)^k}{2^k} \sum_{\tau \in \mathbb{S}_k} (-1)^{\gamma(\tau)} D_\tau(t) \tag{6.8}$$

Note that the last two trees above are conjugate and that they have the same weights. This is true in general, since if τ and $\tilde{\tau}$ are conjugate then they have the same number of white vertices, $\gamma(\tau) = \gamma(\tilde{\tau})$. Therefore, conjugate trees translate into (scaled) symmetric triple products and we stay safely within the Lie algebra.

Absolute convergence of the Cayley expansion (6.8) was proved in Iserles (1999b) for $t \in (0, t^*)$, provided that $\int_0^t \|A(\xi)\| \, d\xi < 2$, a result that can be somewhat improved for certain norms and Lie algebras. However, as far as convergence in norm with respect to $\| \cdot \|_2$ is concerned, the Magnus-expansion condition (4.13) of Moan (2000) remains valid in the present setting.

As in Magnus expansions, it makes sense to truncate the series (6.8) *by power*,

$$\Delta(t) \approx \sum_{m=0}^{p-1} \sum_{\tau \in \mathbb{G}_m} \frac{(-1)^{\beta(\tau)+\gamma(\tau)}}{2^{\beta(\tau)}} D_\tau(t), \tag{6.9}$$

where $\beta(\tau) + 1$ is the number of integrals in D_τ (in other words, $\tau \in \mathbb{S}_{\beta(\tau)}$), while $\mathbb{G}_m$ stands for the set of trees of *power m*,

$$\tau \in \mathbb{G}_m \iff D_\tau(t) = \mathcal{O}(t^{m+1})$$

for all sufficiently smooth matrix functions A. The mechanism that allows

m to exceed $\beta(\tau)$ is subtly different from that of Magnus expansions (4.14) since, except for the second tree in the expansion (6.9), we never encounter an instance of $D_\tau(t) = \int[D_{\tau_1}, D'_{\tau_1}]$. Instead, we say that a tree is *basic* if it has no black nodes with two children (equivalently, if the corresponding expansion term contains no commutators). The first few basic trees are

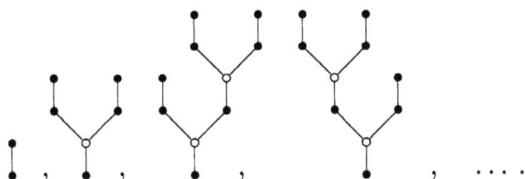

Each basic tree has an even number of vertices and it is easy to verify that if $\tau \in \mathbb{S}_{2m}$ and τ is basic then

$$A(t) = A_0 + \mathcal{O}(t) \qquad \Rightarrow \qquad D_\tau(t) = cA_0^{2m+1}t^{2m+1} = \mathcal{O}\!\left(t^{2m+2}\right)$$

where $c \neq 0$ is scalar. In other words, $\tau \in \mathbb{G}_{2m}$ and nothing is gained. However, as soon as we form the tree

$$\tau = \;\begin{matrix}\tau_1 \\ \vee \\ \bullet\end{matrix}\; ,$$

where τ_1 is basic, it is trivial to notice that $\tau \in \mathbb{G}_{2m+2}$, a 'gain' of one unit in power. Needless to say, this gain is inherited each time τ features as a component of a larger tree.

Truncating by power economizes on the number of components: it has been proved in Iserles (1999b) that

$$\limsup_{k \to \infty}(\#\mathbb{S}_k)^{1/k} = 3 \qquad \text{and} \qquad \limsup_{m \to \infty}(\#\mathbb{G}_m)^{1/m} = 2.69805\ldots$$

(in either case there is substantial saving in comparison with the Magnus expansion; *cf.* (4.15)). However, a very important feature of Magnus expansions is unfortunately lost: *The Cayley expansion* (6.9), *truncated by power, is no longer time-symmetric!* (We should perhaps emphasize that the rôle of time symmetry survives when the exponential is replaced with the Cayley transform: it still implies even order.) In other words, if we truncate the Cayley expansion (with all the integrals evaluated exactly) so that $p = 3$, say, in (6.9), the order will be just three.

As in (4.16)–(4.18), we conclude by presenting Cayley expansions in standard notation, as follows, rather than in a tree terminology.

$$\Delta(t) = \int_0^t A(\xi)\,\mathrm{d}\xi \quad \dots\dots\dots\dots\dots\dots\dots\dots\dots\dots\dots\dots\text{order 2}$$

$$-\tfrac{1}{2}\int_0^t\int_0^{\xi_1}[A(\xi_2),A(\xi_1)]\,\mathrm{d}\boldsymbol{\xi}$$

$$+\tfrac{1}{4}\int_0^t\int_0^{\xi_1}\int_0^{\xi_2}[[A(\xi_3),A(\xi_2)],A(\xi_1)]\,\mathrm{d}\boldsymbol{\xi} \quad \dots\dots\dots\dots\dots\text{order 3}$$

$$-\tfrac{1}{4}\int_0^t\int_0^{\xi_1}\int_0^{\xi_1}A(\xi_2)A(\xi_1)A(\xi_3)\,\mathrm{d}\boldsymbol{\xi} \quad \dots\dots\dots\dots\dots\dots\text{order 4}$$

$$-\tfrac{1}{8}\int_0^t\int_0^{\xi_1}\int_0^{\xi_2}\int_0^{\xi_3}[[[A(\xi_4),A(\xi_3)],A(\xi_2)],A(\xi_1)]\,\mathrm{d}\boldsymbol{\xi} \quad \dots\dots\dots\text{order 5}$$

$$+\tfrac{1}{8}\int_0^t\int_0^{\xi_1}\int_0^{\xi_2}\int_0^{\xi_2}[A(\xi_3)A(\xi_2)A(\xi_4),A(\xi_1)]\,\mathrm{d}\boldsymbol{\xi}$$

$$+\tfrac{1}{8}\int_0^t\int_0^{\xi_1}\int_0^{\xi_2}\int_0^{\xi_1}[A(\xi_3),A(\xi_2)]A(\xi_1)A(\xi_4)\,\mathrm{d}\boldsymbol{\xi}$$

$$+\tfrac{1}{8}\int_0^t\int_0^{\xi_1}\int_0^{\xi_1}\int_0^{\xi_3}A(\xi_2)A(\xi_1)[A(\xi_4),A(\xi_3)]\,\mathrm{d}\boldsymbol{\xi}$$

$$+\tfrac{1}{16}\int_0^t\int_0^{\xi_1}\int_0^{\xi_2}\int_0^{\xi_3}\int_0^{\xi_4}[[[[A(\xi_5),A(\xi_4)],A(\xi_3)],A(\xi_2)],A(\xi_1)]\,\mathrm{d}\boldsymbol{\xi}$$

$$-\tfrac{1}{16}\int_0^t\int_0^{\xi_1}\int_0^{\xi_2}\int_0^{\xi_3}\int_0^{\xi_1}[[A(\xi_4),A(\xi_3)],A(\xi_2)]A(\xi_1)A(\xi_5)\,\mathrm{d}\boldsymbol{\xi}$$

$$-\tfrac{1}{16}\int_0^t\int_0^{\xi_1}\int_0^{\xi_1}\int_0^{\xi_3}\int_0^{\xi_4}A(\xi_2)A(\xi_1)[[A(\xi_5),A(\xi_4)],A(\xi_3)]\,\mathrm{d}\boldsymbol{\xi} \quad \text{order 6}$$

$$+\cdots.$$

The above expansion underscores the importance of time symmetry in reducing the number of terms: compare the order-6 truncation with (4.18). We hasten to reassure the disappointed reader that not all is lost: time symmetry and even order will be regained in the next subsection.

6.3. Quadrature of the Cayley expansion and hierarchical algebras

In principle, the terms in the Cayley expansion (6.9) can be approximated exactly like 'Magnus integrals', since they are all consistent with (5.1): integrals of a multilinear form $\boldsymbol{L}$ over a polytope $\mathcal{S}$. The theory of Sections 5.1–5.2 is robust enough to cater for symmetric triple products, not just commutators.

Thus, to obtain a third-order method we truncate by power,

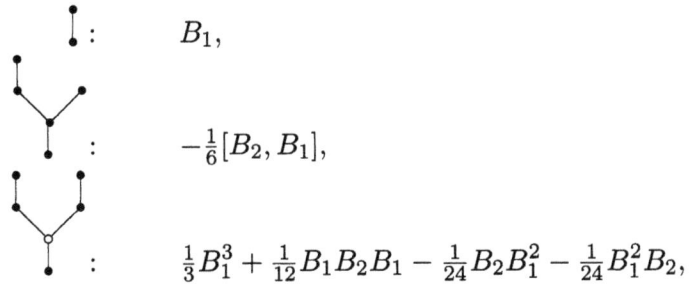

$$\Delta(t) \;\approx\; \mid \;-\tfrac{1}{2}\,\Upsilon\; -\tfrac{1}{4}\,\Upsilon\;, \tag{6.10}$$

and replace integrals by quadrature using a self-adjoint basis. More specifically, we evaluate hA at the Gauss–Legendre points $(\tfrac{1}{2} \pm \tfrac{\sqrt{3}}{6})h$, denote these function values by A_1, A_2 and let $B_1 = \tfrac{1}{2}(A_1 + A_2)$, $B_2 = \sqrt{3}(A_2 - A_1)$. The relevant fourth-order quadratures are

$$\mid \;:\qquad B_1,$$

$$\Upsilon \;:\qquad -\tfrac{1}{6}[B_2, B_1],$$

$$\Upsilon \;:\qquad \tfrac{1}{3}B_1^3 + \tfrac{1}{12}B_1 B_2 B_1 - \tfrac{1}{24}B_2 B_1^2 - \tfrac{1}{24}B_1^2 B_2,$$

but we can throw away the last three terms with complete impunity: after all, we want a third-order method! The outcome is

$$B_1 + \tfrac{1}{12}[B_2, B_1] - \tfrac{1}{12}B_1^3. \tag{6.11}$$

Just to be on the safe side, we expand the solution, only to find that, lo and behold, the order of (6.11) is *four*. Not the order of (6.10), we hasten to say: the miracle has occurred just as the integrals have been replaced by quadrature!

This is not a serendipitous coincidence. *Quadrature recovers time symmetry, thereby boosting the order of an odd-order truncation* (6.9) (Iserles 1999b). Thus, herewith for example a sixth-order method, where B_0, B_1, B_2 have been obtained from order-six Gauss–Legendre quadrature:

$$B_1 + \tfrac{1}{12}B_3 + \tfrac{1}{12}[B_2, B_1] - \tfrac{1}{12}B_1^3 - \tfrac{1}{240}[B_3, B_2] - \tfrac{1}{240}[[B_3, B_1], B_1]$$

$$- \tfrac{1}{240}[[B_2, B_1], B_2] - \tfrac{1}{48}B_1 B_3 B_1 - \tfrac{3}{320}[B_2, B_1^3] + \tfrac{7}{960}B_1[B_2, B_1]B_1 \tag{6.12}$$

$$+ \tfrac{1}{960}[[[B_2, B_1], B_1], B_1] + \tfrac{1}{120}B_1^5.$$

Having hopefully learnt something from our analysis of discretized Magnus expansions in Section 5, our next question is whether all the terms in (6.12) are necessary or can we perhaps replace some with linear combinations of other terms. In other words, we wish to repeat here the discussion from Section 5.3, except that in the present situation we should reckon with two operations: commutation and the symmetric triple product, the latter characteristic of quadratic Lie algebras. Wishing to derive the dimension of linear spaces of *graded free algebras*, along the lines of Section 5.3, a natural temptation is to express symmetric triple products in terms of commutators, but this soon leads to mushrooming complexity. A more effective approach is described in Iserles and Zanna (2000).

Let $(\mathfrak{g}, +)$ be an Abelian group over a field of zero characteristic and introduce a countable family of m-ary operations

$$[\![\,\cdot\,,\dots,\,\cdot\,]\!]_m : \overbrace{\mathfrak{g} \times \mathfrak{g} \times \cdots \times \mathfrak{g}}^{m \text{ times}} \to \mathfrak{g}, \qquad m \in \mathbb{N},$$

which is subject to the following three axioms.

(1) Alternate symmetry: for all $F_1, F_2, \dots, F_m \in \mathfrak{g}$

$$[\![F_1, F_2, \dots, F_m]\!]_m + (-1)^m [\![F_m, F_{m-1}, \dots, F_1]\!]_m = O.$$

(2) Multilinearity: $[\![F_1, F_2, \dots, F_m]\!]_m$ is linear in each of its m arguments.

(3) Hierarchy condition: for all $F_1, \dots, F_m, E_1, \dots, E_s \in \mathfrak{g}$ and $1 \le l \le m$ it is true that

$$[\![F_1, \dots, F_{l-1}, [\![E_1, \dots, E_s]\!]_s, F_{l+1}, \dots, F_m]\!]_m$$
$$= [\![F_1, \dots, F_{l-1}, E_1, \dots, E_s, F_{l+1}, \dots, F_m]\!]_{m+s-1}$$
$$- (-1)^s [\![F_1, \dots, F_{l-1}, E_s, \dots, E_1, F_{l+1}, \dots, F_m]\!]_{m+s-1}.$$

The triple $(\mathfrak{g}, +, \{[\![\cdots]\!]_m\}_{m \in \mathbb{N}})$ has been called a *hierarchical algebra* by Iserles and Zanna (2000). It is easy to see that each hierarchical algebra is a Lie algebra (with the commutator defined as $[\,\cdot\,,\,\cdot\,] = [\![\,\cdot\,,\,\cdot\,]\!]_2$, while each quadratic Lie algebra is hierarchical with

$$[\![F_1, \dots, F_m]\!]_m = F_1 F_2 \cdots F_m - (-1)^m F_m F_{m-1} \cdots F_1, \qquad F_1, \dots, F_m \in \mathfrak{g}.$$

We now proceed as in Section 5.3: choose a set $G_1, G_2, \dots, G_\nu$ of generators and define a *free hierarchical algebra* (FHA) similarly to Definition 5.1. We endow the generators with grading ω and extend the grading to FHA in a natural manner, that is,

$$\omega([\![H_{i_1}, H_{i_2}, \dots, H_{i_r}]\!]_r) = \sum_{k=1}^{r} \omega(H_k).$$

One should not take the analogy with FLAs too far, since FHAs require a subtly different approach. At the heart of the discussion of Section 5.3 is the fact that, using for example the Hall basis, we can express every element of an FLA as a linear combination of *primitive* terms of the form

$$[G_{i_1}, [G_{i_1}, [\dots, [G_{i_{r-1}}, G_{i_r}] \cdots]]].$$

In the case of FHAs, Iserles and Zanna (2000) prove that the primitive 'building blocks' can be chosen to have the form

$$[\![G_{i_1}, G_{i_2}, \dots, G_{i_r}]\!]_r.$$

The method of proof is constructive, repeatedly using the three axioms. Anticipating future discussion, we exemplify it with one of the terms in the sixth-order Cayley expansion (6.12), assuming that $\nu \geq 2$:

$$[[[[G_2, G_1]]_2, G_1]]_2, G_1]]_2 \overset{\text{Axiom 3}}{=} [[[G_2, G_1]]_2, G_1, G_1]]_3 - [[G_1, [[G_2, G_1]]_2, G_1]]_3$$
$$\overset{\text{Axiom 3}}{=} ([[G_2, G_1, G_1, G_1]]_4 - [[G_1, G_2, G_1, G_1]]_4)$$
$$- ([[G_1, G_2, G_1, G_1]]_4 - [[G_1, G_1, G_2, G_1]]_4)$$
$$\overset{\text{Axiom 1}}{=} [[G_2, G_1, G_1, G_1]]_4 - 3[[G_1, G_2, G_1, G_1]]_4.$$

Let $\mathfrak{g}_m$ be the set of all the grade-m elements in the FHA $\mathfrak{g}$. As for FLAs, we can express $\mathfrak{g}$ as a direct sum of $\mathfrak{g}_m$ for $m \in \mathbb{N}$. The dimension of each $\mathfrak{g}_m$ has been characterized in Iserles and Zanna (2000).

Theorem 6.1. Let $\mathfrak{g}$ be the graded FHA generated by $S = \{G_1, \ldots, G_\nu\}$, with grades $\omega_1, \ldots, \omega_\nu$ respectively. Denote by $\lambda_1, \ldots, \lambda_r$ the roots of the rth degree polynomial

$$p(z) = 1 - \sum_{i=1}^{\nu} z^{\omega_i}, \qquad r = \max_{1 \leq i \leq \nu} \omega_i,$$

and assume that they are all distinct. Then

$$\bar{\sigma}_{2m} = \tfrac{1}{2} \sum_{l=1}^{r} \frac{\lambda_l^{-m-1}}{p'(\lambda_l)} \left\{ 2 - \lambda_l^{-m} - \tfrac{1}{2}[p(\lambda_l^{1/2}) + p(-\lambda_l^{1/2})] \right\},$$
$$\bar{\sigma}_{2m+1} = \tfrac{1}{2} \sum_{l=1}^{r} \frac{\lambda_l^{-m-3/2}}{p'(\lambda_l)} \left\{ -\lambda_l^{-m-1/2} + \tfrac{1}{2}[p(\lambda_l^{1/2}) - p(-\lambda_l^{1/2})] \right\},$$

(6.13)

where $\bar{\sigma}_m = \dim \mathfrak{g}_m$, $m \in \mathbb{N}$.

The proof is long and technical. Its main step is in demonstrating that

$$\sum_{m=1}^{\infty} t^m \dim \mathfrak{g}_m = \tfrac{1}{2} \left[\frac{1}{p(t)} + \frac{p(t)}{p(t^2)} \right].$$

Comparing Table 6.1 with Tables 5.1–5.2 demonstrates that the dimension

Table 6.1. *Dimensions (6.13) of a graded FHA for $\nu = 3$ in two cases: $\omega_i \equiv 1$ and $\omega_i = i$*

	m	1	2	3	4	5	6	7	8	9	10
$\omega_i \equiv 1$	$\bar{\sigma}_m$	3	3	18	36	135	351	1134	3240	9963	29403
$\omega_i = i$	$\bar{\sigma}_m$	1	1	3	3	8	11	25	39	80	134

of graded subspaces of FHA grows larger than in the FLA case. This is hardly a surprise. A Lie algebra is closed with respect to just one binary operation, commutation, in addition to the usual linear-space operations. A hierarchical algebra, however, is closed with respect to a countable number of operations! On the face of it, there are infinitely more ways of forming terms in FHA. Fortunately, the hierarchy conditions mean that the operations are interconnected and the growth in dimension is not as bad as we might have expected.

The ratio of the dimensions $\bar{\rho}_m$ and $\bar{\sigma}_m$ from (5.12) and (6.13), respectively, can be determined asymptotically. The dominant zero of the polynomial p, λ_1, say, is in $(1, \infty)$ and simple. Iserles and Zanna (2000) proved that

$$\frac{\bar{\sigma}_m}{\bar{\rho}_m} = -\frac{\lambda_1 m}{2p'(\lambda_1)}[1 + o(1)], \qquad m \gg 1.$$

The method of proof of Theorem 6.1 is constructive and it naturally leads to a basis and to algorithmic means of its construction. We refer the reader to Iserles and Zanna (2000) for details, here just presenting the results for $\nu = 2$ and the grades $\omega(G_i) = i$. The basis of $\mathfrak{g}_m$ is denoted by $\mathcal{B}_m$.

$\mathcal{B}_1$: $\{[\![G_1]\!]_1\}$,

$\mathcal{B}_2$: $\{[\![G_2]\!]_1\}$,

$\mathcal{B}_3$: $\{[\![G_3]\!]_1, [\![G_1, G_2]\!]_2, [\![G_1, G_1, G_1]\!]_3\}$,

$\mathcal{B}_4$: $\{[\![G_1, G_3]\!]_2, [\![G_1, G_1, G_2]\!]_3, [\![G_1, G_2, G_1]\!]\}$,

$\mathcal{B}_5$: $\{[\![G_2, G_3]\!]_2, [\![G_1, G_2, G_2]\!]_3, [\![G_1, G_1, G_3]\!]_3, [\![G_2, G_1, G_2]\!]_3,$

  $[\![G_1, G_3, G_1]\!]_3, [\![G_1, G_1, G_1, G_2]\!]_4, [\![G_1, G_1, G_2, G_1]\!]_4,$

  $[\![G_1, G_1, G_1, G_1, G_1]\!]_5\}$.

We conclude by going back to the sixth-order Cayley expansion (6.12) and representing it in the FHA basis. As before, we let $G_i = B_i$ and $\omega_i = i$. The outcome, after long but straightforward algebra, is

$[\![B_1]\!]_1$ ....grade 1

$+ \frac{1}{12}[\![B_3]\!]_1 - \frac{1}{12}[\![B_1, B_2]\!]_2 - \frac{1}{24}[\![B_1, B_1, B_1]\!]_3$ ....grade 3

$+ \frac{1}{240}[\![B_2, B_3]\!]_2 + \frac{1}{240}[\![B_1, B_2, B_2]\!]_3 - \frac{1}{240}[\![B_1, B_1, B_3]\!]_3 - \frac{1}{240}[\![B_2, B_1, B_2]\!]_3$

$- \frac{1}{160}[\![B_1, B_3, B_1]\!]_3 + \frac{1}{120}[\![B_1, B_1, B_1, B_2]\!]_4 - \frac{1}{240}[\![B_1, B_1, B_2, B_1]\!]_4$

$+ \frac{1}{240}[\![B_1, B_1, B_1, B_1, B_1]\!]_5$ ....grade 5

Note that, thanks to time symmetry, only odd-grade elements enter the expansion.

6.4. Canonical coordinates of the second kind I: A naive approach

The main idea of the present section is to consider alternatives to the standard exponential map expm : $\mathfrak{g} \to \mathcal{G}$, which we can write in the form

$$\mathfrak{g} \ni \sum_{k=1}^{d} \theta_k C_k \sim (\theta_1, \theta_2, \ldots, \theta_d) \to \operatorname{expm}\left(\sum_{k=1}^{d} \theta_k C_k\right) \in \mathcal{G}, \qquad (6.14)$$

where $d = \dim \mathfrak{g}$ and $C = \{C_1, C_2, \ldots, C_d\}$ is a basis of the Lie algebra. The
map (6.14) induces (at least locally, near the identity) a coordinate system
in the Lie group which has been termed by Varadarajan (1984) the *canonical
coordinates of the first kind*. An alternative to (6.14) (which, incidentally,
explains why we have insisted on writing it in such a strange form) are the
canonical coordinates of the second kind (CCSK):

$$\mathfrak{g} \ni \sum_{k=1}^{d} \theta_k C_k \sim (\theta_1, \theta_2, \ldots, \theta_d) \to \mathrm{e}^{\theta_1 C_1} \mathrm{e}^{\theta_2 C_2} \cdots \mathrm{e}^{\theta_d C_d} \in \mathcal{G}. \qquad (6.15)$$

Why should we consider (6.15)? On the face of it, we have replaced a single
exponential with d exponentials (and the whole exercise becomes really interesting when $d \gg 1$!), hardly a sensible point of departure. However, as
long as C is appropriately chosen, the computation of each $\operatorname{expm}(\theta_k C_k)$ can
be exceedingly cheap and, moreover, the approach lends itself naturally to
the exploitation of sparsity: as long as we can expect that there should be
no component in the C_k direction, say (or that it is suitably small), we can
drop the relevant exponential from the product.[2] A useful analogy is the distinction between Householder reflections and Givens rotations in numerical
algebra.

It is possible to approach the issue of CCSK within the context of this
survey from two distinctive points of view. Although ultimately they are
closely related, they follow different philosophies, the first 'naive' and the
other more mathematically sophisticated. This subsection is devoted to the
more 'naive' approach, which associates CCSK with *splittings*.

Splitting methods have a rich history throughout numerical analysis of
differential equations and they are exceedingly useful in geometric integration, for instance in the computation of Hamiltonian systems (Sanz Serna
and Calvo 1994, Yoshida 1990) and in the recovery of integrals and conservation laws (McLachlan, Quispel and Robidoux 1998). Yet, the splitting of
a flow into components *corresponding to elements of a basis* allows a significant enhancement of the technique. For simplicity, let us assume that we
are solving the linear Lie-group equation (4.2), namely $Y' = A(t)Y$, $t \geq 0$.

---

[2] We return to this point in far greater detail in Section 8.

We express the solution in the form

$$Y(t) = e^{\theta_1(t)C_1} e^{\theta_2(t)C_2} \cdots e^{\theta_d(t)C_d} Y_0, \qquad t \geq 0, \tag{6.16}$$

where $\theta_1, \theta_2, \dots, \theta_d$ are scalar functions. It has been proved by Wei and Norman (1964) that such functions always exist locally (and, in the case of solvable Lie algebras or for 2×2 real matrices, globally). The exact derivation of $\theta_1, \theta_2, \dots, \theta_d$ is, needless to say, impossible in general, otherwise we could have written down the solution of (4.2) explicitly! Instead, we replace the θ_ks with *polynomials*, which are chosen so as to match suitable order conditions at $t = 0$.

Letting $t = 0$ in (6.16), we note that $\theta_k(0) = 0$, $k = 1, 2, \dots, d$. To obtain more useful order conditions we differentiate Y. Some brief algebra confirms that

$$\begin{aligned}
A(t) &= Y'(t)Y^{-1}(t) \\
&= \sum_{k=1}^{d} \theta_k'(t) e^{\theta_1(t)C_1} \cdots e^{\theta_{k-1}(t)C_{k-1}} C_k e^{-\theta_{k-1}(t)C_{k-1}} \cdots e^{-\theta_1(t)C_1} \\
&= \sum_{k=1}^{d} \theta_k'(t) \mathrm{Ad}_{\mathrm{expm}[\theta_1(t)C_1]} \cdots \mathrm{Ad}_{\mathrm{expm}[\theta_{k-1}(t)C_{k-1}]} C_k.
\end{aligned} \tag{6.17}$$

Letting $t = 0$ in (6.17) we obtain the first-order condition

$$A(0) = \sum_{k=1}^{d} \theta_k'(0) C_k.$$

Recalling that $A(0) \in \mathfrak{g}$, we can expand it in the elements of C and this yields $\theta_k'(0)$ explicitly.

Higher-order conditions can be obtained by differentiating (6.17) and massaging the formulae with a great deal of (fairly unpleasant) algebra. Thus, for example,

$$\begin{aligned}
A' = & \sum_{k=1}^{d} \theta_k'' \mathrm{Ad}_{e^{\theta_1 C_1}} \cdots \mathrm{Ad}_{e^{\theta_{k-1} C_{k-1}}} C_k \\
& + \sum_{k=1}^{d} \sum_{l=1}^{k-1} \theta_k' \theta_l' \mathrm{Ad}_{e^{\theta_1 C_1}} \cdots \mathrm{Ad}_{e^{\theta_l C_l}} [C_l, \mathrm{Ad}_{e^{\theta_{l+1} C_{l+1}}} \cdots \mathrm{Ad}_{e^{\theta_{k-1} C_{k-1}}} C_k]
\end{aligned}$$

and, letting $t = 0$, we have

$$\sum_{k=1}^{d} \theta''(0) C_k = A'(0) - \sum_{k=1}^{d} \sum_{l=1}^{k-1} \theta_k'(0) \theta_l'(0) [C_l, C_k].$$

Recall that C is a basis of $\mathfrak{g}$, hence there exist scalars $c_{k,l}^j$ such that

$$[C_k, C_l] = \sum_{j=1}^{d} c_{k,l}^j C_j, \qquad k, l = 1, 2, \dots, d.$$

They are called the *structure constants* of $\mathfrak{g}$ and play an important rôle in the theory of Lie algebras (Olver 1995, Varadarajan 1984). Using structure constants and observing that $A'(0) \in \mathfrak{g}$ can be expanded in elements of C, we obtain

$$\sum_{k=1}^{d} \theta_k''(0) C_k = A'(0) + \sum_{j=1}^{d} \sum_{k=1}^{d} \sum_{l=1}^{k-1} \theta_k'(0) c_{k,l}^j \theta_l'(0) C_j, \qquad (6.18)$$

hence second-order conditions.

Typically, the dimension d is quite large, for example $\dim \mathfrak{so}(N) = \frac{1}{2}(N-1)N$ and $\dim \mathfrak{sl}(N) = N^2 - 1$. Thus, in principle it might be costly to evaluate $\theta_k''(0)$, $k = 1, 2, \dots, d$ in (6.18). Higher-order conditions are substantially costlier still. Yet, the cost can be reduced a very great deal by the right choice of the basis C.

The most suitable basis C is provided by a *root-space decomposition* of the (non-nilpotent) Lie algebra $\mathfrak{g}$. Deferring our discussion of this construct to the next subsection, we describe in a more nontechnical setting the special case of $\mathfrak{so}(N)$. To this end we choose the basis

$$C = \{C_{k,l} = e_k e_l^{\mathrm{T}} - e_l e_k^{\mathrm{T}} : 1 \le k < l \le N\},$$

where $e_j \in \mathbb{R}^N$ is the jth unit vector. Note that

$$B \in \mathfrak{so}(N) \qquad \Rightarrow \qquad B = \sum_{k=1}^{N-1} \sum_{l=k+1}^{N} b_{k,l} C_{k,l}$$

and that $U = \mathrm{expm}(t C_{k,l})$ is a rigid rotation in the (k, l) plane: it coincides with the identity matrix, except for

$$\begin{bmatrix} u_{k,k} & u_{k,l} \\ u_{l,k} & u_{l,l} \end{bmatrix} = \begin{bmatrix} \cos t & \sin t \\ -\sin t & \cos t \end{bmatrix}.$$

Therefore, multiplying a matrix with $\mathrm{expm}(t C_{k,l})$ is cheap. Moreover, it is easy to verify that

$$[C_{r,s}, C_{k,l}] = \begin{cases} C_{k,s}, & r = l, \ s \ne k, \\ C_{l,r}, & r \ne l, \ s = k, \\ C_{r,k}, & r \ne k, \ s = l, \qquad r < s, \ k < l, \\ C_{s,l}, & r = k, \ s \ne l, \\ \mathbf{O}, & \text{otherwise}, \end{cases}$$

where we identify $C_{i,j}$ with $-C_{j,i}$ for $i > j$. The condition (6.18) simplifies to

$$\theta''_{k,l}(0) = a'_{k,l}(0) - \sum_{i=1}^{N} a_{k,i}(0)a_{i,l}(0), \qquad 1 \le k < l \le N,$$

where $A(t) = \sum_{k=1}^{N-1} \sum_{l=k+1}^{N} a_{k,l}(t)C_{k,l}$. This requires $\mathcal{O}(N^3)$ flops altogether, in comparison with $\mathcal{O}(N^6)$ if sparsity of structure constants is disregarded. Similar constructions can be applied in other Lie algebras.

In principle, we can go on to derive $\theta_k^{(i)}(0)$ for $i = 0, 1, \ldots, p$, but this procedure, even while utilizing root-space decomposition, becomes progressively more expensive for larger orders p. Herewith we present a device which, to our knowledge, is new and which allows one to obtain order p while computing one less derivative. Observing that (6.16) is sensitive to the ordering of the basis, the main idea is to alternate the order of elements of C while time-stepping the numerical method. Thus, suppose that $t_m = mh$ and

$$Y_{2n+1} = e^{\theta_{2n,1}(t_{2n+1})C_1} e^{\theta_{2n,2}(t_{2n+1})C_2} \cdots e^{\theta_{2n,d}(t_{2n+1})C_d} Y_{2n},$$

$$Y_{2n+2} = e^{\theta_{2n+1,d}(t_{2n+2})C_d} e^{\theta_{2n+1,d-1}(t_{2n+2})C_{d-1}} \cdots e^{\theta_{2n+1,1}(t_{2n+2})C_1} Y_{2n+1},$$

where $\theta_{m,k}$ are p-degree polynomials. Without loss of generality, we assume that $\theta_{m,k}$ are consistent with order p for $m = 0, 1, \ldots, 2n$. Let

$$X(t) = e^{\theta_{2n+1,d}(t)C_d} e^{\theta_{2n+1,d-1}(t)C_{d-1}} \cdots e^{\theta_{2n+1,1}(t)C_1} Y_{2n+1};$$

hence $Y_{2n+1} = X(t_{2n+1})$ and $Y_{2n+2} = X(t_{2n+2})$. Repeatedly multiplying by inverted exponentials, we obtain

$$Y_{2n+1} = e^{-\theta_{2n+1,1}(t)C_1} e^{-\theta_{2n+1,2}(t)C_2} \cdots e^{-\theta_{2n+1,d}(t)C_d} X(t).$$

Assuming that the $\theta_{2n+1,k}$ are chosen consistently with order p and letting $t = t_{2n}$ we deduce that

$$Y_{2n+1} = e^{-\theta_{2n+1,1}(t_{2n})C_1} e^{-\theta_{2n+1,2}(t_{2n})C_1} \cdots e^{-\theta_{2n+1,d}(t_{2n})C_d} Y_{2n} + \mathcal{O}(h^{p+1}).$$

In other words, we may take $\theta_{2n+1,k}(t_{2n}) = -\theta_{2n,k}(t_{2n+1})$, $k = 1, 2, \ldots, d$. This, together with the values of $\theta_{2n+1,k}^{(i)}(t_{2n+1})$, $i = 0, 1, \ldots, p-1$, is just right to determine the $\theta_{2n+1,k}$s consistently with order p.

6.5. Canonical coordinates of the second kind II: Admissible bases

The technique of canonical coordinates of the second kind can be enhanced a great deal at the cost of increased mathematical sophistication. The point of departure for our discussion, presently based on the important paper of Owren and Marthinsen (1999a), is the equation (6.17). We rewrite it in the form

$$\sum_{k=1}^{d} \theta_k' \mathrm{Ad}_{\mathrm{expm}(\theta_1 C_1)} \mathrm{Ad}_{\mathrm{expm}(\theta_2 C_2)} \cdots \mathrm{Ad}_{\mathrm{expm}(\theta_{k-1} C_{k-1})} C_k = A(t). \qquad (6.19)$$

Here $\theta_1, \theta_2, \dots, \theta_d$ are known and we seek the scalars $\theta_1', \theta_2', \dots, \theta_d'$.

Suppose that we have the means to solve (6.19) for arbitrary inputs $\theta_1, \theta_2, \dots, \theta_d$ (in (6.17) we needed just $\boldsymbol{\theta} = \mathbf{0}$). This yields a *differential equation*

$$\boldsymbol{\theta}' = \boldsymbol{g}(\boldsymbol{\theta}, A(t)), \quad t \geq 0, \qquad \boldsymbol{\theta}(0) = \mathbf{0}. \qquad (6.20)$$

Once the solution of (6.20) is known (or adequately approximated), we can use it to advance the solution of the Lie-group equation (4.2) through the CCSK representation (6.16). Moreover, the argument extends at once to nonlinear Lie-group equations $Y' = A(t, Y)Y$, where $A : \mathcal{G} \times \mathbb{R}^+ \to \mathfrak{g}$. Again, we represent the solution in the CCSK form (6.16), except that now

$$\boldsymbol{g} = \mathrm{dccsk}_{\boldsymbol{\theta}}^{-1} A = \boldsymbol{g}(\boldsymbol{\theta}, A(t, e^{\theta_1 C_1} e^{\theta_2 C_2} \cdots e^{\theta_d C_d} Y_0))$$

in the *dccskinv equation* (6.20).

Applying a Runge–Kutta method, say, to (6.20) results in a time-stepping scheme that is guaranteed to respect Lie-group structure. In effect, the only difference between the RK-MK methods of Section 3 and these methods is that, in place of canonical coordinates of the first kind and the dexpinv equation, they utilize canonical coordinates of the second kind and the dccskinv equation. All this motivates a thorough discussion of the problem of how to invert an equation of the form (6.19).

Letting $F_k = \mathrm{Ad}_{\mathrm{expm}(\theta_k C_k)}$, $v_k = \theta_k'$, $k = 1, 2, \dots, d$, we commence by writing (6.20) as

$$\sum_{k=1}^{d} v_k F_1 F_2 \cdots F_{k-1} C_k = A.$$

Let $\mathcal{P}_l$ be a projection on the trailing $d - l$ coordinates,

$$\mathcal{P}_l \sum_{k=1}^{d} c_k C_k = \sum_{k=l+1}^{d} c_k C_k,$$

and set $\hat{F}_l = I - \mathcal{P}_l + \mathcal{P}_l A$, $l = 1, 2, \dots, d$. We can easily verify that

$$\hat{F}_l C_k = \begin{cases} F_l C_k, & l < k, \\ C_k, & l \geq k. \end{cases}$$

Owren and Marthinsen (1999a) say that $\boldsymbol{C}$ is an *admissible ordered basis* (AOB) if for every $\theta_1, \theta_2, \dots, \theta_d$ it is true that

$$F_1 F_2 \cdots F_k \mathcal{P}_k = \hat{F}_1 \hat{F}_2 \cdots \hat{F}_k \mathcal{P}_k, \qquad k = 1, 2, \dots, d - 1. \qquad (6.21)$$

Provided that C is an AOB, it is simple to prove that our equation can be rewritten in the form

$$\sum_{k=1}^{d} v_k \hat{F}_1 \hat{F}_2 \cdots \hat{F}_{d-1} C_k = \hat{F}_1 \hat{F}_2 \cdots \hat{F}_{d-1} F = A, \qquad (6.22)$$

where $F = \sum_{k=1}^{d} v_k C_k$. Later we will see that, in a number of important cases, AOB implies that each $\hat{F}_k$ can be inverted very cheaply.

At first glance, the AOB condition (6.21) is exceedingly demanding. Surprisingly, it is often achievable but we need to introduce a little bit more Lie-algebra theory before being in a position to describe exactly how. The following brief extract should ideally be supplemented by perusing a Lie-algebra monograph: the book by Varadarajan (1984) is a good place to start.

- A subalgebra $\mathfrak{h}$ of a Lie algebra $\mathfrak{g}$ is an *ideal* if $[\mathfrak{h}, \mathfrak{g}] \subseteq \mathfrak{h}$.

- The Lie algebra $\mathfrak{g}$ is *solvable* if there exists $m \in \mathbb{Z}^+$ such that $\mathfrak{g}^{(m)} = \{0\}$, where $\mathfrak{g}^{(0)} = \mathfrak{g}$ and $\mathfrak{g}^{(i+1)} = [\mathfrak{g}^{(i)}, \mathfrak{g}^{(i)}] \subseteq \mathfrak{g}^{(i)}$.

- The *radical* of $\mathfrak{g}$, denoted by $\operatorname{Rad} \mathfrak{g}$, is the maximal solvable ideal in $\mathfrak{g}$. We say that the Lie algebra is *semisimple* if $\operatorname{Rad} \mathfrak{g} = \{0\}$.

  If definitions have become hazy by now, let us just point out that all specific Lie algebras in this survey (and in known applications within its framework) are semisimple, inclusive of $\mathfrak{sl}(N)$, $\mathfrak{so}(N)$ and $\mathfrak{sp}(N)$. (All these three Lie algebras are, as a matter of fact, *simple*: their only ideals are $\{0\}$ and the algebra itself.)

- An element in $\mathfrak{g}$ is *semisimple* if all the roots of its minimal polynomial are distinct: in a matrix representation it means that the element can be diagonalized.

- A subalgebra is *toral* if all its elements are semisimple. It is easy to see that every toral algebra must be abelian: in a matrix representation we can restate this by saying that the elements of the subalgebra share all eigenvectors, hence they commute.

- Unless $\mathfrak{g}$ is nilpotent, it possesses a nonzero *maximal toral subalgebra*. Such subalgebra, which we denote by $\mathfrak{h}$, is unique up to an isomorphism. If $\mathfrak{g}$ is a simple algebra, $\mathfrak{h}$ is also known (subject to an equivalent definition) as a *Cartan subalgebra*.

- Suppose that $\mathfrak{g}$ is a linear space over $\mathbb{C}$. We denote by $\mathfrak{h}^*$ the *dual space* of a maximal toral subalgebra. It consists of all linear functionals $\mathfrak{h} \to \mathbb{C}$. The nonzero functional $\alpha \in \mathfrak{h}^*$ is a *root* if there exists $f \in \mathfrak{g} \backslash \{0\}$ such that

$$[h, f] = \alpha(h)f, \qquad h \in \mathfrak{h}.$$

In a matrix representation, $\alpha(H)$ is an eigenvalue of the commutator operator generated by $H \in \mathfrak{h}$, while F can be 'translated' into its eigenvector.

- Denote the set of all roots of $\mathfrak{g}$ by Φ. It is possible to prove that $\mathfrak{g}$ can be subjected to the *root-space decomposition*

$$\mathfrak{g} = \mathfrak{h} \oplus \bigoplus_{\alpha \in \Phi} \mathfrak{g}_\alpha, \tag{6.23}$$

where $\mathfrak{g}_\alpha = \{f \in \mathfrak{g} : [h, f] = \alpha(h)f, \ h \in \mathfrak{h}\} \neq \{0\}$.

- The decomposition (6.23) motivates the choice of a *Chevalley basis* of the Lie algebra $\mathfrak{g}$: we choose one basis vector for each one-dimensional subspace $\mathfrak{g}_\alpha$, $\alpha \in \Phi$, and combine it with an arbitrary basis of $\mathfrak{h}$.

- There exists an integer $k^* \geq 1$ such that $\mathrm{ad}_h^{k^*+1} = 0$ for every $h \in \mathfrak{g}_\alpha$, $\alpha \in \Phi$.

Many of the above concepts can be illustrated briefly with an example, and we choose $\mathfrak{sl}(N, \mathbb{C})$. It is semisimple (as a matter of fact, we have already mentioned that it is a simple Lie algebra). Using the standard representation of $\mathfrak{sl}(N, \mathbb{C})$ as matrices of zero trace, we can easily identify a maximal toral subalgebra $\mathfrak{h}$ with diagonal zero-trace matrices. Setting $E_{k,l} = e_k e_l^{\mathrm{T}}$, $k, l = 1, 2, \ldots, N$, we may choose the basis $\{E_{k,k} - E_{k+1,k+1} : k = 1, 2, \ldots, N-1\}$ for $\mathfrak{h}$. Moreover, given

$$\mathfrak{h} \ni H = \sum_{k=1}^{N} h_k E_{k,k}, \qquad \sum_{k=1}^{N} h_k = 0,$$

and letting $h_0 = 0$, we verify easily that

$$[H, E_{r,s}] = (h_r - h_{r-1} - h_s + h_{s-1})E_{r,s}, \qquad r, s = 1, 2, \ldots, N, \quad r \neq s.$$

Hence we identify the root $\alpha(H) = h_r - h_{r-1} - h_s + h_{s-1}$ and construct our basis by placing there $E_{r,s}$ for every $r \neq s$ and appending to this the above basis of $\mathfrak{h}$. Note that this results in $N^2 - 1$ terms, matching exactly the dimension of $\mathfrak{sl}(N, \mathbb{C})$.

To determine k^* we compute

$$\mathrm{ad}_{E_{r,s}}^2 E_{k,l} = \begin{cases} E_{r,s}, & k = s, \ l = r, \\ O, & \text{otherwise}, \end{cases} \quad r \neq s, \ k \neq l \quad \Rightarrow \quad \mathrm{ad}_{E_{r,s}}^3 E_{k,l} = O$$

and $\mathrm{ad}_{E_{r,s}}^2(E_{k,k} - E_{k+1,k+1}) = O$. Therefore $\mathrm{ad}_{E_{r,s}}^3 = O$, $r \neq s$, and we deduce that $k^* = 2$.[3]

---

[3] There are easier ways to determine k^* but they require more Lie-algebra theory.

Theorem 6.2.   Let $\{\rho_1, \rho_2, \ldots, \rho_{d_*}\}$, where $d_* = d - \dim \mathfrak{h}$, be the set of roots Φ of a semisimple non-nilpotent Lie algebra $\mathfrak{g}$. Suppose that the Chevalley basis C is ordered so that the basis of $\mathfrak{h}$ comes last. Then this basis is AOB if

$$k\rho_i + \rho_j = \rho_m, \quad m < i < j \leq d_*, \, 1 \leq k \leq k^* \quad \Rightarrow \quad \rho_m + \rho_n \notin \Phi \cup \{0\}$$
(6.24)

for all $n = m+1, m+2, \ldots, i-1$. Moreover, in that case

$$\hat{F}_k^{-1} = I + \sum_{l=1}^{k^*} (-1)^l \frac{\theta_k^l}{l!} \mathrm{ad}_{C_k}^l \mathcal{P}_k, \qquad k = 1, 2, \ldots, d-1.$$
(6.25)

Returning to $\mathfrak{so}(N, \mathbb{C})$, it has been proved in Owren and Marthinsen (1999a) that conditions of Theorem 6.2 are satisfied as long as super-diagonal elements are ordered lexicographically by rows in front of the elements underneath the diagonal, which are ordered lexicographically by columns: thus, the ordered basis is

$$\{E_{k,l} : 1 \leq k < l \leq N\} \cup \{E_{k,l} : 1 \leq l < k \leq N\}$$
$$\cup \, \{E_{k,k} - E_{k+1,k+1} : 1 \leq k \leq N-1\}.$$

We omit the largely technical proof. Likewise, it is possible, using Theorem 6.2, to identify AOB of $\mathfrak{sp}(N, \mathbb{C})$ and $\mathfrak{so}(N, \mathbb{C})$. This, however, is probably of less importance than in the case of $\mathfrak{so}(N, \mathbb{C})$, since the increase in dimension due to the replacement of $\mathbb{R}$ with $\mathbb{C}$ leads to a significant increase in the volume of computations. In the present stage of the development of Lie-group methods it is fair to say, we believe, that the Cayley-transform-based techniques from Sections 6.1–6.2 are the method of choice for quadratic Lie groups, while CCSK should be used with the special linear group.

7. Adjoint methods

In the previous sections we have encountered a number of numerical integrators for Lie groups. Although such methods produce solutions that stay on a given Lie group $\mathcal{G}$ *by design*, it is of interest to study how well such schemes respect other qualitative features of the underlying equations: the retention of a symplectic form, Lie–Poisson structure, conservation of energy, time symmetry, time reversibility, *et cetera*. Given the novelty of the proposed schemes, many of the above features and their implications on the 'quality' of the solution are still under investigation (Engø and Faltinsen 1999, Faltinsen 2000). For this reason, we shall focus here just on *time symmetry* for Lie-group methods, which is at present one of the few features that are better understood, deriving adjoint and self-adjoint Lie-group methods.

Before proceeding further, let us recall that the *flow* Φ of the differential equation in $\mathbb{R}^N$

$$\boldsymbol{y}' = \boldsymbol{f}(t, \boldsymbol{y}), \quad t \geq t_0, \qquad \boldsymbol{y}(t_0) = \boldsymbol{y}_0,$$

defined as

$$\Phi(t, t_0, \boldsymbol{y}_0) = \boldsymbol{y}(t),$$

obeys the following conditions:

(i) $\Phi(t_0, t_0, \boldsymbol{y}_0) = \boldsymbol{y}_0$,
(ii) $\Phi(t + \tau, t_0, \boldsymbol{y}_0) = \Phi(\tau, t, \Phi(t, t_0, \boldsymbol{y}_0))$,

provided that the above function $\boldsymbol{f}$ is Lipschitz with respect to $\boldsymbol{y}$ (Hairer et al. 1993). In particular, the second condition implies that

$$\Phi(-\tau, t + \tau, \Phi(\tau, t, \boldsymbol{y}(t))) = \boldsymbol{y}(t),$$

a condition that in literature is mostly known as *time symmetry* or *self-adjointness* of the exact flow Φ.

7.1. Adjoint methods in the classical setting

Numerical methods usually approximate the flow Φ by a discrete flow, say Ψ, so that

$$\boldsymbol{y}_{n+1} = \Psi(t_n + h, t_n, \boldsymbol{y}_n), \qquad n \in \mathbb{Z}^+,$$

approximates the exact solution $\boldsymbol{y}(t_n + h)$ to given order p. Although numerical integrators for ODEs always obey the condition $\Psi(t_n, t_n, \boldsymbol{y}_n) = \boldsymbol{y}_n$, they usually fail to satisfy condition (ii). However, its weaker variant, *time symmetry*, is easier to impose and numerical methods such that

$$\Psi(-h, t_n + h, \Psi(h, t_n, \boldsymbol{y}_n)) = \boldsymbol{y}_n, \qquad n \in \mathbb{Z}^+,$$

are usually called *self-adjoint* or *time-symmetric* methods. If a method is not self-adjoint, its *adjoint* Ψ^* is defined as the map

$$\Psi^*(-h, t_n + h, \Psi(h, t_n, \boldsymbol{y}_n)) = \boldsymbol{y}_n.$$

In shorthand notation, we write Ψ_h for $\Psi(h, t_n, \cdot)$, so that Ψ_h^* denotes the adjoint method of Ψ_h and moreover $\Psi_h^* \circ \Psi_h = \text{Id}$, the identity map, or equivalently $\Psi_{-h} = \Psi_h^{-1}$, if and only if the method is self-adjoint.

The theory of adjoint and self-adjoint numerical methods for ODEs in $\mathbb{R}^N$ is well established and we refer the reader to Hairer et al. (1993) for further reading.

One might question why self-adjointness of a numerical integration scheme is desirable. It turns out that for numerical integration schemes that are self-adjoint it is possible to develop a theory analogous to the KAM theory for Hamiltonian and time-reversible problems (Moser 1973), which usually implies better approximation of the solution and slower accumulation of error over long integration intervals (Estep and Stuart 1995, Reich 1996).

With regard to Lie-group methods and time symmetry, we have already shown in Section 4 that the Magnus expansion truncated by power is time-symmetric when applied to linear Lie-group differential equations $Y' = A(t)Y$, provided that the underlying quadrature is based on quadrature nodes in $[0, 1]$ which are symmetric with respect to $\frac{1}{2}$. A similar result applies also to the RK-MK methods, provided that the underlying Runge–Kutta scheme is self-adjoint.

However, it can be easily verified by means of numerical experiments that the above-mentioned Lie-group methods are not self-adjoint for nonlinear problems. Hence, using self-adjoint methods to solve a Lie-algebra differential equation is not sufficient to derive self-adjoint Lie-group methods!

In this section we shall be discussing a more general procedure, derived by Zanna, Engø and Munthe-Kaas (1999), that allows us to construct self-adjoint Lie-group methods for linear and nonlinear problems alike and for all types of coordinate maps $\phi : \mathfrak{g} \to \mathcal{G}$ that we may use to represent the solution. In lifting the Lie-group equation from $\mathcal{G}$ to $\mathfrak{g}$, we make an implicit choice of a coordinate map, which has to be taken into account in the construction of self-adjoint methods.

7.2. Coordinate maps centred at arbitrary points

Schematically, the first step in the development of Lie-group schemes introduced in Sections 3–6 is the choice of a smooth map, say $\phi : \mathfrak{g} \to \mathcal{G}$, such that $\phi(\mathbf{O}) = \mathbf{I}$, the identity of the Lie group G, and $\phi'(\mathbf{O}) = \mathbf{I}$ (more precisely $\mathrm{T}\phi(\mathbf{O}, B) = B$, where $B \in \mathfrak{g}$). In other words, ϕ is a diffeomorphism mapping a neighbourhood of $\mathbf{O} \in \mathfrak{g}$ into a neighbourhood of $\mathbf{I}$ in $\mathcal{G}$. Thus,

$$\phi(B) = \mathrm{expm}(B), \qquad B \in \mathfrak{g},$$

is an example of such a map, which in (6.14) we have termed *canonical coordinates of the first kind*. Similarly, we might consider *canonical coordinates of the second kind* (6.15), namely

$$\phi(B) = \mathrm{ccsk}(B) = \mathrm{expm}(\beta_1 B_1)\, \mathrm{expm}(\beta_2 B_2) \cdots \mathrm{expm}(\beta_d B_d), \qquad B \in \mathfrak{g},$$

where $B = \sum_{i=1}^{d} \alpha_i B_i$, α_is being real coefficients, the B_is are basis elements of the algebra $\mathfrak{g}$, and the β_is real functions of $\alpha_1, \alpha_2, \ldots, \alpha_d$. Yet another example of this kind of map is the *Cayley transform* (6.3), which in the current formalism reads

$$\phi(B) = (\mathbf{I} - \tfrac{1}{2}B)^{-1}(\mathbf{I} + \tfrac{1}{2}B), \qquad B \in \mathfrak{g},$$

that maps $\mathfrak{g}$ into $\mathcal{G}$ whenever $\mathcal{G}$ is a quadratic group and $\mathfrak{g}$ is its corresponding quadratic algebra.

Secondly, the Lie-group differential equation

$$Y' = A(t, Y)Y, \quad t \geq 0, \qquad Y(0) = Y_0 \in \mathcal{G}, \tag{7.1}$$

is *lifted* by means of the inverse of the map $\mathrm{d}\phi$ to an ordinary differential equation in $\mathfrak{g}$. Thus, for coordinates of the first kind, one has $\mathrm{d}\phi^{-1} = \mathrm{dexp}^{-1}$, the *dexpinv equation* (3.2) that we have already encountered time and again in the course of the present article. Similarly, we have derived the expressions dcay^{-1} and dccsk^{-1} in Section 6.

Finally, the $\mathrm{d}\phi^{-1}$ equation is solved in $\mathfrak{g}$ with either a Runge–Kutta method or with a Magnus or a Cayley-type expansion. Assuming that an approximation $Y_n \in \mathcal{G}$ has already been derived, we typically solve

$$\Theta' = \mathrm{d}\phi_\Theta^{-1}(A(t, \phi(\Theta)Y_n)), \qquad \Theta(t_n) = O, \quad t \in [t_n, t_{n+1}],$$

where $t_{n+1} = t_n + h$. The choice of the initial condition $\Theta(t_n) = O$ is equivalent to 'centring' the coordinate map ϕ at Y_n. Instead, coordinates centred at any point $X \in \mathcal{G}$ can be obtained inverting the map $B \in \mathfrak{g} \mapsto \phi(B)X \in \mathcal{G}$. In the general case we write

$$X = \phi(C)^{-1}Y_n,$$

for some $C \in \mathfrak{g}$ to be specified later. Note that $\phi(C)^{-1}$ is the inverse (in $\mathcal{G}$) of the group element $\phi(C)$. Before proceeding further, we observe that both canonical coordinates of the first kind and the Cayley transform obey the relation

$$\phi(B)^{-1} = \phi(-B),$$

for all $B \in \mathfrak{g}$. For canonical coordinates of the second kind one has instead

$$\mathrm{ccsk}(B)^{-1} = \mathrm{expm}(-\beta_d B_d)\,\mathrm{expm}(-\beta_{d-1}B_{d-1}) \cdots \mathrm{expm}(-\beta_1 B_1).$$

We seek a solution of (7.1) of the form

$$Y(t) = \phi(\Theta(t))\phi(C)^{-1}Y_n \tag{7.2}$$

for $t \in [t_n, t_{n+1}]$. Differentiating in the usual fashion we obtain a differential equation for $\Theta(t)$

$$\Theta' = \mathrm{d}\phi_\Theta^{-1}(A(t, Y)), \qquad t \in [t_n, t_{n+1}], \tag{7.3}$$

where Y is as in (7.2), in tandem with the initial condition

$$\Theta(t_n) = C. \tag{7.4}$$

Thus, changing the centre of coordinate map does not affect the differential equation obeyed by Θ, just its initial condition.

7.3. The adjoint of Lie-group methods

Assume that the Lie-algebra differential equation (7.3), with the initial condition (7.4), is computed with a numerical method Ψ_h, and denote by Ψ_h^* its adjoint in the classical sense of Section 7.1, namely

$$(\Psi_{-h}^* \circ \Psi_h)B = B$$

for all $B \in \mathfrak{g}$. The corresponding Lie-group method is such that

$$Y_{n+1} = \tilde{\Psi}_h Y_n = \phi(\Theta_{h,n+1})\phi(C_{h,n})^{-1}Y_n,$$
$$\Theta_{h,n+1} = \Psi(h, t_n, C_{h,n}),$$

where $C_{h,n}$ is the initial condition of Θ in the interval $[t_n, t_{n+1}]$, and we allow it to depend on the interval of integration and on the step-size h. In order to obtain the adjoint of the Lie-group method $\tilde{\Psi}_h$, we need not just use Ψ_h^* in $\mathfrak{g}$, but also make sure that the coordinate map employed while stepping forward with the method $\tilde{\Psi}_h$ is the same as when stepping backward with the adjoint method $\tilde{\Psi}_h^*$. Define a pair of methods $\tilde{\Psi}$ and $\tilde{\Psi}^*$ on $\mathcal{G}$ as

$$\tilde{\Psi}(t_n + h, t_n, Y_n) = \phi(\Theta_{h,n+1})\phi(C_{h,n})^{-1}Y_n,$$
$$\tilde{\Psi}^*(t_n + h, t_n, Y_n) = \phi(\Theta_{h,n+1}^*)\phi(C_{h,n}^*)^{-1}Y_n,$$
$$C_{-h,n+1}^* = \Theta_{h,n+1}, \tag{7.5}$$

where $\Theta_{h,n+1} = \Psi(h, t_n, C_{h,n})$ and $\Theta_{h,n+1}^* = \Psi^*(h, t_n, C_{h,n}^*)$.

Theorem 7.1. (Zanna et al. 1999) The method $\tilde{\Psi}^*$ is the Lie-group adjoint of $\tilde{\Psi}$. Moreover, $(\tilde{\Psi}^*)^* = \tilde{\Psi}$.

Proof. With the same notation as above, we have

$$(\tilde{\Psi}_{-h}^* \circ \tilde{\Psi}_h)Y_n = \phi(\Theta_{-h,n+2}^*)\phi(C_{-h,n+1}^*)^{-1}\phi(\Theta_{h,n+1})\phi(C_{h,n})^{-1}Y_n.$$

Because of (7.5), one has $\phi(C_{h,n})^{-1}Y_n = \phi(C_{-h,n+1}^*)Y_{n+1}$, from which we deduce that

$$\phi(C_{-h,n+1}^*)^{-1}\phi(\Theta_{h,n+1}) = I.$$

Furthermore $\Theta_{h,n+1}$ and $\Theta_{-h,n+2}^*$ are solutions of the same differential equation (7.3) whereby the initial condition of Θ^* is the endpoint of Θ and Θ^* is obtained by means of Ψ_{-h}^*, where Ψ_h^* is the adjoint of Ψ_h in the classical sense. Thus $(\tilde{\Psi}_{-h}^* \circ \tilde{\Psi}_h)Y_n = Y_n$ and the assertion follows. □

We have not yet specified what $C_{h,n}$ is. In general we let it be a function of the step-size of integration h and of the current stage values F_i of a Runge–Kutta method Ψ,

$$C_{h,n} = \vartheta(F_1, F_2, \ldots, F_\nu),$$

ν being the number of the stages of the scheme (see Appendix A for notation). Thus,

$$C^*_{-h,n+1} = \vartheta^*(F_1^*, F_2^*, \ldots, F_\nu^*),$$

therefore the functions ϑ and ϑ^* obey the fundamental *adjointness condition*

$$\vartheta^*(F_1^*, F_2^*, \ldots, F_\nu^*) = \vartheta(F_1, F_2, \ldots, F_\nu) + \sum_{i=1}^{\nu} b_i F_i,$$

for the centre of the coordinate map.

The following result, which can be found in Zanna et al. (1999), characterizes self-adjoint Lie-group methods.

Theorem 7.2. Assume that Ψ is self-adjoint on $\mathfrak{g}$. Then $\tilde{\Psi}$ is self-adjoint on G provided that

$$\vartheta(F_\nu, F_{\nu-1}, \ldots, F_1) = \vartheta(F_1, F_2, \ldots, F_\nu) + \sum_{i=1}^{\nu} b_i F_i. \tag{7.6}$$

7.4. Geodesic- and flow-symmetric coordinate maps

Basically, there are two distinct ways to generate coordinate maps such that (7.6) is obeyed. One way to achieve this goal is to choose ϑ so that the value $Y_{n+1/2} = \phi(C_{h,n})^{-1}Y_n$ is a 'midpoint in space' between Y_n and Y_{n+1}, which will generate what we call the *geodesic midpoint method*. An alternative is to choose ϑ so that $Y_{n+1/2} = \phi(C_{h,n})^{-1}Y_n$ is instead a 'midpoint in time', thus generating a *flow midpoint*. The situation is schematically represented in Figure 7.1.

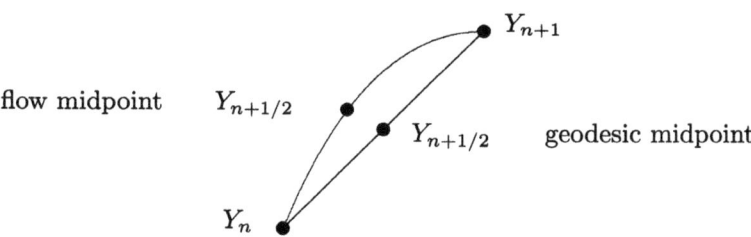

Fig. 7.1. Representation of the geodesic- and flow-symmetric midpoint

In the first instance we say that the coordinate map is *geodesic-symmetric*, while in the second case we say that the coordinate map is *flow-symmetric*.

Geodesic-symmetric coordinate maps are always defined and correspond to the choice

$$C_{h,n} = \vartheta(F_1, F_2, \ldots, F_\nu) = -\tfrac{1}{2}\sum_{i=1}^{\nu} b_i F_i.$$

Flow-symmetric coordinates are instead more naturally defined for methods based on collocation. If $\ell_i(x)$, $i = 1,\ldots\nu$, denote the familiar cardinal polynomials of Lagrangian interpolation, already introduced in Section 5.1; we set

$$w_i = \int_0^{\frac{1}{2}} \ell_i(\tau)\,\mathrm{d}\tau, \qquad i = 1, 2, \ldots, \nu, \tag{7.7}$$

and the choice

$$C_{h,n} = \vartheta(F_1, F_2, \ldots, F_\nu) = -\sum_{i=1}^{\nu} w_i F_i$$

corresponds to the flow midpoint.

There are other choices of functions ϑ that obey (7.6) and it is possible to show that the set of such functions is convex. See Zanna et al. (1999) for further examples of coordinate maps that yield self-adjoint schemes.

To conclude this section, we illustrate with a numerical experiment the benefits of using the geodesic- and flow-symmetric coordinates instead of classical coordinates centred at Y_n. We consider the *Euler equations* for a rigid body, that is,

$$\boldsymbol{y}' = \boldsymbol{y} \times M\boldsymbol{y}, \quad t \geq 0, \qquad \boldsymbol{y}(0) = \boldsymbol{y}_0, \tag{7.8}$$

where $\boldsymbol{y} \in \mathbb{R}^3$ (we assume that $\|\boldsymbol{y}_0\|_2 = 1$), the symbol '$\times$' denotes the classical vector product on $\mathbb{R}^3$ and $M = \mathrm{diag}\,(m_1, m_2, m_3)$ is a diagonal matrix. This system has the Hamiltonian function $H(\boldsymbol{y}) = \tfrac{1}{2}(m_1 y_1^2 + m_2 y_2^2 + m_3 y_3^2)$ and obeys $\|\boldsymbol{y}\|_2 = 1$. It can be represented by means of Lie-group action of SO(3) on $\mathbb{R}^3$ by representing the solution $\boldsymbol{y}(t)$ as $Q(t)\boldsymbol{y}_n$, $n = 0, 1, 2, \ldots$, with $Q \in \mathrm{SO}(3)$, $t \in [t_n, t_{n+1}]$. Hence, in each interval $[t_n, t_{n+1}]$ we solve the differential equation

$$Q'(t) = A(\boldsymbol{y}(t))Q(t), \quad t \geq t_n, \qquad Q(t_n) = I, \tag{7.9}$$

where

$$A(\boldsymbol{y}(t)) = -\begin{bmatrix} 0 & -m_3 y_3 & m_2 y_2 \\ m_3 y_3 & 0 & -m_1 y_1 \\ -m_2 y_2 & m_1 y_1 & 0 \end{bmatrix}.$$

In this numerical experiment, $m_1 = 1$, $m_2 = \tfrac{1}{3}$ and $m_3 = \tfrac{1}{5}$ and $h = t_{n+1} - t_n = \tfrac{1}{10}$, while the initial condition is a random 3-vector with unit norm. We remark that such action automatically obeys the homogeneous-space condition $\|\boldsymbol{y}\|_2 = 1$ whenever a Lie-group method is applied to (7.9).

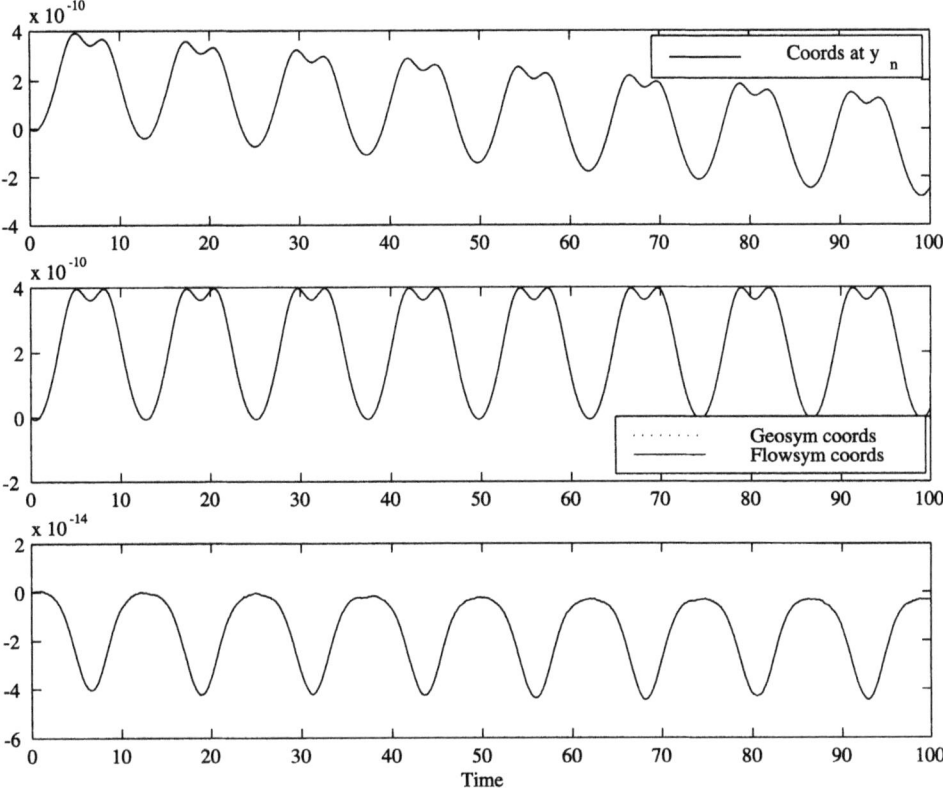

Fig. 7.2. Error in the Hamiltonian versus time for an order-four RK-MK method based on Gaussian nodes. The top plot corresponds to coordinates centred at $\boldsymbol{y}_n$ while the second plot corresponds to geodesic- and flow-symmetric coordinates. Although generally the two latter choices would correspond to different error, in this case the errors are very similar. The bottom plot corresponds to the difference between the errors in geodesic- and flow-symmetric coordinates

We compare an RK-MK method of order four based on Gauss–Legendre quadrature, using coordinates centred at $\boldsymbol{y}_n$ with geodesic- and flow-symmetric coordinates. The error in the Hamiltonian function, evaluated as

$$\mathrm{err}_H = H(\boldsymbol{y}_n) - H(\boldsymbol{y}_0),$$

is displayed in Figure 7.2. A similar comparison is displayed in Figure 7.3 for a method based on Magnus expansion of order four. The outcome reveals that methods employing geodesic- and flow-symmetric coordinates display better error propagation.

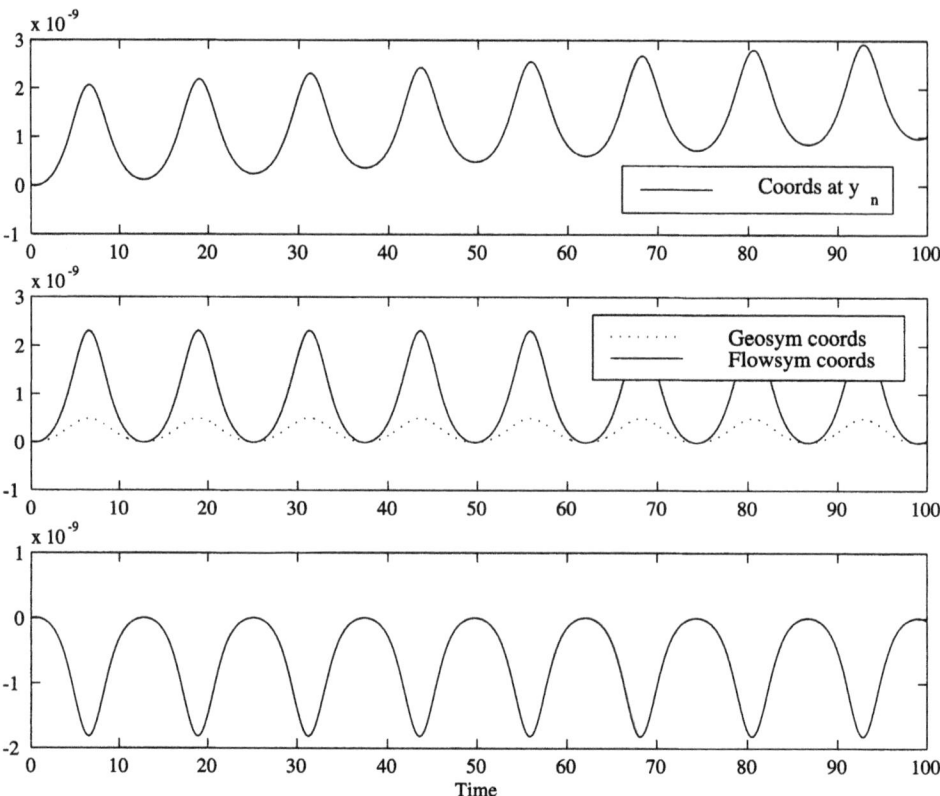

Fig. 7.3. Error in the Hamiltonian versus time for an order-four Magnus method based on Gaussian nodes. The top plot corresponds to coordinates centred at y_n while the second plot corresponds to geodesic- and flow-symmetric coordinates. The bottom plot corresponds to the difference between the errors in geodesic- and flow-symmetric coordinates

Precise details of numerical schemes used in the above example can be found in Appendix A.

8. Computation of exponentials

8.1. Six dubious ways to compute the exponential of a matrix in a Lie algebra

Most (but by no means all) Lie-group methods require repeated calculation of exponentials or, in a more realistic setting, an approximation of exponentials. In principle, this is a well-tried and familiar task in numerical analysis and can be accomplished in one of several ways: rational approx-

imants (Baker 1975, Iserles and Nørsett 1991), Krylov-subspace methods
(Hochbruck and Lubich 1997), Schur decomposition (Golub and Van Loan
1996) and so on. Although such methods have occasionally attracted healthy
scepticism (Moler and Van Loan 1978), it is fair to say that they have a distinguished track record across numerical analysis. However, as we have
already commented in Section 6, our present task is subject to a crucial restriction: *Our approximant must map the Lie algebra $\mathfrak{g}$ to the Lie group $\mathcal{G}$!*

Low-dimensional algebras are easy and often we can evaluate the exponential explicitly. In particular, the following two cases are of practical
importance.

- Firstly, given $A = \begin{bmatrix} a & b \\ c & -a \end{bmatrix} \in \mathfrak{sl}(2)$, we can easily establish that

$$e^A = \cosh \omega \, \boldsymbol{I} + \frac{\sinh \omega}{\omega} A, \qquad \text{where} \qquad \omega = \sqrt{a^2 + bc}. \qquad (8.1)$$

  This will be of use in Section 11, in our discussion of the application of
  Magnus expansions to the calculation of Sturm–Liouville spectra.

- Secondly, the exponential of

$$A = \begin{bmatrix} 0 & a & b \\ -a & 0 & c \\ -b & -c & 0 \end{bmatrix} \in \mathfrak{so}(3)$$

  is

$$\boldsymbol{I} + \frac{\sin \sigma}{\sigma} A + \frac{1 - \cos \sigma}{\sigma^2} A^2,$$

  where $\sigma = \sqrt{a^2 + b^2 + c^2}$. Given the number of spatial dimensions in
  our universe, it will come as little surprise that many useful equations,
  for instance (7.8), can be formulated in $\mathfrak{so}(3)$. We will return to the
  above expression, known as the *Rodrigues formula*, in Appendix B.

These, however, are the exceptions.

An exact formula being unavailable, an appealing alternative is to compute the exponential to machine accuracy. This, however, is neither affordable nor always reliable. The MATLAB function **expm** computes the exponential by scaling and squaring a diagonal Padé approximant: the procedure
is very expensive for large dimensions and the outcome often falls short of
machine accuracy and is subject to fast error accumulation.

When neither an explicit formula nor computation to machine accuracy
are feasible, we must resort to approximation. Any such procedure must
conform with two conditions. The outcome lies in the correct Lie group $\mathcal{G}$
and it departs from the exact exponential only to an extent consistent with
the order of the Lie-algebra method. A standard means of approximation is
to replace e^z with a function $r(z)$, analytic in a neighbourhood of the origin.

The action of such a function can be extended from $\mathbb{C}$ to $\mathfrak{gl}(N)$, hence to any matrix Lie algebra, by elementary means. Our two desiderata can now be reformulated by requiring that $r(z) = \mathrm{e}^z + \mathcal{O}(z^{p+1})$, where $p \geq 1$ is the order of the Lie-algebra time-stepping procedure, and $r(\mathfrak{g}) \subseteq \mathcal{G}$.

As we have already mentioned in Section 6, the above conditions might be much too restrictive. As has been proved by Kang and Zai-jiu (1995), the only analytic function that maps $\mathfrak{sl}(N)$ to $\mathrm{SL}(N)$ for every $N \in \mathbb{N}$ and takes zero to identity is $r(z) = \mathrm{e}^{\alpha z}$ for $\alpha \in \mathbb{R}$. Requiring consistency means that we must choose the exact exponential! On the other hand, in a *quadratic* Lie algebra we are faced with an abundance of riches: given an arbitrary odd function f, analytic about the origin, it is true that $\mathrm{e}^{f(\mathfrak{g})} \subseteq \mathcal{G}$ (Celledoni and Iserles 1998). In particular, this is the case with all *diagonal Padé approximants*,

$$r(z) = \frac{p_m(z)}{p_m(-z)}, \qquad \text{where} \qquad p_m(z) = \sum_{k=0}^{m} \binom{m}{k} \frac{(2m-k)!}{(2m)!} z^k, \quad m \in \mathbb{N}.$$

Yet, all this is of lesser utility since, arguably, the method of choice for quadratic Lie algebras rests upon the use of the Cayley transform (*i.e.*, the diagonal Padé approximant with $m = 1$) as an alternative action, thereby avoiding altogether the need to approximate the exponential function!

Yet another option is to evaluate the exponential with a *Krylov-subspace method*. Assuming for simplicity that we wish to approximate $\mathrm{e}^A v$, where $A \in \mathfrak{gl}(N)$ and $v \in \mathbb{R}^N$, such techniques choose the approximant from the space $\mathcal{K}_{N,M} = \mathrm{span}\,\{v, Av, \ldots A^{M-1}v\}$. Surprisingly small values of M produce remarkably good and affordable approximants (Hochbruck and Lubich 1997). Yet there is absolutely nothing in this approach to guarantee that the outcome resides in the correct Lie group.

The last (and perhaps the most obvious) candidate for our list of alternatives to the matrix exponential is *projection*. For example, to travel from $\mathfrak{sl}(N)$ to $\mathrm{SL}(n)$, we may employ a diagonal Padé approximant. The outcome, $V = r(A)$, say, cannot be expected to reside in the special linear group. However, replacing V with $V/(\det V)^{1/N}$ produces an element in $\mathrm{SL}(n)$. Unfortunately, experience tells us that this procedure is prone to instability (*cf.* Section 11.4).

8.2. Splitting methods

Let $A \in \mathfrak{g}$. In the spirit of Celledoni and Iserles (1998), we wish to approximate

$$\mathrm{e}^{tA} \approx R(tA) = \mathrm{e}^{tB_1}\mathrm{e}^{tB_2} \cdots \mathrm{e}^{tB_s}, \tag{8.2}$$

where the matrices $\boldsymbol{B} = \{B_1, B_2 \ldots, B_s\}$ are subject to the following requirements.

(1) Each B_l resides in $\mathfrak{g}$.
(2) It is cheap to evaluate $\mathrm{expm}(tB_l) \in \mathcal{G}$ exactly for each l.
(3) It is cheap to multiply exponentials in (8.2).
(4) The error is suitably small, $R(tA) = \mathrm{e}^{tA} + \mathcal{O}(t^{p+1})$.

We choose the *splitting* $\boldsymbol{B}$ so that it consists of low-rank matrices.

Let K and L be two $N \times r$ matrices, where $r \geq 1$ is small, and assume that $C = B_l = KL^{\mathrm{T}}$. Then

$$\mathrm{e}^{tC} = I + tKD^{-1}(\mathrm{e}^{tD} - I)L^{\mathrm{T}},$$

where $D = L^{\mathrm{T}}K$ (Celledoni and Iserles 1998). Note that D is just $r \times r$ and the cost of evaluating e^{tC} exactly is modest for small values of r.

As an example, let us consider $\mathfrak{g} = \mathfrak{so}(N)$. We let $r = 2$, $s = N - 1$, set $B^{[0]} = A = [\boldsymbol{b}_1^{[0]}, \boldsymbol{b}_2^{[0]}, \dots, \boldsymbol{b}_N^{[0]}]$, and choose $B_1 = \boldsymbol{b}_1^{[0]}\boldsymbol{e}_1^{\mathrm{T}} - \boldsymbol{e}_1\boldsymbol{b}_1^{[0]\mathrm{T}} \in \mathfrak{so}(N)$, where $\boldsymbol{e}_k \in \mathbb{R}^N$ is the kth unit vector. Letting $B^{[1]} = B^{[0]} - B_1$, we observe that its first row and column vanish. We continue in a manner similar to LU factorization, letting $B^{[i]} = B^{[i-1]} - B_i$ and

$$B_i = \boldsymbol{b}_i^{[i-1]}\boldsymbol{e}_i^{\mathrm{T}} - \boldsymbol{e}_i\boldsymbol{b}_i^{[i-1]\mathrm{T}} \in \mathfrak{so}(N), \qquad i = 1, 2, \dots, N - 1.$$

We refer to Celledoni and Iserles (1998) for precise estimation of cost, implementation details and a similar example for $\mathfrak{sl}(N)$, as well as for an example of a splitting, again with $r = 2$, that eliminates two rows and columns of $B \in \mathfrak{so}(N)$ at a time. Although the number of exponentials in (8.2) is generally quite large for low-rank splittings, the underlying linear algebra carries a reasonable price tag.

The main disadvantage of low-rank splitting methods is the quality of approximation: in general, we can expect order $p = 1$. The second-order condition is

$$\sum_{k=1}^{s-1} \sum_{l=k+1}^{s} [B_k, B_l] = O$$

and it is easy to verify that it is satisfied when (8.2) is the *Strang splitting*: $s = 2\tilde{s} + 1$ and $B_{\tilde{s}+i} = B_{\tilde{s}-i}$, $i = 1, 2, \dots, \tilde{s}$. In principle, it is easy to convert any low-rank matrix so that it becomes a Strang splitting by first approximating $\frac{1}{2}A$ with a *first-order* splitting (8.2), next approximating the same matrix with the same splitting but with the matrices B_l arranged in reverse order, and finally 'aggregating' the middle two terms. The outcome,

$$S(tA) = \mathrm{e}^{tB_1/2}\mathrm{e}^{tB_2/2} \cdots \mathrm{e}^{tB_{s-1}/2}\mathrm{e}^{tB_s}\mathrm{e}^{tB_{s-1}/2} \cdots \mathrm{e}^{tB_2/2}\mathrm{e}^{tB_1/2}, \qquad (8.3)$$

costs twice as much as (8.2) but it has a crucial advantage: it is not just second-order but also *time-symmetric*. This renders (8.3) amenable to the application of the *Yoshida device* (Sanz Serna and Calvo 1994, Yoshida

1990). Thus, the function

$$S(\alpha tA)S((1-2\alpha)tA)S(\alpha tA), \qquad \text{where} \qquad \alpha = \frac{2}{3} + \frac{\sqrt[3]{2}}{3} + \frac{\sqrt[3]{4}}{6},$$

approximates e^{tA} to order four. A similar procedure can be used to increase the order further in increments of two.

A most welcome feature of low-rank splittings is that they can be implemented to take advantage of sparsity. Provided that A is banded, say, all the computations can be confined to the relevant band and sparsity is inherited as we are 'mopping up' rows and columns as in the $\mathfrak{so}(N)$ algorithm above.

8.3. Canonical coordinates of the second kind

Our point of departure is similar to the reasoning behind the CCSK representation (6.16). Again, $C = \{C_1, C_2, \dots, C_d\}$ is a basis of the Lie algebra $\mathfrak{g}$ and we seek polynomials $\theta_1, \theta_2, \dots, \theta_d$ so that

$$e^{\theta_1(t)C_1} e^{\theta_2(t)C_2} \cdots e^{\theta_d(t)C_d} = e^{tA} + \mathcal{O}(t^{p+1}). \tag{8.4}$$

C being a basis, there exist scalars $a_1, a_2, \dots, a_d$ so that $A = \sum_{k=1}^{d} a_k C_k$. Letting $\theta_k(t) = a_k t$, $k = 1, 2, \dots, d$, gives us a first-order splitting (8.2). With greater generality, we set $\theta_k(0) = 0$, $\theta_k'(0) = a_k$, $k = 1, 2, \dots, d$, to guarantee $p \geq 1$.

To obtain higher-order conditions in (8.4) we proceed as in Section 6.4. Differentiation and further algebra produce, analogously to (6.17), the equation

$$A = \sum_{k=1}^{d} a_k C_k = \sum_{k=1}^{d} \theta_k'(t) \mathrm{Ad}_{\,\mathrm{expm}[\theta_1(t)C_1]} \cdots \mathrm{Ad}_{\,\mathrm{expm}[\theta_1(t)C_1]} C_k + \mathcal{O}(t^p).$$

$$\tag{8.5}$$

We go on differentiating (8.5) and setting $t = 0$. This yields order conditions, which need to be unscrambled by further algebra, exploiting the *structure constants* (*cf.* Section 6.4 for the definition) of C. The outcome is

$$p \geq 2: \quad \theta_k''(0) = \sum_{l=1}^{d} \sum_{j=1}^{l-1} a_l c_{l,j}^k a_j, \qquad k = 1, 2, \dots, d,$$

$$p \geq 3: \quad \theta_k'''(0) = 2 \sum_{l=1}^{d} \sum_{j=1}^{l-1} c_{l,j}^k [\theta_l''(0) a_j + a_l \theta_j''(0)]$$

$$+ 2 \sum_{l=1}^{d} \sum_{j=1}^{l-1} \sum_{i=1}^{j-1} \sum_{m=1}^{d} c_{l,j}^m c_{i,m}^k a_l a_j a_i$$

$$+ \sum_{l=1}^{d} \sum_{j=1}^{d} \sum_{i=1}^{l-1} c_{l,i}^j c_{i,j}^k \beta_l \beta_i^2, \qquad k = 1, 2, \dots, d$$

(Celledoni and Iserles 1999).

The cost of this procedure is, at first glance, prohibitive. We might just about get away with computing order-two conditions at the cost of $\mathcal{O}(d^3)$ operations, but a price tag of $\mathcal{O}(d^5)$ operations for order three, to say nothing of higher orders, is out of the question. This naive impression is misleading, since we are absolutely free to exploit, along the lines of Sections 6.4–6.5, sparsity in structure constants in a serendipitously chosen basis $\boldsymbol{C}$.[4] Again, Chevalley bases present the best choice, as well as leading to very easy computation of exponentials in (8.4). For example, choosing the basis $\{E_{k,l}-E_{l,k} : 1 \leq k < l \leq N\}$ of $\mathfrak{so}(N)$, an order-two approximation is attained by letting

$$\theta_{k,l}(t) = a_{k,l}t + \tfrac{1}{2}\sum_{i=1}^{N} a_{k,i}a_{i,l}t^2, \qquad 1 \leq k < l \leq N,$$

an $\mathcal{O}(N^3) = \mathcal{O}(d^{3/2})$ procedure. In the case of $\mathfrak{sl}(N)$ the approach advocated in Celledoni and Iserles (1999) is to choose the ordered basis $\boldsymbol{C} = \boldsymbol{C}_1 \cup \boldsymbol{C}_2$, where

$$\boldsymbol{C}_1 = \{E_{k,l} : k \neq l\}, \quad \boldsymbol{C}_2 = \{E_{k,k} - E_{k+1,k+1} : k = 1, 2, \ldots, N-1\},$$

where each set is ordered lexicographically. Let

$$A = \sum_{\substack{k,l=1 \\ k\neq l}}^{N} a_{k,l}E_{k,l} + \sum_{k=1}^{N-1} b_k(E_{k,k} - E_{k+1,k+1})$$

and denote the coefficients corresponding to terms in $\boldsymbol{C}_1$ and $\boldsymbol{C}_2$ by $\theta_{k,l}$ and η_k respectively. The second-order conditions are

$$\theta_{k,l}(t) = a_{k,l}t + \tfrac{1}{2}\left[\sum_{i=1}^{k-1} a_{k,i}a_{i,l} + a_{k,l}(b_{k-1} - b_k + b_l - b_{l-1})\right.$$
$$\left. - \sum_{i=k+1}^{N} a_{k,i}a_{i,l}\right]t^2, \qquad k \neq l,$$

$$\eta_k(t) = b_k t - \tfrac{1}{2}\sum_{i=1}^{k}\sum_{j=k+1}^{N} a_{i,j}a_{j,i}t^2, \qquad k = 1, 2, \ldots, N-1,$$

again just $\mathcal{O}(N^3) = \mathcal{O}(d^{3/2})$ operations.

An intriguing aspect of approximants based on CCSK is that they might well be ideally suitable to handling sparsity in the matrix A. Although this

---

[4] Sparsity in structure constants has no connection whatsoever with sparsity (or otherwise) of the matrix A.

issue is by no means fully understood, there are enough encouraging pointers to justify a brief discussion. One mechanism that exploits sparsity is that the latter can be taken into account in the evaluation of the θ_ks because of the association between terms in a Chevalley basis and the entries of A. For example, the second-order conditions for a *tridiagonal* matrix $A \in \mathfrak{sl}(N)$ reduce to

$$
\theta_{k,l}(t) = \begin{cases}
\frac{1}{2}a_{k,k-1}a_{k-1,k-2}t^2, & l = k-2, \\
a_{k,k-1}t - \frac{1}{2}a_{k,k-1}(b_k - 2b_{k-1} + b_{k-2})t^2, & l = k-1, \\
a_{k,k+1}t + \frac{1}{2}a_{k,k+1}(b_{k+1} - 2b_k + b_{k-1})t^2, & l = k+1, \\
-\frac{1}{2}a_{k,k+1}a_{k+1,k+2}t^2, & l = k+2, \\
0, & |k-l| \neq 1,
\end{cases}
$$

$$
\eta_k(t) = b_k t - \frac{1}{2}a_{k,k+1}a_{k+1,k}t^2.
$$

This entails just $\mathcal{O}(N)$ operations and, equally importantly, just $\mathcal{O}(N)$ terms survive in the product (8.4). The cost scales with the number of nonzero elements in the matrix, rather than with dimension, a hallmark of a good method for sparse matrices.

Another mechanism is of more tentative value, yet we believe that it deserves mentioning. Even if the matrix A is sparse, its exponential is dense. However, it is possible to prove, at least in the case of *banded matrices*, that most of the entries are exceedingly small and that the loci of large elements are predictable (Iserles 1999a). As an example, we have averaged 1000 exponentials of 100×100 matrices with the cruciform sparsity pattern displayed in Figure 8.1(a) and with random elements uniformly distributed in $(-1, 1)$ and normalized so that $\max_{k,l}|a_{k,l}| = 1$. Let W be the average of all the exponentials. The matrix is dense, yet most of the elements of W are tiny! Thus, Figure 8.1(b) displays the sparsity pattern of W without all its entries that are smaller than 10^{-6} in magnitude. Although the cruciform shape 'swells', most of the matrix consists of zero entries. This observation is affirmed in Figure 8.1(c), where we have plotted the matrix $\log_{10}|W|$. The vertical axis tells the magnitude of the entries in terms of significant (decimal) digits. The decay outside the original cruciform shape is evident. Using upper bounds from Iserles (1999a), it is possible to say how rapidly entries decay for banded A but computer experiments (and, indeed, Figure 8.1) indicate that similar behaviour takes place for more exotic sparsity patterns.

Choosing a Chevalley basis, members of C mostly correspond to elements of A. Given a tolerance $\varepsilon > 0$ and knowing which elements of e^A are bound to be smaller than ε in magnitude, we are free to remove them altogether from the product (8.4). The outcome departs entry by entry from the exact exponential by at most ε and, by design, it resides in the Lie group $\mathcal{G}$.

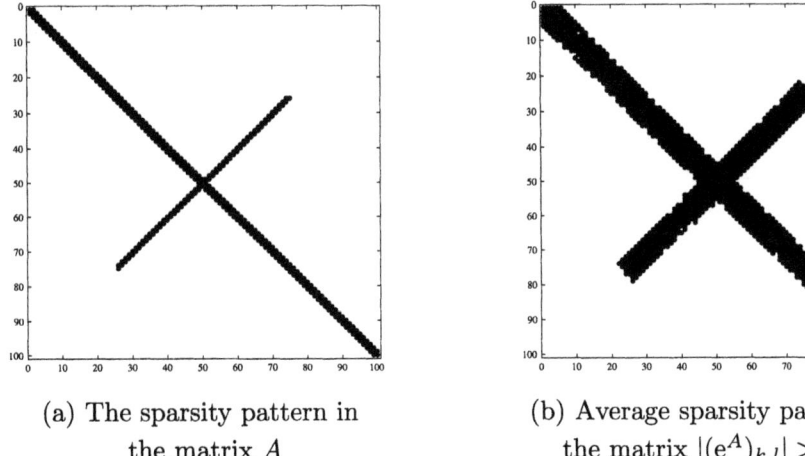

(a) The sparsity pattern in
the matrix A

(b) Average sparsity pattern in
the matrix $|(e^A)_{k,l}| > 10^{-6}$

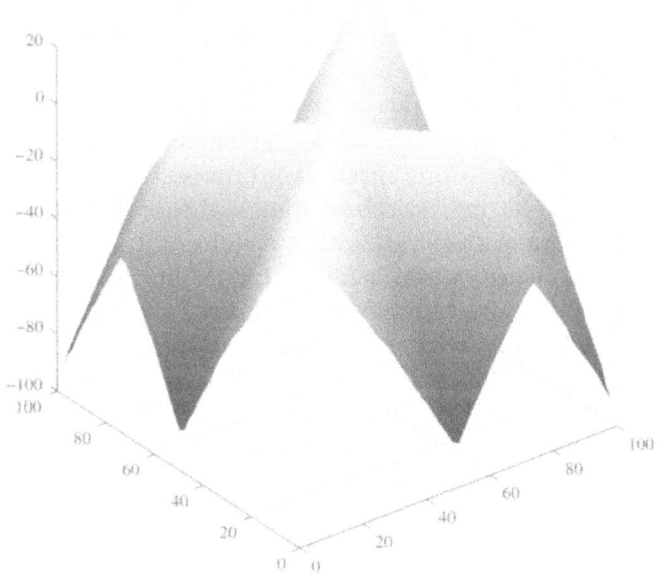

(c) The 'shape' of the matrix $\log_{10} |(e^A)_{k,l}|$

Fig. 8.1. How large is the exponential of a sparse matrix?

The sparsity pattern of a matrix exponential of a sparse matrix after the excision of small entries is at present unknown. (Banded matrices are an exception.) Moreover, full implications of this phenomenon to the subject matter of this section are far from clear. Having said this, our analysis emphasizes a very important and welcome feature of methods based upon canonical coordinates of the second kind, applied in conjunction with Chevalley bases: the connection between basis elements and entries of the matrix is a powerful tool, which we can bring to bear with pointwise precision upon the exploitation of sparsity.

9. Stability and backward error analysis

How stable are Lie-group methods? The combined wisdom of half a century of computational analysis of ODEs is that numerical algorithms for initial-value problems are to all intents and purposes useless unless they exhibit favourable stability properties. Indeed, much of the narrative of modern numerical analysis of ODEs is the tale of stability, linear and nonlinear alike, culminating in a profound understanding of the subject. We refer the reader to the monograph of Stuart and Humphries (1996) for a comprehensive review of this important subject area.

The word 'stability' has so far been conspicuously absent from our exposition. A partial reason is ignorance: much remains to be done in the realm of stability investigations in a Lie-group setting. Interesting results abound which cannot yet be fitted into a general theory. Thus, we can learn from computation that merely projecting a solution into the right manifold often leads to instabilities, while intrinsic Lie-group methods exhibit much more favourable behaviour. Much of the advance in stability theory for classical numerical ODEs was concerned with identifying appropriate *stability models*: broad enough to provide insight about many differential systems of interest, yet sufficiently focused and narrowly defined to be amenable to rigorous analysis. Such are the linear model, the monotone model and the many more advanced models from Stuart and Humphries (1996), originating in the theory of nonlinear dynamical systems. This chapter in the narrative of Lie-group methods cannot yet be written, except for the observation that some Lie-group solvers, for instance RK-MK, Magnus and Fer expansions with exactly calculated exponentials compute the solution of linear equations with *constant coefficients* exactly: in that case there is no need for stability analysis!

The last few years have seen the emergence of an alternative stability theory, mainly within the context of symplectic integration of Hamiltonian ODEs and discretization of dynamical systems. In addition to asking 'How near is the numerical solution to the exact one and how influenced is it by small perturbations?', the new breed of stability researchers also poses a different query: 'What is it that our numerical method solves *exactly*? And how far apart is it from the equation that we wish to solve?' This is precisely the question of *backward error analysis* that J. H. Wilkinson made into the centrepiece of modern numerical linear algebra. Arguably, it is just as relevant in the ODE setting and it has already led to impressive new insights (Benettin and Giorgilli 1994, Hairer 1994, Hairer and Lubich 1997, Neishtadt 1984, Reich 1996).

In this section we report briefly on recent work of Faltinsen (1998), who has generalized backward error analysis to a Lie-group setting.

Discussion of stability is meaningless without the concept of *distance*. Thus, given a Lie group $\mathcal{G}$, we seek $d : \mathcal{G} \times \mathcal{G} \to \mathbb{R}^+$, which is consistent with the standard axioms of a metric in an Euclidean space and, in addition, compatible with the topology of $\mathcal{G}$. A good way to ensure compatibility is to require that d is *left invariant*, *i.e.*, that

$$d(ZX, ZY) = d(X, Y) \qquad X, Y, Z \in \mathcal{G}. \tag{9.1}$$

Ideally, we would have liked d to be *bi-invariant*: satisfying both (9.1) and the condition $d(XZ, YZ) = d(X, Y)$ for all $X, Y, Z \in \mathcal{G}$. This, however, is not always possible. On the other hand, according to the Birkhoff–Kakutani theorem (Birkhoff 1936), every Lie group $\mathcal{G}$ admits a left-invariant, *almost right-invariant* metric which, in addition to (9.1), obeys $d(XZ, YZ) \le \rho(Z)\, d(X, Y)$, where the function ρ is finite. Note that existence and uniqueness of a bi-invariant Riemannian metric is assured when the Lie group is compact and connected (Boothby 1975, p. 244).

Assumption 9.1. The metric d is left-invariant and almost right-invariant.

An example is the *geodesic metric* in $O(N)$ (which, as a matter of fact, is bi-invariant): $d(X, Y) = \|\boldsymbol{\eta}\|_2$, where $e^{i\eta_1}, e^{i\eta_2}, \ldots, e^{i\eta_N}$ are the eigenvalues of $X^{\mathrm{T}}Y$. (Note that the spectrum of elements in $O(N)$ lives on the complex unit circle, hence the η_ks are real.) Other examples are more complicated to derive explicitly, but this is of little consequence since we do not require d in a closed form.

We are concerned with the solution of the Lie-group equation

$$Y' = A(t, Y)Y, \quad t \ge t_0, \qquad Y(t_0) = Y_0 \in \mathcal{G}, \tag{9.2}$$

where $A : [t_0, \infty) \times \mathcal{G} \to \mathfrak{g}$, by a Lie-group method. The flow corresponding to (9.2) is denoted by $\Phi_{t,t_0,A}$, therefore $Y(t) = \Phi_{t,t_0,A}(Y_0)$.

Assumption 9.2. The matrix function A is real analytic and there exist constants $\alpha, \beta, t^* > 0$ such that

$$\|A(t, \hat{Y})\hat{Y}\|_{\mathrm{F}} \le \alpha, \qquad t \in [0, t^*], \quad \hat{Y} \in \mathcal{B}_\beta(Y_0),$$

where $\| \cdot \|_{\mathrm{F}}$ is the *Frobenius norm* and $\mathcal{B}_\beta(Y_0) = \{\hat{Y} \in \mathcal{G} : d(Y_0, \hat{Y}) \le \beta\}$.

We are interested in Lie-group solvers that lift the solution to the corres-ponding Lie algebra $\mathfrak{g}$, whether once or repeatedly in the course of every time-step. All the methods that we have described in this survey: Crouch–Grossman and Runge–Kutta–Munthe-Kaas schemes, Magnus, Fer and Cay-ley expansions and methods based on canonical coordinates of the second kind fit this framework. An example of a method that is outside the scope of the theory of this section is *projection*. For example, in the case $\mathcal{G} = \mathrm{O}(N)$ we might time-step from Y_n to Y_{n+1}, say with an arbitrary ODE method which produces a new value $\bar{Y}_{n+1}$. We subject $\bar{Y}_{n+1}$ to a *polar decomposi-tion* and retain the orthogonal part as our new Y_{n+1} (Higham 1997). We hasten to acknowledge that this is a perfectly valid procedure, except that it is outside the scope of our present discussion.

The *map* induced by the numerical method will be denoted by $\Psi_{h,t_n,A}$, hence $Y_{n+1} = \Psi_{h,t_n,A}(Y_n)$, $n \in \mathbb{Z}^+$.

Assumption 9.3.   The Lie-group method is accurate to order $p \geq 1$, *i.e.*,

$$d(\Phi_{h,t_0,A}(Y_0), \Psi_{h,t_0,A}(Y_0)) = \mathcal{O}(h^{p+1}).$$

What is it that $\Psi_{h,t_n,A}$ solves exactly? In linear algebra this is precisely the question of backward error analysis. In so far as ODEs are concerned, however, the situation is slightly more complicated.

Theorem 9.1. (Faltinsen 1998)   Subject to Assumptions 9.1–9.3, there exists a matrix function $A_h : [t_0, \infty) \times \mathcal{G} \to \mathfrak{g}$ such that

$$\|A_h(t, X) - A(t, X)\|_{\mathrm{F}} = \mathcal{O}(h^p) \tag{9.3}$$

and

$$d(\Phi_{h,t_0,A_h}(Y_0), \Psi_{h,t_0,A}(Y_0)) = \mathcal{O}(e^{-\gamma/h}), \tag{9.4}$$

where $\gamma > 0$ is a constant.

Note that Φ_{h,t_0,A_h} evolves on the very same Lie group $\mathcal{G}$ and, because of (9.3), the *modified equation* $Y_h' = A_h(t, Y_h)Y_h$ 'approximates' the ODE (9.2). The above theorem argues that a single step of the numerical method *departs to an exponentially small extent* from the exact solution of a nearby equation! It is reminiscent of similar results in symplectic integration and its method of proof generalizes the work of Reich (1996) to a Lie-group setting.

A good approximation across a single step does not tell us much. Ideally, we wish to extend the scope of Theorem 9.1 to $[t_0, \infty)$ or, at the very least, to a large number of steps. This, however, requires further conditions. As in backward error analysis for symplectic integration, the verification of such conditions in a nonlinear case might be difficult and require bespoke analysis for different methods. Matters simplify a great deal, though, for linear equations $Y' = A(t)Y$. Suppose that the exact solution is $Y(t) =$

expm$[\Theta(t)]Y_0$, $t \geq t_0$. Let $\mu \in \mathbb{R}$ be the least real number such that

$$\|\mathrm{Ad}_{\mathrm{expm}\Theta} B\|_\mathrm{F} \leq c\,\mathrm{e}^{\mu(t-t_0)}\|B\|_\mathrm{F}, \qquad t \geq t_0$$

for some $c > 0$ which may depend on Θ. Then (9.4) can be extended to a longer interval. Specifically, it is true that

$$d(\Phi_{mh,t_0,A_h}(Y_0), \Psi_{mh,t_0,A}(Y_0)) = \mathcal{O}\big(\mathrm{e}^{-\gamma^*/h}\big),$$

where $\gamma^* > 0$ and $m \leq M(h)$, where

$$M(h) = \begin{cases} \mathcal{O}(1), & \mu > 0, \\ \mathcal{O}(h^{-p}), & \mu = 0, \\ \infty, & \mu < 0. \end{cases}$$

It is possible to prove that $\mathfrak{g} = \mathfrak{so}(N)$ implies that the operator $\mathrm{Ad}_{\mathrm{expm}\Theta}$ is itself skew symmetric and $\mu = 0$. This implies that the scope of backward error analysis is quite significant at least in this case.

Another interesting phenomenon originally identified in the analysis of Hamiltonian problems is *linear error growth* (Estep and Stuart 1995). Provided that the exact solution of a Hamiltonian differential system is periodic with period $T > 0$ and a pth-order numerical method satisfies convenient requirements (*e.g.*, reversibility or conservation of Hamiltonian energy), it is possible to prove that

$$\|\boldsymbol{y}_{nT/h} - \boldsymbol{y}(nT)\| \leq cnh^p, \tag{9.5}$$

where $c > 0$ (for simplicity we assume that T/h is integer). This is a very important feature of successful long-term integration, since lesser methods typically produce quadratic error growth and the solution is unlikely to remain periodic for long.

Does (9.5) remain valid in a Lie-group setting? Unfortunately, not always. In a recent paper, though, Engø and Faltinsen (1999) proved that solving a *Lie–Poisson system* with a method that is both a Lie-group action and conserves Hamiltonian energy results in linear error growth. As an example, let us recall the Euler equations for a rigid body (7.8). This is a Lie–Poisson system which conserves the Hamiltonian energy $H(\boldsymbol{y}) = \frac{1}{2}(m_1 y_1^2 + m_2 y_2^2 + m_3 y_3^2)$, as well as evolving on the unit sphere in $\mathbb{R}^3$. Therefore $\boldsymbol{y}(t)$ lives on a 'circle' (an intersection of a sphere and an ellipsoid), a feature shared by energy-conserving methods based on group actions, and is periodic there. We have already seen in Figures 7.2 and 7.3 that self-adjoint methods do well in recovering a periodic solution.

In Figure 9.1 we have integrated the rigid body equations (7.8) with three RK-MK methods: Forward Euler, Runge–Kutta–Heun and the trapezoidal rule, all applied in the Lie algebra. Only the latter method conserves energy (Engø and Faltinsen 1999). The results have been displayed as point-plots of $\boldsymbol{y}_n$ in three dimensions. First note that all the solution trajectories evolve

(a) Forward Euler   (b) Runge–Kutta–Heun   (c) Trapezoidal rule

Fig. 9.1. Rigid body equations (7.8), as solved by three different Lie-group
methods with $h = \frac{1}{10}$, integrating for 100000 steps

on the unit sphere: unsurprising, since we are using Lie-group methods, yet beyond the reach of most classical algorithms. Secondly, forward Euler is no respecter of periodicity and its trajectory spirals to a fixed point. Runge–Kutta–Heun is much better, yet more careful examination demonstrates how the error accumulates and periodicity is lost. The solution is qualitatively correct for a while, but long integration leads to false dynamics in this case also. The energy-conserving trapezoidal rule, though, produces a trajectory which to all intents and purposes is periodic.

10. Implementation, error control and *DiffMan*

10.1. *Implementation and error control of Lie-group solvers*

Practical implementation of Lie-group solvers requires much more than merely programming a numerical method. We must address ourselves to issues like error control and variable-step implementation. As is perhaps natural in a new subject, implementation details have so far received less attention than theoretical issues, a situation that is likely to be remedied in the next few years.

In this section we survey the little that is presently known about implementational issues, commencing with the welcome observation that variable-step procedures do not interfere with the retention of Lie-group structure. This is important, since it is known that another important geometric-integration technique, symplectic solution of Hamiltonian systems, loses many of its most favourable features unless implemented with (essentially) constant step-size (Sanz Serna and Calvo 1994).

There are two levels of discretization in Lie-group solvers and each should be monitored in a variable-step implementation and contribute to the estimate of local error:

(1) the error committed in the evaluation of the coordinate map;   and

(2) the error incurred in the solution of the Lie-algebraic equation.

In so far as the coordinate map is concerned, the situation is simple. The Cayley map (6.3) requires an inversion of a matrix. Unless its size is large, this can be accomplished by direct methods, otherwise it requires iteration. In the first case the only source of error is round-off and this issue is well understood by classical numerical algebra. In the second case the error is determined by the termination criteria and the issue is, again, transparent. The use of techniques based on coordinates of the second kind (6.15) generates round-off error only, since each individual exponential can be evaluated easily in exact arithmetic. Unfortunately, the situation is different with the most important coordinate map, expm. To date, there are no efficient means to monitor the error of methods from Section 8. The last statement refers not just to approximation methods but also to 'exact' calculation of the exponential: it will be seen in Section 11.5 that the MATLAB function expm is a stumbling block to high-precision long-term integration of some systems.[5] It is entirely conceivable that good techniques to monitor the error in expm should depend on the Lie group in question: intuitively, a compact object like $O(N)$ is easier to handle than $SL(N)$, say.

Estimation of the error in Lie algebra presents a different challenge, more akin to classical error-control theory for numerical ODEs.

A fair share of quality software for 'classical' ODEs is based on multistep methods. There exist multistep Lie-group solvers (Faltinsen, Marthinsen and Munthe-Kaas 1999), but their implementation requires travelling forwards and backwards between the group and the algebra in a manner which is, arguably, too cumbersome for the use of the 'automatic differentiation' techniques of Gear (1971), the cornerstone of most multistep ODE software.

In principle, error control of explicit RK-MK can be accomplished in a straightforward manner, using *embedded Runge–Kutta schemes* (Hairer et al. 1993). The sole difference to the classical framework is that, instead of estimating the error in the original configuration space, we do so in the algebra. An embedded RK scheme has the Butcher tableau

$$
\begin{array}{c|cccc}
c_1 & a_{1,1} & a_{1,2} & \cdots & a_{1,\nu} \\
c_2 & a_{2,1} & a_{2,2} & \cdots & a_{2,\nu} \\
\vdots & \vdots & \vdots & & \vdots \\
c_\nu & a_{\nu,1} & a_{\nu,2} & \cdots & a_{\nu,\nu} \\
\hline
 & b_1 & b_2 & \cdots & b_\nu \\
 & \bar{b}_1 & \bar{b}_2 & \cdots & \bar{b}_\nu
\end{array}
$$

---

[5] In fairness to MATLAB, in our experience expm performed better than alternatives.

and the approximants

$$\boldsymbol{y}_{n+1} = \boldsymbol{y}_n + h \sum_{l=1}^{\nu} b_l \boldsymbol{f}_l$$

$$\bar{\boldsymbol{y}}_{n+1} = \boldsymbol{y}_n + h \sum_{l=1}^{\nu} \bar{b}_l \boldsymbol{f}_l$$

(compare with (3.3)!) are of order p and $\bar{p} \geq p+1$ respectively. The higher-order approximant is used for error control, $\boldsymbol{y}_{n+1} - \boldsymbol{y}(t_{n+1}) \approx \boldsymbol{y}_{n+1} - \bar{\boldsymbol{y}}_{n+1}$. Applying a similar trick in the Lie algebra readily yields an estimate of the local error at relatively little extra cost for low-order methods. If the order is high, typically the embedded RK scheme requires an increasingly large number of additional stages to evaluate $\bar{\boldsymbol{y}}_{n+1}$ and the cost mounts. This is a problem common to high-order RK methods in the classical setting also.

We have neglected in our discussion of RK-MK one important source of error: in practical applications the dexp^{-1} operator (2.46) is truncated consistently with the order of the method, typically replaced with

$$\mathrm{dexp}_A^{-1}(C, p) = \sum_{j=0}^{p-1} \frac{\mathrm{B}_j}{j!} \mathrm{ad}_A^j C.$$

This carries an error which we are forbidden to neglect, but can estimate easily from the leading term of $\mathrm{dexp}_A^{-1} C - \mathrm{dexp}_A^{-1}(C, p)$,

$$\frac{|\mathrm{B}_q|}{q!} \|\mathrm{ad}_A^q C\|, \qquad \text{where} \qquad q = \begin{cases} 1, & p = 1, \\ 2\lfloor (p+1)/2 \rfloor, & p \geq 2. \end{cases}$$

Much more challenging is the error control of Magnus expansions, the subject of Iserles, Marthinsen and Nørsett (1999). Again, there are two discretization steps: truncation by power (4.14) of the Magnus expansion and the replacement of integrals by quadrature. Later we restrict the discussion to the linear case $Y' = A(t)Y$.

To estimate the error in the leading truncation term, we commence by assuming that the leading terms in the expansion of the matrix A are known,

$$A(t) = C_0 + C_1 t + C_2 t^2 + \cdots.$$

The error term in $\int_0^t C_\tau(\xi)\,\mathrm{d}\xi$ (cf. (4.5) for the definition of C_τ) can be evaluated from the tree τ in the following manner. We label each leaf of the tree by a few leading terms of the expansion and prune the tree according to the original composition rules from Section 4.1.

(1) If two leaves share a parent, they are excised and the parent is labelled by the commutator of their labels.

(2) If a leaf is the only child of a parent then it is eliminated and its parent is labelled with the integral of its label.

By the end of this procedure we throw away all the terms except for the leading one. An example will clarify this procedure:

C_0+C_1t

C_0+C_1t C_0 C_0

$C_0t+\frac{1}{2}C_1t^2$ C_0+C_1t C_0t C_0

$\frac{1}{2}[C_0,C_1]t^2$ C_0t C_0

$\Rightarrow$ $\Rightarrow$

$\frac{1}{6}[C_0,C_1]t^3$ C_0 C_0t

C_0t $-\frac{1}{6}[C_0,[C_0,C_1]]t^3$

$-\frac{1}{6}[C_0,[C_0,[C_0,C_1]]]t^4$

$\Rightarrow$ $\Rightarrow$ $\Rightarrow$

$$\Rightarrow \quad -\tfrac{1}{30}[C_0,[C_0,[C_0,C_1]]]t^5.$$

This procedure has been automated in Iserles et al. (1999), using a `Maple` program.

Often it is easy to find derivatives of A explicitly but in general-purpose software one needs to approximate them by finite differences. The scaled function values $A_1, A_2, \ldots, A_\nu$ are a perfectly good starting point for this elementary calculation. Note that very few derivatives are required, for instance just C_0 and C_1 for the power-5 tree in the above example, and the approximants need not be very precise. If we are using the adjoint basis, we might just as well use (appropriately scaled) $h^{-l}B_l$ in place of C_{l-1}: the difference between expansions at 0 and $\frac{1}{2}h$ is subsumed into higher-order terms.

The computation of quadrature error is more challenging and the results of Iserles et al. (1999) are of a more tentative nature. This is hardly surprising, since even the estimation of the error of univariate Gauss–Legendre quadrature is difficult (Davis and Rabinowitz 1984). Indeed, perhaps paradoxically, the univariate quadrature is the most problematic in our setting also. Of course, we may evaluate each integral by two quadratures, for instance Gauss–Legendre and Gauss–Lobatto, or perhaps consecutive Gauss–Legendre rules, but this doubles the cost. The remedy, proposed in Iserles et al. (1999), is again to use the derivatives of A: recall that in many instances they are easy to evaluate explicitly.

Straightforward, yet messy, calculation, expanding everything in sight into Taylor series in h, demonstrates that the leading error terms in the first four integrals in (4.10), using two-point Gauss–Legendre quadrature, are

$$E_1 = -\tfrac{1}{180}C_4 h^5,$$
$$E_2 = -(\tfrac{1}{240}[C_0, C_3] + \tfrac{1}{1080}[C_1, C_2])h^5,$$
$$E_3 = \tfrac{1}{1080}([C_0, [C_0, C_2]] + [C_1, [C_0, C_1]])h^5,$$
$$E_4 = -(\tfrac{1}{270}[C_0, [C_0, C_2]] + \tfrac{1}{720}[C_1, [C_0, C_1]])h^5,$$

respectively. Note that only E_1 depends on the fourth derivative.

A Matlab program incorporating variable time-stepping strategy with the above error estimators for the fourth-order Magnus method has been applied in Iserles et al. (1999) to a large number of test problems, in each case behaving predictably and producing error consistent with the specified tolerance. Yet much work remains to be done in this subject area, not just to estimate quadrature error without the need for higher derivatives, but also to develop the right strategies to the methods of Blanes et al. (1999) from Section 5.4.

10.2. The DiffMan package

DiffMan (Engø et al. 1999) is an object-oriented Matlab toolbox for solving differential equations on manifolds. The package embodies most of the algorithms discussed in this survey. This software is in the public domain and can be obtained from the *DiffMan* home page at **www.ii.uib.no/diffman**. Some of the numerical examples of Section 11 are distributed with *DiffMan*. Since the package is still undergoing intensive development, we do not wish to elaborate excessively upon the details of *DiffMan* and of its usage. Instead, we refer the reader to the latest version of the *DiffMan* user manual, which can be found at the above-mentioned home page.

The basic philosophy behind *DiffMan* is the idea of *coordinate-free numerics* (Munthe-Kaas and Haveraaen 1996), a research programme devoted to the study of the rôle of abstract formulations, independent of particular representations, in computational and applied mathematics. We have touched briefly upon this topic in Section 3, emphasizing the importance of using abstractions as a tool for organizing object-oriented software.

The *DiffMan* package is built upon *classes* modelling continuous mathematical structures. There are currently three main classes: *domains, fields* and *flows*. Domains consist of Lie algebras, Lie groups and homogeneous manifolds. Fields are structures built over manifolds. Currently the only fields are vector fields, but more general tensor fields (and possibly the even more general fibre bundles) are likely to be included in future. Numerical algorithms are incorporated in the class of flows on manifolds.

The *DiffMan* package displays the importance of integrating symbolic and numerical techniques in the same software. For example, a particular Lie algebra implemented is the *free Lie algebra* class. It is capable of symbolic computations which remain valid in *any* particular finite-dimensional Lie algebra. Hence, numerical algorithms may be developed using this package, and the resulting expressions may be evaluated later, substituting data from a concrete Lie algebra. Abstract concepts play a central rôle, characteristic not just of pure mathematics but also of modern non-numerical software.

Another important insight that has emerged from this research is the observation that *some* numerical algorithms require detailed knowledge of representations of objects, while others can be completely formulated using general, coordinate-independent operations. This distinction is most easily seen in the area of solving linear algebraic equations. For example, *Gaussian elimination* requires detailed knowledge of matrices in terms of components (there is a major difference between sparse and dense Gaussian elimination!), while other algorithms such as *conjugate gradients* just require matrix-vector products. Algorithms that can be formulated in terms of general, coordinate-independent operations are much more flexible in their use than algorithms tied to particular representations. The fact that Runge–Kutta methods can indeed be phrased as coordinate-free algorithms has many implications that we are only now beginning to understand.

11. Applications

11.1. The Lagrange top

Many systems evolving in physical space can be modelled by motions consisting of rotations, like the rigid body that we have already encountered in (7.8), or translations, or combinations thereof, in which case we say that the motion is described by the *special Euclidean group* SE(3).

The *Lagrange top* is an important instance in which we make use of SE(3) action. It differs from the rigid body equations (7.8) because of the presence of gravity.

The equations of motion can be described in either *space* or *body coordinates*. In this section we briefly discuss both, considering space coordinates first. We denote by $\mathbf{\Pi}_s$ the angular momentum of the Lagrange top, by $\mathbf{\Omega}_s$ its angular velocity, by I_s its inertia tensor, and by $\boldsymbol{\lambda}_s$ the unit vector pointing towards the centre of mass of the Lagrange top. The equations of motion are

$$\mathbf{\Pi}_s' = Mgl\boldsymbol{k} \times \boldsymbol{\lambda}_s,$$
$$I_s' = [I_s, \hat{\mathbf{\Omega}}_s],$$
$$\boldsymbol{\lambda}_s' = \mathbf{\Omega}_s \times \boldsymbol{\lambda}_s,$$

where M is the mass of the top, g is the gravitational constant, l is the distance from the origin of the space-coordinate system to the centre of

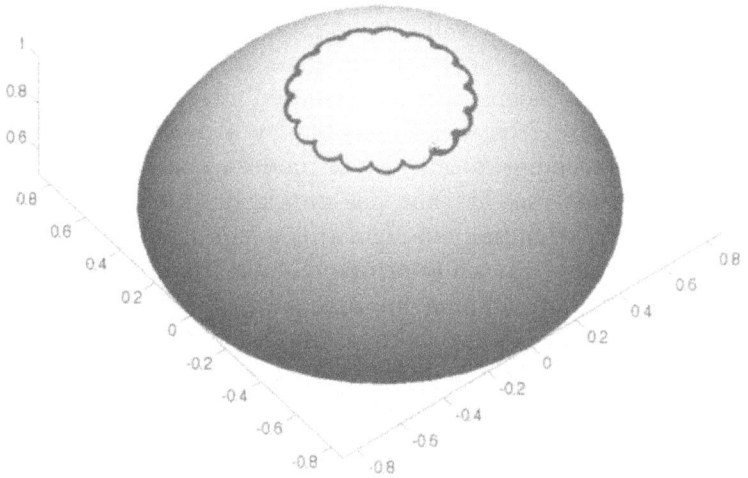

Fig. 11.1. Precession and nutations for $t \in [0, 10]$ of the axis of a Lagrange top

mass of the Lagrange top and $\mathbf{k}$ is the unit vector along the z-axis. The angular velocity $\mathbf{\Omega}_{\mathrm{s}}$ is related to the angular momentum $\mathbf{\Pi}_{\mathrm{s}}$ by the identity

$$\mathbf{\Omega}_{\mathrm{s}} = I_{\mathrm{s}}^{-1} \mathbf{\Pi}_{\mathrm{s}}.$$

We refer the reader to Arnold (1989) and Goldstein (1980) for background, theory and notation.

It is well known that the axis of the top displays a very special motion composed of a *precession* about the vertical axis z and *nutations*, nodding up and down between two bounding angles θ_1 and θ_2 as displayed in Figure 11.1 for a top with initial angular velocity $\mathbf{\Omega}_{\mathrm{b}} = [0, 0, 100]^{\mathrm{T}}$ about its figure axis, $l = \sqrt{3}/2$, and mass $M = 20/(gl)$. The inertia tensor in body coordinates is $I_{\mathrm{b}} = \mathrm{diag}(1, 1, 1/5)$ and the initial position of the axis of the top is rotated at an angle $\theta = -\pi/10$ with respect to the z-axis.

The same motions can be described in body coordinates, in terms of two vectors $\mathbf{\Pi}_{\mathrm{b}}$ and $\mathbf{\Gamma}_{\mathrm{b}}$ in $\mathbb{R}^3$, the angular momentum and the gravity vector in body coordinates respectively (Marsden and Ratiu 1994). The equations simplify to

$$\begin{aligned} \mathbf{\Pi}_{\mathrm{b}}' &= \mathbf{\Pi}_{\mathrm{b}} \times \mathbf{\Omega}_{\mathrm{b}} + Mgl\mathbf{\Gamma}_{\mathrm{b}} \times \mathbf{\chi}, \\ \mathbf{\Gamma}_{\mathrm{b}}' &= \mathbf{\Gamma}_{\mathrm{b}} \times \mathbf{\Omega}_{\mathrm{b}}, \end{aligned} \tag{11.1}$$

where $\mathbf{\chi}$ is the unit vector on the figure axis of the top. Note that in body coordinates the inertia tensor is constant. Also, in this case, $\mathbf{\Omega}_{\mathrm{b}} = I_{\mathrm{b}}^{-1} \mathbf{\Pi}_{\mathrm{b}}$. Although the vector $\mathbf{\Gamma}_{\mathrm{b}}$ is parallel to the z-axis of the space-coordinate system, its motion does not correspond to the motion of the figure axis $\mathbf{\lambda}_{\mathrm{s}}$

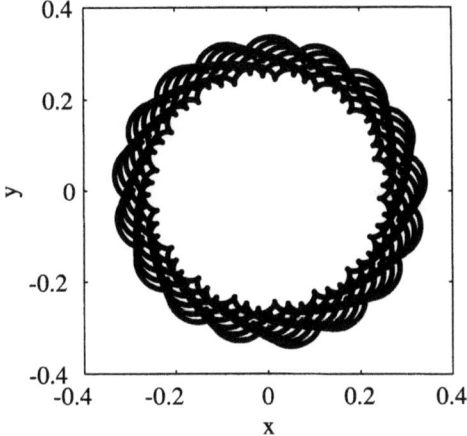

Fig. 11.2. Numerical precession and nutations projected on the
(x, y)-plane for $t \in [0, 100]$ of the axis of a Lagrange top. The equations
are solved in body coordinates with the MATLAB routine ode45

of the Lagrange top. To reconstruct the motion of the axis we integrate the
configuration matrix $R \in \mathrm{SO}(3)$, so that

$$\mathbf{\Omega}_\mathrm{s} = R\mathbf{\Omega}_\mathrm{b},$$

and, in general, the same type of transformation of body coordinates to
space coordinates holds for all relevant vectors. The configuration matrix R
obeys the differential equation

$$R' = -R\widehat{\mathbf{\Omega}}_\mathrm{b}. \tag{11.2}$$

Once the configuration matrix R is known, the position of the axis of the
top can be recovered by means of the transformation

$$\boldsymbol{\lambda}_\mathrm{s} = R \begin{bmatrix} 0 \\ 0 \\ 1 \end{bmatrix}.$$

In our first experiment, the equations (11.1)–(11.2) were solved with the
MATLAB routine ode45 with variable step-size and error control. Since
in body coordinates the gravity vector $\mathbf{\Gamma}_\mathrm{b}$ rotates very fast, the MATLAB
routine employs a very small step-size in the integration interval $[0, 10]$,
ranging from $h_\mathrm{min} \approx 0.5 \times 10^{-6}$ to $h_\mathrm{max} \approx 0.00214$. The average value is
$h_\mathrm{mean} = 0.0018$ and the standard deviation $\sigma = 1.972 \times 10^{-4}$. The motion of
the axis is quite similar to the one observed in Figure 11.1. However, when
the integration is performed over longer time, the numerical approximation
of $\boldsymbol{\lambda}_\mathrm{s}$ is no longer on the unit sphere and the amplitude of precessions shrinks
(see Figure 11.2).

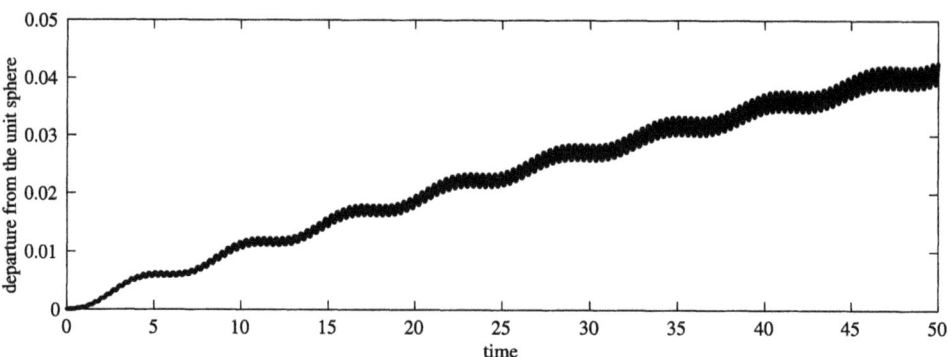

Fig. 11.3. Departure from unity for 'unit vector' along the axis
of the top for $t \in [0, 100]$. The equations are solved in body
coordinates with the MATLAB routine ode45

Routine ode45 does not preserve much of the geometry of the underlying problem, and many of the conserved quantities of the problem (Casimirs and first integrals – see Marsden and Ratiu (1994)) are not conserved (see Figure 11.3).

To illustrate the advantages of the methods discussed in this article, we solve numerically the equations of the Lagrange top (11.1)–(11.2) using the *coadjoint action* of SE(3) on the dual of the algebra $\mathfrak{se}(3)^*$ (Marsden and Ratiu 1994),

$$\Lambda\Big((R, \boldsymbol{\Omega}_{\mathrm{b}}), (\boldsymbol{\Pi}_{\mathrm{b}}, \boldsymbol{\Gamma}_{\mathrm{b}})\Big) = (R\,\boldsymbol{\Pi}_{\mathrm{b}} + \boldsymbol{\Omega}_{\mathrm{b}} \times R\,\boldsymbol{\Gamma}_{\mathrm{b}}, R\,\boldsymbol{\Gamma}_{\mathrm{b}}).$$

Using numerically the coadjoint action implies that all the Casimirs, in this case the projection of the angular momentum on the gravity axis, $\boldsymbol{\Pi}_{\mathrm{b}}{\cdot}\boldsymbol{\Gamma}_b$, and the norm of the gravity vector $\|\boldsymbol{\Gamma}_{\mathrm{b}}\|$, are automatically preserved to machine accuracy (Engø and Faltinsen 1999, Zanna et al. 1999). In Figure 11.4 we plot the precessions and nutations in $[0, 100]$ of the above Lagrange top, numerically solved with the explicit fourth-order RK-MK method based on the Butcher tableau

$$\begin{array}{c|cccc}
0 & & & & \\
\frac{1}{2} & \frac{1}{2} & & & \\
\frac{1}{2} & 0 & \frac{1}{2} & & \\
1 & 0 & 0 & 1 & \\
\hline
 & \frac{1}{6} & \frac{1}{3} & \frac{1}{3} & \frac{1}{6}
\end{array}$$

(*cf.* Appendix A), with step-size $h = \frac{1}{100}$. We refer to this numerical scheme as RKMK4.

Comparing Figure 11.4 with Figure 11.2, we observe that the tip of the

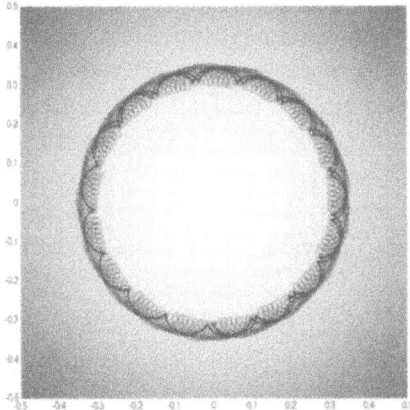

Fig. 11.4. Numerical precessions and nutations projected on
the (x, y)-plane for $t \in [0, 100]$ of the axis of a Lagrange top.
The equations are solved in body coordinates (employing the
coadjoint action) with the Lie-group scheme RKMK4

axis remains on the unit sphere (with five times as large step-size as the
average one used by ode45), as we expected from theory. There is only a
very slight variation in the angles θ_1 and θ_2, bounding the nutations. This
comes as no surprise: Lie group-type methods perform very well for systems
with oscillatory behaviour, as will be further discussed in Section 11.5. Note
that this scheme is neither time-reversible nor energy preserving. Using self-
adjoint schemes, such as those discussed in Section 7, or energy-preserving
schemes, such as those of Engø and Faltinsen (1999), it is possible to further
improve other geometrical features of the numerical solution.

We remark that, for this type of action, solved by Lie-group schemes based
on canonical coordinates of the first kind, the integrator requires numerous
evaluations of exponentials and dexp^{-1} of 3×3 skew-symmetric matrices,
which can be computed exactly and very rapidly using the formulae described in Appendix B.

11.2. Sturm–Liouville problems

Numerous problems in applied mathematics require the solution of *Sturm–
Liouville problems*: finding λ and y so that

$$\mathcal{L}y = -y'' + q(t)y = \lambda y, \qquad t \in (0, \alpha), \tag{11.3}$$

with the sufficiently smooth *potential* $q \in C^m(0, \alpha)$ and the boundary conditions

$$y(0)\boldsymbol{b}_0 + y(\alpha)\boldsymbol{b}_\alpha = \boldsymbol{0}, \tag{11.4}$$

where rank $[\boldsymbol{b}_0, \boldsymbol{b}_\alpha] = 2$. Examples range from fluid flow to Schrödinger spectra to geophysics to NMR imaging and beyond.

It is well known from classical functional analysis that the *spectrum* $\sigma(\mathcal{L})$, that is, the set of all λ that solve (11.3), is real, countable and accumulates at $+\infty$. Its computation has attracted significant effort since the very dawn of numerical analysis and led to an impressive array of methods and software. We refer the reader to Pryce (1993) and to references therein.

The problem (11.3) can always be formulated in SL(2),

$$\boldsymbol{y}' = \begin{bmatrix} 0 & 1 \\ q(t) - \lambda & 0 \end{bmatrix} \boldsymbol{y}, \quad t \in (0, \alpha), \qquad \text{where} \qquad \boldsymbol{y} = \begin{bmatrix} y \\ y' \end{bmatrix}, \qquad (11.5)$$

and this makes it a 'natural' for Lie-group methods. The main idea is to approximate the fundamental solution of (11.5) by a product of exponentials and impose the boundary conditions (11.4). This technique has been introduced and thoroughly analysed by Moan (1998) and it constitutes a far-fetched generalization of *Preuss–Fulton methods* (Preuss 1973). The outcome is a scalar nonlinear equation for the spectral parameter λ. For example, suppose that (11.4) simplifies to $y(0) = y(\alpha) = 0$ and our approximation to the fundamental solution of (11.5) is

$$Y(t) = \text{expm} \begin{bmatrix} a & b \\ c & -a \end{bmatrix}.$$

It follows from (8.1) that

$$Y(t) = \begin{bmatrix} \cosh\omega + a\frac{\sinh\omega}{\omega} & b\frac{\sinh\omega}{\omega} \\ c\frac{\sinh\omega}{\omega} & \cosh\omega - a\frac{\sinh\omega}{\omega} \end{bmatrix}, \qquad \text{where} \qquad \omega = \sqrt{a^2 + bc}.$$

Note that a, b, c, ω are functions of both t and λ. We impose the boundary conditions and the outcome is the equation

$$b(\alpha, \lambda)\frac{\sinh\omega(\alpha, \lambda)}{\omega(\alpha, \lambda)} = 0. \qquad (11.6)$$

Suppose for example (and a very trivial example it is!) that Y is the second-order truncation of the Magnus expansion,

$$Y(t) = \text{expm} \begin{bmatrix} 0 & t \\ \int_0^t q(\xi)\,d\xi & 0 \end{bmatrix}.$$

Therefore (11.6) becomes

$$\alpha\frac{\sin\sqrt{\alpha^2\lambda - \alpha\int_0^\alpha q(\xi)\,d\xi}}{\sqrt{\alpha^2\lambda - \alpha\int_0^\alpha q(\xi)\,d\xi}} = 0$$

and the approximate eigenvalues are

$$\lambda_m \approx \frac{m^2 \pi^2 + \alpha \int_0^\alpha q(\xi) \, d\xi}{\alpha^2}, \qquad m \in \mathbb{N}.$$

Adding the next term to the Magnus expansion somehow yields a better approximation,

$$\lambda_m \approx \frac{m^2 \pi^2 + \alpha \int_0^\alpha q(\xi) \, d\xi - \frac{1}{4} \left[\int_0^\alpha (2\xi - \alpha) q(\xi) \, d\xi \right]^2}{\alpha^2}, \qquad m \in \mathbb{N}.$$

We can continue in this vein, possibly replacing integrals by quadrature, except that the outcome of this naive approach is very poor. Magnus expansions are an excellent means to approximate the fundamental solution *locally* and we can hardly expect a single exponential to approximate the solution well in the entire interval $(0, \alpha)$. The situation is further exacerbated when α is large or when (as is the case in many instances of practical interest) we desire to approximate a large number of eigenvalues. It is possible to show that ω is a polynomial in λ, of a degree that grows with the order of the Magnus expansion. Bearing in mind Theorem 4.1, we can thus hardly expect convergence for large λ. A superior alternative is to partition the interval into small subintervals, where convergence and adequate precision can be assured. The details of this procedure are reported in Moan (1998).

Let us restrict the discussion for the sake of simplicity to a fourth-order method. Letting

$$q_{n,1} = q\big((n + \tfrac{1}{2} - \tfrac{\sqrt{3}}{6})h\big), \qquad q_{n,2} = q\big((n + \tfrac{1}{2} + \tfrac{\sqrt{3}}{6})h\big),$$

where $h = \alpha/n^*$, we set

$$\Theta_n(\lambda) = \begin{bmatrix} -\frac{\sqrt{3}}{12} h^2 (q_{n,1} - q_{n,2}) & h \\ \frac{1}{2} h (q_{n,1} + q_{n,2}) - h\lambda & \frac{\sqrt{3}}{12} h^2 (q_{n,1} - q_{n,2}) \end{bmatrix}, \qquad n = 0, 1, \dots, n^* - 1.$$

The fourth-order approximant to the solution of (11.5) at α is $Y^\lambda \boldsymbol{y}(0)$, where

$$Y^\lambda = e^{\Theta_{n^*-1}(\lambda)} \cdots e^{\Theta_1(\lambda)} e^{\Theta_0(\lambda)}.$$

To force the boundary conditions $y(0) = y(\alpha) = 0$, say, we seek λ so that $Y_{1,2}^\lambda = 0$. This nonlinear equation can be solved by *Newton–Raphson iteration* which, eigenvalues being simple, converges quadratically near the solution. Moan (1998) recommends a procedure based on the Newton–Raphson method being applied to an augmented equation, thereby simplifying the computation of the derivative.

The outcome is a very efficient method for the computation of Sturm–Liouville problems and it can easily be generalized to more elaborate eigenvalue problems and boundary conditions. However, if the interval in question is long or a large number of eigenvalues is desired, efficiency suffers. This is in particular the case if high-order Magnus methods are used, since the degree of elements of Θ_n as polynomials in λ grows. As a matter of fact, the error in a pth-order method grows as $\mathcal{O}(h^{p+1}\lambda^{p/2-1})$. To overcome this phenomenon, which occurs also in other methods for the computation of Sturm–Liouville problems, Moan (1998) introduced an interesting device which might be relevant to other applications of Lie-group methods. We observe first that, given matrix values in a self-adjoint basis $B_0, B_1, \ldots, B_{\nu-1}$ (*cf.* Section 5), it follows at once from (5.7) that only B_0 depends upon λ.

Rather than presenting the construction from Moan (1998), we introduce a conceptually similar idea of Iserles (2000). Our point of departure is a decomposition of the Magnus expansion (4.5) into *streamers*: partial sums of the form

$$\mathcal{H}_{\tau^{[0]}}(t) = \sum_{k=0}^{\infty} \alpha(\tau^{[k]}) \int_0^t \mathcal{C}_{\tau^{[k]}}(\xi)\,\mathrm{d}\xi,$$

where

$$\tau^{[k+1]} = \begin{array}{c}\bullet \\ \diagdown\!\!\diagup \end{array}\!\!\!^{\tau^{[k]}}, \qquad k \in \mathbb{Z}^+. \tag{11.7}$$

Let $\tau = \tau^{[0]}$ be in the form (4.8). It is possible to show that

$$\alpha(\tau^{[k]}) = \frac{\mathrm{B}_{s+k}}{(s+k)!}\hat{\alpha}(\tau), \qquad \text{where} \qquad \hat{\alpha}(\tau) = \prod_{i=1}^{s} \alpha(\tau_i)$$

(*cf.* (4.9)). Therefore we can write the streamer as

$$\mathcal{H}_{\tau^{[0]}}(t) = \hat{\alpha}(\tau^{[0]}) \int_0^t \sum_{k=0}^{\infty} \frac{\mathrm{B}_{s+k}}{(s+k)!}\mathrm{ad}_{\int_0^\xi A(\eta)\,\mathrm{d}\eta}^k \mathcal{C}_{\tau^{[0]}}(\xi)\,\mathrm{d}\xi. \tag{11.8}$$

We say that a tree is *primitive* if it cannot be written in the form (11.7). The set of primitive trees in $\mathbb{F}_m$ is denoted by $\mathbb{F}_m^{\mathrm{p}}$, whereby we might replace the truncated Magnus expansion (4.14) with the *Magnus streamer expansion*

$$\Xi_p(t) = \sum_{m=0}^{p-1} \sum_{\tau \in \mathbb{F}^{\mathrm{p}}} \mathcal{H}_\tau(t). \tag{11.9}$$

Like (4.14), this is a pth-order approximant, except that each tree therein is accompanied by an infinitely long streamer and there is one less level of truncation in the Magnus expansion (4.5).

Using (11.9) makes sense only if there exists a good method to evaluate the streamer (11.8) without truncating the expansion. This is the case if the Lie algebra is of sufficiently small dimension.

Let $\dim \mathfrak{g} = d$ and let $C = \{C_1, C_2, \ldots, C_d\}$ be a basis of the algebra. Then there exists a natural map $\pi : \mathfrak{g} \to \mathbb{R}^d$ defined by

$$\pi(B) = \theta \qquad \text{where} \qquad B = \sum_{k=1}^{d} \theta_k C_k.$$

The action of the ad operator is a linear transformation; thus $\pi(\mathrm{ad}_D B) = \mathscr{C}_D \pi(B)$, where $\mathscr{C}_D \in \mathfrak{gl}(d)$ is a *commutator matrix*.[6] It follows from (11.8) and the definition of Bernoulli numbers that

$$\pi(\mathcal{H}_\tau(t)) = \hat{\alpha}(\tau) \int_0^t \sum_{k=s}^{\infty} \frac{\mathrm{B}_k}{k!} \mathscr{C}_{D(\xi)}^{k-s} \pi(\mathcal{C}_\tau(\xi)) \, \mathrm{d}\xi$$

$$= \hat{\alpha}(\tau) \int_0^t \mathscr{C}_{D(\xi)}^{-s} \left\{ \frac{\mathscr{C}_{D(\xi)}}{\mathrm{e}^{\mathscr{C}_{D(\xi)}} - I} - \sum_{k=0}^{s-1} \frac{\mathrm{B}_k}{k!} \mathscr{C}_{D(\xi)}^{k} \right\} \pi(\mathcal{C}_\tau(\xi)) \, \mathrm{d}\xi.$$

where $D(t) = \int_0^t A(\eta) \, \mathrm{d}\eta$. Provided that d is small, typically we can obtain the commutator matrix and its exponential explicitly. This, together with quadrature, leads to an explicit formula for the calculation of the streamer that does not require any truncation of the series.

A similar idea, reported in Moan (1998), ameliorates the deterioration in accuracy for large λ. For example, the error of sixth-order Magnus streamer is $\mathcal{O}(h^7 \lambda)$, hence it behaves like a fourth-order method when $h^2 \lambda \approx 1$, when the standard sixth-order Magnus, which carries an error of $\mathcal{O}(h^7 \lambda^2)$, reduces to an order-two method.

11.3. Charged particles in a magnetic field

The motion of charged particles moving in a magnetic field was first studied numerically by Carl Størmer (1907) as a part of his work on explaining the origin of northern lights (aurora borealis).

A particle (of unit mass and unit charge) is moving in a magnetic field b according to the equations

$$y'(t) = v,$$
$$v'(t) = b(t, y) \times v,$$

where v is the velocity, y is the position and b is the magnetic field. We attempt to obtain a simpler system by assuming that $b = \text{const}$, neglecting the dependence of b on space and time. Written in a matrix form, this yields

$$\begin{bmatrix} y \\ v \end{bmatrix}' = \begin{bmatrix} 0 & I \\ 0 & \hat{b} \end{bmatrix} \begin{bmatrix} y \\ v \end{bmatrix}, \tag{11.10}$$

---

[6] Note that the usual definition of a commutator matrix is as an object in $\mathfrak{gl}(N^2)$ for $\mathfrak{g} \subseteq \mathfrak{gl}(N)$.

where $\widehat{\boldsymbol{b}}$ is a 3×3 skew-symmetric matrix given by the hat map (B.1) in Appendix B. We compute the bracket of the vector fields in (11.10). Since they are linear, we find that

$$\left[\begin{bmatrix} 0 & \boldsymbol{I} \\ 0 & \widehat{\boldsymbol{b}} \end{bmatrix}, \begin{bmatrix} 0 & \boldsymbol{I} \\ 0 & \widehat{\boldsymbol{d}} \end{bmatrix} \right] = \begin{bmatrix} 0 & \widehat{\boldsymbol{d}} - \widehat{\boldsymbol{b}} \\ 0 & \widehat{\boldsymbol{b} \times \boldsymbol{d}} \end{bmatrix}.$$

Hence, the vector fields do *not* form a Lie algebra. The remedy is to enlarge the family of 'simple' equations into

$$\begin{bmatrix} \boldsymbol{y} \\ \boldsymbol{v} \end{bmatrix}' = \begin{bmatrix} 0 & B \\ 0 & \widehat{\boldsymbol{b}} \end{bmatrix} \begin{bmatrix} \boldsymbol{y} \\ \boldsymbol{v} \end{bmatrix} = \begin{bmatrix} B\boldsymbol{v} \\ \boldsymbol{b} \times \boldsymbol{v} \end{bmatrix}, \tag{11.11}$$

where B is a general 3×3 matrix. This results in the commutator

$$\left[\begin{bmatrix} 0 & B \\ 0 & \widehat{\boldsymbol{b}} \end{bmatrix}, \begin{bmatrix} 0 & C \\ 0 & \widehat{\boldsymbol{c}} \end{bmatrix} \right] = \begin{bmatrix} 0 & B\widehat{\boldsymbol{c}} - C\widehat{\boldsymbol{b}} \\ 0 & \widehat{\boldsymbol{b} \times \boldsymbol{c}} \end{bmatrix}.$$

It is worth remarking on the similarity between this bracket and the brackets in Example 2.8, Example 2.10 and in Section 11.1. These are all special instances of so-called *semidirect product* brackets. Semidirect products are one of the most fundamental ways of constructing Lie algebras and Lie groups from simpler algebras and groups.

The solution of the simplified equation (11.11) is given in terms of initial conditions $\boldsymbol{y}_0$ and $\boldsymbol{v}_0$ as

$$\begin{bmatrix} \boldsymbol{y}(t) \\ \boldsymbol{v}(t) \end{bmatrix} = \mathrm{expm} \begin{bmatrix} 0 & tB \\ 0 & t\widehat{\boldsymbol{b}} \end{bmatrix} \begin{bmatrix} \boldsymbol{y}_0 \\ \boldsymbol{v}_0 \end{bmatrix}.$$

Repeated matrix multiplications yield

$$\mathrm{expm} \begin{bmatrix} 0 & B \\ 0 & \widehat{\boldsymbol{b}} \end{bmatrix} = \begin{bmatrix} \boldsymbol{I} & B\widehat{\boldsymbol{b}}^{-1}(\mathrm{expm}(\widehat{\boldsymbol{b}}) - \boldsymbol{I}) \\ 0 & \mathrm{expm}(\widehat{\boldsymbol{b}}) \end{bmatrix},$$

a formula that can be computed exactly and efficiently using (B.6) and (B.10).

In a numerical experiment we have taken $\boldsymbol{b}(t, \boldsymbol{y}) = \boldsymbol{b}(\boldsymbol{y})$ as the *magnetic dipole field*

$$\boldsymbol{b}(\boldsymbol{y}) = \frac{3\boldsymbol{e}_y^{\mathrm{T}} \boldsymbol{m} \boldsymbol{e}_y - \boldsymbol{m}}{\|\boldsymbol{y}\|^3},$$

where $\boldsymbol{m} = [0, 0, 1]^{\mathrm{T}}$ is the magnetic dipole vector. Figure 11.5 displays the orbit of a single particle with initial conditions $\boldsymbol{y}_0 = [0, -\frac{5}{2}, 0]^{\mathrm{T}}$, $\boldsymbol{v}_0 = 12 \cdot 10^{-3}[0, 0, 1]^{\mathrm{T}}$. The time interval is $[0, 10000]$. The particle is moving back and forth in the *van Allen belt* around the Earth. This system is Hamiltonian, the Hamiltonian energy being $H = \|\boldsymbol{v}\|^2$. The preservation of H also follows trivially from the observation that the acceleration is always orthogonal to $\boldsymbol{v}$.

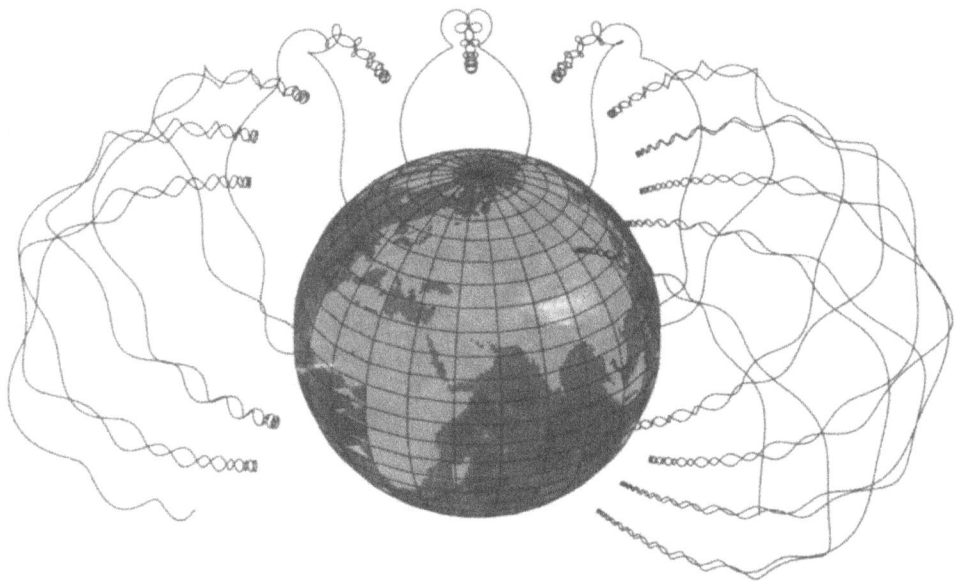

Fig. 11.5. Motion of a charged particle in the Earth's magnetic field

We wish to compare step-sizes in classical integrators and RK-MK-type
methods for this problem. The classical integrator used in the study is the
ode45 code of MATLAB, based on DOPRI5(4), an embedded (4,5) pair of
Runge–Kutta methods due to Dormand and Prince. The code uses variable step-size control. The RK-MK method we use is also based on the
DOPRI5(4) method and we have employed the step-size controller of *Diff-
Man*. The interval of integration ranged from 0 to 500, corresponding to a
motion of the particle from the equator up towards the North Pole, bouncing back once towards the equator. We varied the tolerance of the step-size
controllers and measured the relative error of the answer at the endpoint.
For each simulation we have counted the number of steps taken. Table 11.1
summarizes the results. We see that the Lie-group method is much more
accurate than the classical integrator for the same step-size. Figure 11.6
displays the time-step selection as a function of time for the two methods
for the simulations given in the first line of Table 11.1. The dip in the middle
corresponds to the part of the trajectory where the particle is bouncing back.
Both methods must reduce the step-size in this region, but the classical one
does so relatively more than the Lie-group method. Note that, whereas the
Lie-group integrator preserves $\|v\|$ exactly, the ode45 integrator does so only
up to the order of the method.

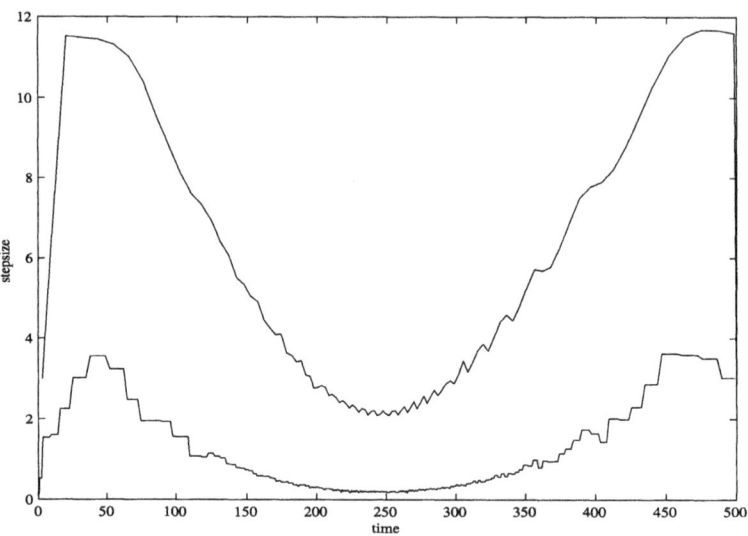

Fig. 11.6. Time-step selection for RK-MK (upper) and classical RK (lower)

We conclude that, for this problem, where both commutators and exponentials can be computed quickly, the total cost of a Lie-group integrator is significantly less than that of a classical integrator. The time interval of integration is relatively small in this example: the importance of preserving geometric properties becomes increasingly more crucial for very long integration intervals.

11.4. Toda-lattice equations

Imagine a regular lattice of N particles, all of unit mass, and assume that each particle interacts with just its nearest neighbours, subject to an exponential interaction potential. The outcome is the *Toda lattice*, governed by

Table 11.1. *Global error and the number of steps for classical RK and a Lie-group integrator*

Classical RK		RK-MK	
Error	Number of steps	Error	Number of steps
$7 \cdot 10^{-3}$	836	$5 \cdot 10^{-3}$	104
$2 \cdot 10^{-4}$	2132	$4 \cdot 10^{-4}$	142
$2 \cdot 10^{-6}$	5356	$1 \cdot 10^{-6}$	353

the Hamiltonian equations of motion

$$\left.\begin{array}{l} \dfrac{\mathrm{d}q_k}{\mathrm{d}t} = p_k, \\[2mm] \dfrac{\mathrm{d}p_k}{\mathrm{d}t} = \mathrm{e}^{q_{k-1}-q_k} - \mathrm{e}^{q_k-q_{k+1}}, \end{array}\right\} \qquad k = 1, 2, \ldots, N, \qquad (11.12)$$

where q_k and p_k are the generalized coordinates and momenta, respectively (Toda 1981), where we have let $q_0 = -\infty$, $q_{N+1} = \infty$ in open configuration and $q_{N+k} = q_k$ when the particles are arranged in a ring (*cf.* Figure 11.7). Equation (11.12) corresponds to the Hamiltonian potential

$$H(\boldsymbol{p}, \boldsymbol{q}) = \tfrac{1}{2} \sum_{k=1}^{N} p_k^2 + \sum_{k=1}^{N} (\mathrm{e}^{q_k-q_{k+1}} - 1).$$

It is a special case of a *Fermi–Pasta–Ulam* (FPU) flow.

(a) Open configuration   (b) Ring configuration

Fig. 11.7. Different Toda-lattice configurations with $N = 5$

Letting $\alpha_k = \tfrac{1}{2}\mathrm{e}^{(q_k-q_{k+1})/2}$, $\beta_k = \tfrac{1}{2}p_k$, Flaschka (1974) showed that (11.12) can be recast in the *Lax form*

$$Y' = [B(Y), Y], \quad t \geq 0, \qquad Y(0) = Y_0, \qquad (11.13)$$

where

$$Y = \begin{bmatrix} \beta_1 & \alpha_1 & 0 & \cdots & & \alpha_N \\ \alpha_1 & \beta_2 & \alpha_2 & \ddots & & \vdots \\ 0 & \alpha_2 & \ddots & \ddots & & 0 \\ \vdots & \ddots & \ddots & \beta_{N-1} & \alpha_{N-1} \\ \alpha_N & \cdots & & 0 & \alpha_{N-1} & \beta_N \end{bmatrix},$$

with $\alpha_N = 0$ for open lattice configuration, and $B(Y) = Y_- - Y_+ \in \mathfrak{so}(N)$, where M_+ and M_- denote the upper-triangular and the lower-triangular portions of the matrix M, respectively. We immediately identify (11.13) with the *isospectral flow* (1.1).

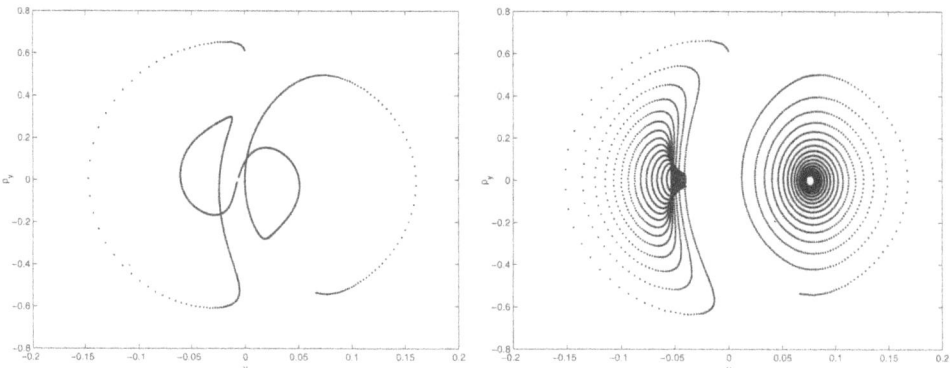

Fig. 11.8. Three-particle lattice solved by the Runge–Kutta method (11.15),
applied (a) directly to the isospectral flow (11.13) (on the left);
and (b) to the orthogonal flow (11.16) (on the right)

A special case of (11.12) that has elicited much interest is three particles
in a ring. In that case, the linear transformation

$$\tilde{p} = Tp, \quad \tilde{q} = Tq, \qquad \text{where} \qquad T = \begin{bmatrix} \frac{\sqrt{6}}{6} & -\frac{\sqrt{6}}{3} & \frac{\sqrt{6}}{6} \\ \frac{\sqrt{2}}{2} & 0 & -\frac{\sqrt{2}}{2} \\ \frac{\sqrt{3}}{3} & \frac{\sqrt{3}}{3} & \frac{\sqrt{3}}{3} \end{bmatrix},$$

in tandem with rescaling and elimination of one of the variables by employing
the linear conservation law $p_1 + p_2 + p_3 \equiv$ const, results in a simplified two-
dimensional Hamiltonian function

$$\bar{H}(\bar{p}, \bar{q}) = \tfrac{1}{2}(\bar{p}_1^2 + \bar{p}_2^2) + \tfrac{1}{24}(e^{2\bar{q}_2 + 2\sqrt{3}\bar{q}_1} + e^{2\bar{q}_2 - 2\sqrt{3}\bar{q}_1} + e^{-4\bar{q}_2}) - \tfrac{1}{8} \quad (11.14)$$

in the new variables $\bar{p}, \bar{q} \in \mathbb{R}^2$ (Zanna 1998).[7] The solution now evolves on
a compact submanifold of $\mathbb{R}^4$, more specifically on a 2-torus. Therefore, a
Poincaré section consists of two closed curves and a good numerical scheme
should retain this important property.

The following calculation has been performed in Zanna (1998) for a variety
of methods which follow a set pattern. First we transform the three-particle
lattice with the initial conditions $p = [1, 1, 0]^T$, $q = 0$, to the Lax form
(11.13). We then solve the isospectral form with the constant step-size $h = \tfrac{1}{10}$ for 10^4 steps, transform the variables to $(\bar{p}, \bar{q})$ and sketch the Poincaré
section. The following methods have been used.

---

[7] One reason why this case is interesting is that truncation of $\bar{H}$ to cubic terms results
in the famous *Hénon–Heiles Hamiltonian*, which is known to be nonintegrable (Berry
1987, Toda 1981).

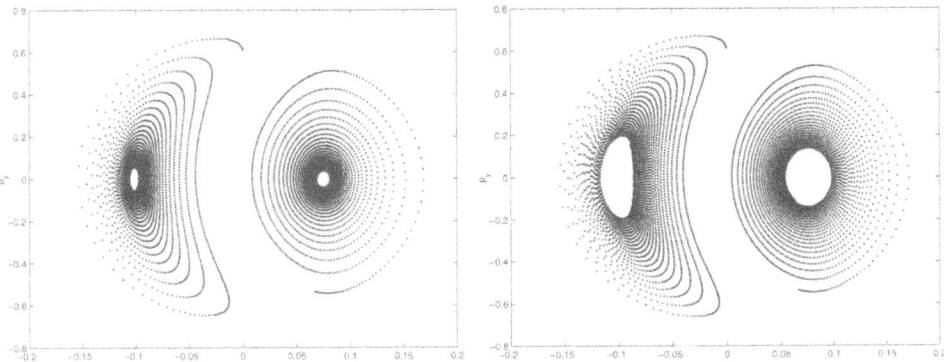

Fig. 11.9. Three-particle lattice solved by the Runge–Kutta method (11.15), applied to the orthogonal flow (11.16) and with polar decomposition. In the plot on the left only the final result is projected, while on the right both the final result and intermediate stages are subjected to this procedure

(1) The three-stage, third-order explicit Runge–Kutta method with the Butcher tableau

$$\begin{array}{c|ccc} 0 & & & \\ \frac{1}{2} & \frac{1}{2} & & \\ 1 & -1 & 2 & \\ \hline & \frac{1}{6} & \frac{2}{3} & \frac{1}{6} \end{array} \qquad (11.15)$$

applied directly to (11.13). Note that this method is not isospectral. The Poincaré section is displayed in Figure 11.8(a) and we note at once that, instead of periodic trajectories, the motion collapses rapidly to a spurious fixed point.

(2) The Runge–Kutta method (11.15) applied to the *orthogonal flow*

$$Q' = B(QY_nQ^{\mathrm{T}})Q, \quad t \geq nh, \qquad Q(nh) = \boldsymbol{I}, \qquad (11.16)$$

translating back to (11.13) with $Y_{n+1} = Q_{n+1}Y_nQ_{n+1}^{\mathrm{T}}$. Note that the numerical approximation $Q_{n+1} \approx Q((n+1)h)$, produced by the method (11.15), does not evolve on SO(3). Figure 11.8(b) demonstrates that the solution again spirals to a fixed point, although less rapidly than in the former case.

(3) The former method can be 'orthogonalized' by *projection*, subjecting Q_{n+1} to polar decomposition and discarding the nonorthogonal part. This can be further enhanced by subjecting the intermediate stages to polar decompositions. Figure 11.9 displays relevant Poincaré sections.

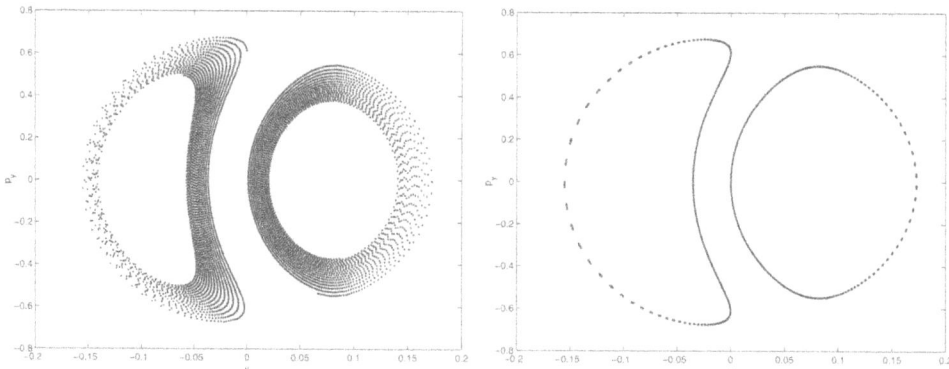

Fig. 11.10. Three-particle lattice solved by Lie-group
methods: (a) the RK-MK method with the tableau (11.15);
and (b) fourth-order Magnus collocation from Section A.3

Evidently, the behaviour improves, yet the solution goes on spiralling
to a spurious fixed point.

(4) We solve (11.16) with the RK-MK method, using the Runge–Kutta
scheme (11.15). The Poincaré section for this computation, the first
with a 'proper' Lie-group method, is displayed in Figure 11.10(a). We
have not managed to get rid of the unwelcome spurious convergence to
a fixed point, yet the trajectory spirals significantly more slowly.

(5) Finally, (11.16) is solved by the fourth-order collocated Magnus method
from Section A.3, applied to the orthogonal flow. The outcome is displayed in Figure 11.10(b): the Poincaré section consists of the qualitatively correct two closed curves, a section across a 2-torus!

Figures 11.8–11.10 display consistent gradual improvement. This is not
evident at all from the retention of the Lie group *per se*, since the qualitative
feature under examination, invariant tori, is of a different flavour. Yet, it
is clear that conservation of Lie-group structure is advantageous and that
purpose-designed Lie-group solvers hold the edge over projection methods.

Note that invariant tori would have been retained by a symplectic method,
applied directly to the Hamiltonian equations induced by the potential
(11.14). This is hardly a surprise: we have, after all, subjected methods
applied in a Lie-group formalism to a 'Hamiltonian test'. It is possible to
devise a test according to 'isospectral' ground rules, for instance by monitoring global error in Lax-equation coordinates. This has been done in Calvo,
Iserles and Zanna (1999), demonstrating a marked advantage of isospectral
methods (originating in Lie-group solvers) over symplectic algorithms.

11.5. Highly oscillatory equations

Highly oscillatory systems of ODEs feature in many applications and their computation currently absorbs much of the overall effort devoted to numerical analysis of ODEs. It is well known that classical solvers perform poorly, and this has motivated many novel and ingenious techniques Petzold, Jay and Yen (1997).

Past experience indicates that Lie-group methods might be a suitable means to solve highly oscillatory systems (Iserles and Nørsett 1999, Iserles et al. 1999). We commence from an example that has already featured in Iserles and Nørsett (1999), the solution of the *Airy equation*

$$y'' + ty = 0, \quad t \geq 0, \qquad y(0) = 1, \quad y'(0) = 1. \tag{11.17}$$

The exact solution of (11.17) can be represented as a linear combination of *Airy functions*

$$y(t) = \tfrac{1}{2}\Gamma(\tfrac{2}{3})[3^{2/3}\mathrm{Ai}(-t) + 3^{1/6}\mathrm{Bi}(-t)], \qquad t \geq 0.$$

It is easy to prove that the trajectory is a bounded function that oscillates like $\sin t^{3/2}$: the frequency increases with time (*cf.* Figure 11.11) (Abramowitz and Stegun 1970, p. 448). This indicates that long-time integration is difficult and this can be confirmed by endeavouring to solve (11.17) with any popular, general-purpose solver. We have solved the Airy equation with two MATLAB routines, **ode113** and **ode15s**, the first for nonstiff and the latter for stiff problems. Both routines employ variable-order methods in tandem with sophisticated error control. Yet, although we have attempted a wide range of possible error tolerances, including the least that the software would accept (a relative tolerance of 2.2×10^{-14}), the pointwise error at $t = 100$

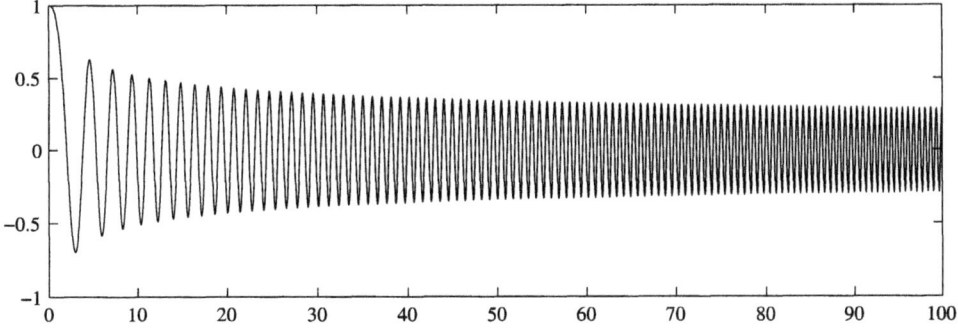

Fig. 11.11. The Airy equation (11.17)

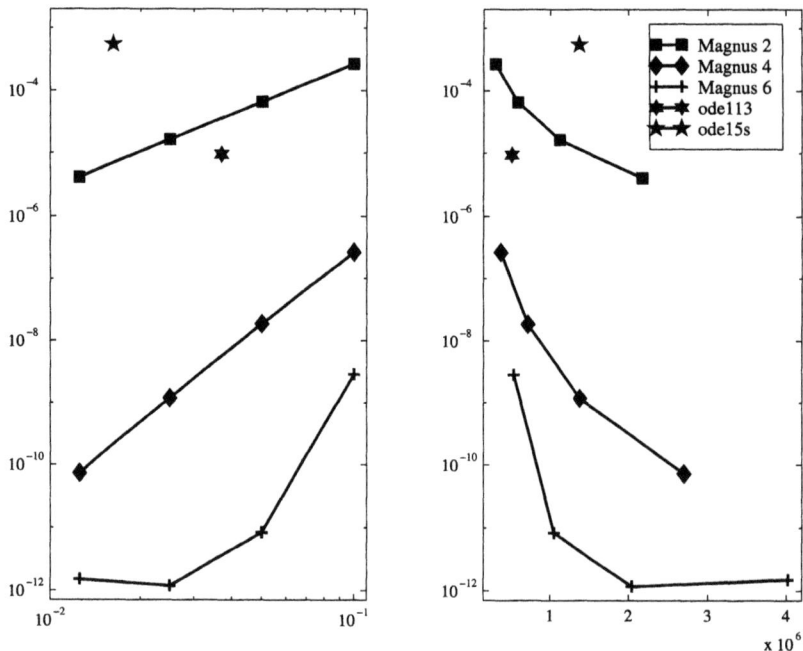

Fig. 11.12. The solution $y(t)$ of the Airy equation (11.17) at $t = 100$

consistently exceeded 9.42×10^{-6} and 5.43×10^{-4} respectively: very poor performance, in particular in the case of the stiff solver ode15s.[8]

We have solved the Airy equation (11.17) with Magnus methods of orders 2, 4 and 6 respectively, taken from (A.9), with the constant step-size sequence $h = \frac{1}{10}, \frac{1}{20}, \frac{1}{40}, \frac{1}{80}$. No attempt has been made to monitor the error or to optimize the solution. In particular, no advantage has been taken of the fact that (11.17) has been rendered as an SL(2) equation and, rather than using (8.1), we have computed the exponential with the MATLAB function expm. Figure 11.12 displays the outcome of our calculations. The plot on the left compares the absolute error at $t = 100$ for different values of h on a doubly logarithmic scale. For comparison purposes we have included the optimal results for ode113 and ode15s, assigning to them h equal to the average step-size. The plot on the right compares the logarithm of the absolute error with the number of flops expanded on the solution: a very imprecise measure, yet a fair reflection of the computational effort.

The main observation is that, unlike the MATLAB routines, Magnus methods perform very well indeed. The deterioration in the accuracy of the sixth-

---

[8] Highly oscillating ODEs should never be confused with stiff ODEs. Our calculation merely confirms this.

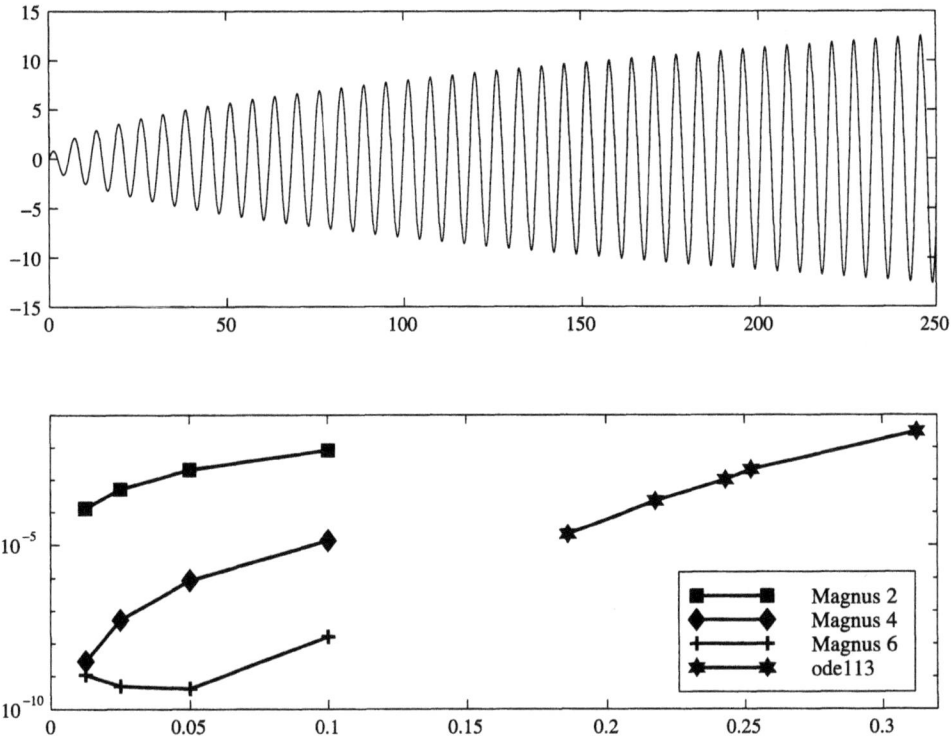

Fig. 11.13. The solution of the Bessel-like equation (11.18)

order method for small step-size can be explained by the accumulation of small errors in the evaluation of the **expm** function. In principle, we could have replaced it by the exact formula (8.1), except that, while nominally reducing the number of flops, this does not enhance the solution because of inexact computation of hyperbolic functions. In all likelihood, the behaviour in Figure 11.12 is as good as one can expect with IEEE computer arithmetic.

Another example of a highly oscillatory ODE is

$$t^2 y'' - ty' + (1 + t^2)y = 0, \qquad t \geq 1, \qquad (11.18)$$
$$y(1) = J_0(1), \quad y'(1) = J_0(1) - J_1(1),$$

whose solution is $y(t) = tJ_0(t)$. Here J_ν is a *Bessel function of the first kind*, whose oscillatory behaviour is well known. Note that we commence integration at $t = 1$, to avoid a regular singularity at the origin. The top plot of Figure 11.13 depicts the exact solution of (11.18). The oscillatory behaviour is evident: as a matter of fact, it is elementary to prove that the solution behaves like $\sqrt{2t/\pi}\sin t$ for $t \gg 1$. The bottom plot displays in a logarithmic scale the maximal error in the interval $[0, 250]$ in second-order, fourth-order and sixth-order Magnus methods for a range of different step-sizes. For comparison purposes, we have also included the same data for the MATLAB routine ode113 for different tolerances, down to the least permitted by the software. As in the case of Figure 11.12, we have assigned to each such computation an average step-size. Note that the step-sizes are significantly larger for MATLAB routines, but so is the error! Even though it uses error control, ode113 appears incapable of driving pointwise accuracy below quite a large threshold. We note in passing that the constant-step-size Magnus methods were (for the largest step-size, $h = \frac{1}{10}$) 244%, 317% and 476% more expensive than ode113, but no attempt has been made to optimize the calculations. As in the case of the Airy equation (11.17), the sixth-order method 'hits the buffers' for small step-sizes and it appears that there exists a limit on attainable accuracy, probably caused by accumulation of round-off error. Yet, this limit is twice the number of significant digits attainable by the general-purpose ode113.

The above comparison might appear as unfair to ode113 which, after all, has not been constructed with highly oscillatory equations in mind. Such criticism misses the point altogether. The whole purpose of geometric integration is to take advantage of structure, rather than employ general-purpose methods! Moreover, our Magnus expansions have been implemented with constant step sizes. There is little doubt (and much evidence in Iserles et al. (1999)) that variable-step implementation of Magnus methods, for instance with the technique from Section 10.1, would have improved the odds drastically in their favour. Of course, there are other bespoke methods for highly oscillatory systems (Petzold et al. 1997). It is not our claim that Lie-group methods are superior, since this has never been investigated comprehensively. Our aim is more modest, namely to argue that such an investigation might be very interesting.

What is the explanation for the remarkable performance of Magnus methods (and other Lie-group methods that we have not implemented in this section) for a highly oscillatory problem? Although this issue constitutes an interesting open research problem, our suspicion is that the conservation of SL(2) structure has absolutely nothing to do with it. Classical numerical methods invariably employ the *ansatz* that locally the solution of an ODE behaves like a *polynomial* in t. Unfortunately, polynomials are a very poor means to approximate rapidly varying and highly oscillating functions. We either require minute step-size or a very high degree of a polynomial – and the latter might lead to ill conditioning. All the Lie-group methods of this paper, however, are based on an entirely different representation of the solution, as an *exponential of a matrix with polynomial entries*. Unlike polynomials, exponentials of matrices with polynomial entries can easily model exponential change, high oscillation and changes in amplitude and frequency. This might well be the mechanism explaining the superior performance of Lie-group methods.

Acknowledgements

The authors wish to thank many of their colleagues and fellow *dexperts* who have read portions of this survey as it was taking shape, in particular Krister Åhlander, Sergio Blanes, Elena Celledoni, Kenth Engø, Stig Faltinsen, Arne Marthinsen, Per Christian Moan, Ranjiva Munasinghe, Brynjulf Owren and Reinout Quispel. Their comments and perceptive remarks have diminished the number of mistakes, typos and infelicities and improved the cohesion and presentation of this paper.

HMK, SPN and AZ wish to acknowledge the support of the Norwegian Research Council through the SYNODE II project (127582/410), while AI gratefully acknowledges the support of the Lars Onsager Fund.

REFERENCES

R. Abraham and J. E. Marsden (1978), *Foundations of Mechanics*, 2nd edn, Addison-Wesley.

M. Abramowitz and I. A. Stegun, eds (1970), *Handbook of Mathematical Functions*, Dover, New York.

V. I. Arnold (1989), *Mathematical Methods of Classical Mechanics*, Vol. 60 of *GTM*, 2nd edn, Springer.

G. A. Baker (1975), *Essentials of Padé Approximants*, Academic Press, New York.

G. Benettin and A. Giorgilli (1994), 'On the Hamiltonian interpolation of near-to-the-identity symplectic mappings with application to symplectic integration algorithm', *J. Statist. Phys.* **74**, 1117–1143.

M. V. Berry (1987), Regular and irregular motion, in *Hamiltonian Dynamical Systems* (R. MacKay and J. Meiss, eds), Adam Hilger, Bristol, pp. 27–53.

G. Birkhoff (1936), 'A note on topological groups', *Compositio Math.* **3**, 427–430.

S. Blanes, F. Casas and J. Ros (1999), Improved high order integrators based on Magnus expansion, Technical Report NA1999/08, DAMTP, University of Cambridge.

S. Blanes, F. Casas, J. A. Oteo and J. Ros (1998), 'Magnus and Fer expansions for matrix differential equations: The convergence problem', *J. Phys. A* **31**, 259–268.

W. M. Boothby (1975), *An Introduction to Differentiable Manifolds and Riemannian Geometry*, Academic Press.

N. Bourbaki (1975), *Lie Groups and Lie Algebras*, Addison-Wesley, Reading, MA.

J. C. Butcher (1963), 'Coefficients for the study of Runge–Kutta integration processes', *J. Austral. Math. Soc.* **3**, 185–201.

M. P. Calvo, A. Iserles and A. Zanna (1997), 'Numerical solution of isospectral flows', *Math. Comput.* **66**, 1461–1486.

M. P. Calvo, A. Iserles and A. Zanna (1999), 'Conservative methods for the Toda lattice equations', *IMA J. Numer. Anal.* **19**, 509–523.

E. Celledoni and A. Iserles (1998), Approximating the exponential from a Lie algebra to a Lie group, Technical Report 1998/NA3, DAMTP, University of Cambridge. To appear in *Math. Comput.*

E. Celledoni and A. Iserles (1999), Methods for the approximation of the matrix exponential in a Lie-algebraic setting, Technical Report 1999/NA3, DAMTP, University of Cambridge. To appear in *IMA J. Numer. Anal.*

M. T. Chu (1993), On a differential equation $\frac{dx}{dt} = [X, k(X)]$ where k is a Toeplitz annihilator, Technical report, North Carolina State University.

M. T. Chu (1998), 'Inverse eigenvalue problems', *SIAM Rev.* **40**, 1–39.

R. Cools (1997), Constructing cubature formulas: the science behind the art, in *Acta Numerica*, Vol. 6, Cambridge University Press, pp. 1–54.

G. Cooper (1987), 'Stability of Runge–Kutta methods for trajectory problems', *IMA J. Numer. Anal.* **7**, 1–13.

P. E. Crouch and R. Grossman (1993), 'Numerical integration of ordinary differential equations on manifolds', *J. Nonlinear Sci.* **3**, 1–33.

P. J. Davis and P. Rabinowitz (1984), *Methods of Numerical Integration*, 2nd edn, Academic Press, Orlando, FL.

P. Deift, T. Nanda and C. Tomei (1983), 'Ordinary differential equations and the symmetric eigenvalue problem', *SIAM J. Numer. Anal.* **20**, 1–22.

L. Dieci, R. D. Russell and E. S. van Vleck (1994), 'Unitary integrators and applications to continuous orthonormalization techniques', *SIAM J. Numer. Anal.* **31**, 261–281.

F. Diele, L. Lopez and R. Peluso (1998), 'The Cayley transform in the numerical solution of unitary differential systems', *Adv. Comput. Math.* **8**, 317–334.

K. Engø (2000), 'On the construction of geometric integrators in the RKMK class', *BIT* **40**, 41–61.

K. Engø and S. Faltinsen (1999), Numerical integration of Lie–Poisson systems while preserving coadjoint orbits and energy, Technical Report No. 179, Dept Comp. Sc., University of Bergen.

K. Engø, A. Marthinsen and H. Z. Munthe-Kaas (1999), DiffMan: an object oriented MATLAB toolbox for solving differential equations on manifolds, Technical Report No. 164, Dept Comp. Sc., University of Bergen.

D. J. Estep and A. M. Stuart (1995), 'The rate of error growth in Hamiltonian-conserving integrators', *Z. Angew. Math. Phys.* **46**, 407–418.

S. Faltinsen (1998), Backward error analysis for Lie-group methods, Technical Report 1998/NA12, DAMTP, University of Cambridge. To appear in *BIT*.

S. Faltinsen (2000), 'Can Lie-group methods be symplectic?' To appear.

S. Faltinsen, A. Marthinsen and H. Z. Munthe-Kaas (1999), Multistep methods integrating ordinary differential equations on manifolds, Technical Report Numerics No. 3/1999, The Norwegian University of Science and Technology, Trondheim.

F. Fer (1958), 'Résolution de l'équation matricielle $\dot{U} = pU$ par produit infini d'exponentielles matricielles', *Bull. Classe des Sci. Acad. Royal Belg.* **44**, 818–829.

H. Flaschka (1974), 'The Toda lattice I', *Phys. Rev. B* **9**, 1924–1925.

C. W. Gear (1971), *Numerical Initial Value Problems in Ordinary Differential Equations*, Prentice-Hall, Englewood Cliffs, NJ.

H. Goldstein (1980), *Classical Mechanics*, Addison-Wesley, Reading, MA.

G. H. Golub and C. F. Van Loan (1996), *Matrix Computations*, 3rd edn, Johns Hopkins University Press, Baltimore.

V. Guillemin and A. Pollack (1974), *Differential Topology*, Prentice-Hall, Englewood Cliffs, NJ.

E. Hairer (1994), Backward analysis of numerical integrators and symplectic methods, in *Scientific Computation and Differential Equations* (K. Burrage, C. Baker, P. van der Houwen, Z. Jackiewicz and P. Sharp, eds), Vol. 1 of *Annals of Numer. Math.*, J. C. Baltzer, Amsterdam, pp. 107–132. Proceedings of the SCADE'93 conference, Auckland, New Zealand, January 1993.

E. Hairer and C. Lubich (1997), 'The life-span of backward error analysis for numerical integrator', *Numer. Math.* **76**, 441–462.

E. Hairer, S. P. Nørsett and G. Wanner (1993), *Solving Ordinary Differential Equations I: Nonstiff Problems*, 2nd revised edn, Springer, Berlin.

F. Harary (1969), *Graph Theory*, Addison-Wesley, Reading MA.

F. Hausdorff (1906), 'Die symbolische Exponentialformel in der Gruppentheorie', *Berichte der Sächsischen Akademie der Wissenschaften* (*Math. Phys. Klasse*) **58**, 19–48.

D. J. Higham (1997), 'Time-stepping and preserving orthonormality', *BIT* **37**, 24–36.

M. Hochbruck and C. Lubich (1997), 'On Krylov subspace approximations to the matrix exponential operator', *SIAM J. Numer. Anal.* **34**, 1911–1925.

A. Iserles (1984), 'Solving linear ordinary differential equations by exponentials of iterated commutators', *Numer. Math.* **45**, 183–199.

A. Iserles (1997), Multistep methods on manifolds, Technical Report 1997/NA13, DAMTP, University of Cambridge.

A. Iserles (1999*a*), How large is the exponential of a bounded matrix?, Technical Report 1999/NA1, DAMTP, University of Cambridge.

A. Iserles (1999*b*), On Cayley-transform methods for the discretization of Lie-group equations, Technical Report 1999/NA4, DAMTP, University of Cambridge.

A. Iserles (2000), 'Fast computation of Magnus series'. To appear.

A. Iserles and S. P. Nørsett (1991), *Order Stars*, Chapman and Hall, London.

A. Iserles and S. P. Nørsett (1999), 'On the solution of linear differential equations in Lie groups', *Phil. Trans Royal Society A* **357**, 983–1020.

A. Iserles and A. Zanna (2000), 'On the dimension of certain graded Lie algebras arising in geometric integration of differential equations', *LMS J. Comput. & Math.* **3**, 44–75.

A. Iserles, A. Marthinsen and S. P. Nørsett (1999), 'On the implementation of the method of Magnus series for linear differential equations', *BIT* **39**, 281–304.

A. Iserles, S. P. Nørsett and A. F. Rasmussen (1998), Time-symmetry and high-order Magnus methods, Technical Report 1998/NA06, DAMTP, University of Cambridge.

G. Julia (1918), 'Mémoire sur l'itération des fonctions rationnelles', *J. Math.* **8**, 47–245.

F. Kang and S. Zai-jiu (1995), 'Volume-preserving algorithmms for source-free dynamical systems', *Numer. Math.* **71**, 451–463.

H. J. Landau (1994), 'The inverse eigenvalue problem for real symmetric Toeplitz matrices', *J. Amer. Math. Soc.* **7**, 749–767.

Y. Liu (1998), Projected Runge–Kutta methods for differential equations on matrix Lie groups, Technical Report 1998/NA1, DAMTP, University of Cambridge.

W. Magnus (1954), 'On the exponential solution of differential equations for a linear operator', *Comm. Pure Appl. Math* **VII**, 649–673.

J. E. Marsden and T. S. Ratiu (1994), *Introduction to Mechanics and Symmetry*, Springer, New York.

R. I. McLachlan, G. R. W. Quispel and N. Robidoux (1998), 'A unified approach to Hamiltonian systems, Poisson systems, gradient systems, and systems with Lyapunov functions and/or first integrals', *Phys. Rev. Lett.* **81**, 2399–2403.

P. C. Moan (1998), Efficient approximation of Sturm–Liouville problems using Lie-group methods, Technical Report 1998/NA11, DAMTP, University of Cambridge.

P. C. Moan (2000), 'On the convergence of Magnus and Cayley expansions'. To appear.

C. Moler and C. F. Van Loan (1978), 'Nineteen dubious ways to compute the exponential of a matrix', *SIAM Rev.* **20**, 801–836.

J. Moser (1973), *Stable and Random Motion in Dynamical Systems*, Princeton University Press.

H. Munthe-Kaas (1995), 'Lie–Butcher theory for Runge–Kutta methods', *BIT* **35**, 572–587.

H. Munthe-Kaas (1998), 'Runge–Kutta methods on Lie groups', *BIT* **38**, 92–111.

H. Munthe-Kaas (1999), 'High order Runge–Kutta methods on manifolds', *Appl. Numer. Math.* **29**, 115–127.

H. Munthe-Kaas and M. Haveraaen (1996), Coordinate free numerics: Closing the gap between 'pure' and 'applied' mathematics?, in *Proceedings of ICIAM–95, Zeitschrift für Angewandte Mathematik und Mechanik (ZAMM)*.

H. Munthe-Kaas and E. Lodden (2000), 'Lie group integrators for parabolic PDEs'. To appear.

H. Munthe-Kaas and B. Owren (1999), 'Computations in a free Lie algebra', *Phil. Trans Royal Society A* **357**, 957–982.

H. Munthe-Kaas and A. Zanna (1997), Numerical integration of differential equations on homogeneous manifolds, in *Foundations of Computational Mathematics* (F. Cucker and M. Shub, eds), Springer, pp. 305–315.

A. I. Neishtadt (1984), 'The separation of motions in systems with rapidly rotating phase', *J. Appl. Math. Mech.* **48**, 133–139.

S. P. Nørsett and G. Wanner (1981), 'Perturbed collocation and Runge–Kutta methods', *Numer. Math.* **38**, 193–208.

P. J. Olver (1995), *Equivalence, Invariants, and Symmetry*, Cambridge University Press.

B. Owren and A. Marthinsen (1999*a*), Integration methods based on canonical coordinates of the second kind, Technical Report Numerics No. 5/1999, Norwegian University of Science and Technology, Trondheim.

B. Owren and A. Marthinsen (1999*b*), 'Runge–Kutta methods adapted to manifolds and based on rigid frames', *BIT* **39**, 116–142.

B. Owren and B. Welfert (1996), The Newton iteration on Lie groups, Technical Report Numerics No. 3/1996, Norwegian University of Science and Technology, Trondheim.

L. R. Petzold, L. O. Jay and J. Yen (1997), Numerical solution of highly oscillatory ordinary differential equations, in *Acta Numerica*, Vol. 6, Cambridge University Press, pp. 437–483.

S. Preuss (1973), 'Solving linear boundary value problems by approximating the coefficients', *Math. Comput.* **27**, 551–561.

J. D. Pryce (1993), *Numerical Solution of Sturm–Liouville Problems*, Oxford University Press, New York.

S. Reich (1996), Backward error analysis for numerical integrators, Technical Report SC 96-21, Konrad-Zuse Zentrum für Informationstechnik, Berlin.

J. M. Sanz Serna and M. P. Calvo (1994), *Numerical Hamiltonian Problems*, Chapman & Hall.

C. Størmer (1907), 'Sur les trajectoires des corpuscules électrisés', *Arch. Sci. Phys. Nat., Genève* **24**, 5–18, 113–158, 221–247.

A. M. Stuart and A. R. Humphries (1996), *Dynamical Systems and Numerical Analysis*, Cambridge Monographs on Applied and Computational Mathematics, Cambridge University Press, Cambridge.

M. Toda (1981), *Theory of Nonlinear Lattices*, Springer, Berlin.

W. F. Trench (1997), 'Numerical solution of the inverse eigenvalue problem for real symmetric Toepliz matrices', *SIAM J. Sci. Comput.* **18**, 1722–1736.

V. S. Varadarajan (1984), *Lie Groups, Lie Algebras, and Their Representations*, GTM 102, Springer.

J. Wei and E. Norman (1964), 'On global representations of the solutions of linear differential equations as a product of exponentials', *Proc. Amer. Math. Soc.* **15**, 327–334.

H. Yoshida (1990), 'Construction of higher order symplectic integrators', *Phys. Lett. A* **150**, 262–268.

N. J. Zabusky and M. D. Kruskal (1965), 'Interaction of solitons in a collisionless plasma and the recurrences of initial states', *Phys. Rev. Lett.* **15**, 240–243.

A. Zanna (1996), The method of iterated commutators for ordinary differential equations on Lie groups, Technical Report 1996/NA12, DAMTP, University of Cambridge.

A. Zanna (1998), On the Numerical Solution of Isospectral Flows, PhD thesis, University of Cambridge, England.

A. Zanna (1999), 'Collocation and relaxed collocation for the Fer and the Magnus expansions', *SIAM J. Numer. Anal.* **36**, 1145–1182.

A. Zanna and H. Munthe-Kaas (1997), Iterated commutators, Lie's reduction method and ordinary differential equations on matrix Lie groups, in *Foundation of Computational Mathematics* (F. Cucker and M. Shub, eds), Springer, pp. 434–441.

A. Zanna, K. Engø and H. Z. Munthe-Kaas (1999), Adjoint and selfadjoint Lie-group methods, Technical Report NA1999/02, DAMTP, University of Cambridge.

A. List of methods

In this appendix we list a few of the methods that have been described in the survey. No attempt has been expanded to explain the methods, beyond references to the relevant earlier material.

A.1. RK-MK methods

All classical RK methods can be translated into Lie-group methods. Assume that the Butcher tableau

$$
\begin{array}{c|cccc}
c_1 & a_{1,1} & a_{1,2} & \cdots & a_{1,\nu} \\
c_2 & a_{2,1} & a_{2,2} & \cdots & a_{2,\nu} \\
\vdots & \vdots & \vdots & & \vdots \\
c_\nu & a_{\nu,1} & a_{\nu,2} & \cdots & a_{\nu,\nu} \\
\hline
& b_1 & b_2 & \cdots & b_\nu
\end{array}
$$

defines a Runge–Kutta method of order p in the classical sense (Hairer et al. 1993) and that ϕ is a map from $\mathfrak{g}$ to $\mathcal{G}$, for instance $\phi = $ expm, the exponential mapping, or $\phi = $ cay, the Cayley mapping for quadratic Lie groups. Then the corresponding order-p RK-MK algorithm for the Lie-group equation $Y' = A(t, Y)Y$ is obtained as

$$
\left.
\begin{aligned}
\Theta_k &= \sum_{l=1}^{\nu} a_{k,l} F_l, \\
F_k &= \mathrm{d}\phi^{-1}(\Theta_k, A_k, p), \\
A_k &= hA(t_n + c_k h, \phi(\Theta_k) Y_n), \\
\Theta &= \sum_{l=1}^{\nu} b_l F_l, \\
Y_{n+1} &= \phi(\Theta) Y_n,
\end{aligned}
\right\}
\qquad k = 1, \ldots, \nu,
\qquad\qquad \text{(A.1)}
$$

for $n \in \mathbb{N}$, and it is explicit provided that the underlying RK scheme is explicit. The function $\mathrm{d}\phi^{-1}(B, C, p)$ is a truncation of $\mathrm{d}\phi_B^{-1}(C)$ to order $p - 1$, which is usually sufficient for a method of order p, given that the error is subsumed in the $\mathcal{O}(h^{p+1})$ term. In some instances, for example when $\mathfrak{g} = \mathfrak{so}(3)$, the function $\mathrm{d}\phi_B^{-1}(C)$ can be evaluated exactly (see Appendix B).

Some popular schemes of the type (A.1) are Lie-group versions of

- forward Euler,

$$
Y_{n+1} = \phi\big(hA(t_n, Y_n)\big) Y_n;
$$

- the implicit midpoint rule,

$$
F_1 = hA(t_n + \tfrac{1}{2}h, \phi(\tfrac{1}{2}F_1)Y_n),
$$
$$
\Theta = F_1,
$$
$$
Y_{n+1} = \phi(\Theta)Y_n;
$$

- the trapezoidal rule,

$$
F_1 = hA(t_n, Y_n),
$$
$$
F_2 = hA(t_n + h, \phi(\tfrac{1}{2}(F_1 + F_2))Y_n),
$$
$$
\Theta = \tfrac{1}{2}(F_1 + F_2),
$$
$$
Y_{n+1} = \phi(\Theta)Y_n;
$$

- Heun's method,

$$
F_1 = hA(t_n, Y_n),
$$
$$
F_2 = hA(t_n + \tfrac{1}{2}h, \phi(\tfrac{1}{2}F_1)Y_n),
$$
$$
\Theta = F_2,
$$
$$
Y_{n+1} = \phi(\Theta)Y_n.
$$

The scheme (A.1) employs coordinates centred at Y_n. Schemes with co-ordinates centred at an arbitrary point can be obtained in a similar manner,

$$
\left.
\begin{aligned}
\Theta_k &= C_{h,n} + \sum_{l=1}^{\nu} a_{k,l} F_l, \\
F_k &= \mathrm{d}\phi^{-1}(\Theta_k, A_k, p), \\
A_k &= hA(t_n + c_k h, \phi(\Theta_k)\phi(C_{h,n})^{-1} Y_n),
\end{aligned}
\right\} \qquad k = 1, \ldots, \nu,
$$

$$
C_{h,n} = \vartheta(F_1, \ldots, F_\nu),
$$

$$
\Theta = C_{h,n} + \sum_{l=1}^{\nu} b_l F_l,
$$
$$
Y_{n+1} = \phi(\Theta)\phi(C_{h,n})^{-1} Y_n,
$$

for $n = 0, 1, 2, \ldots$ and ϑ. In particular, for *geodesic-symmetric coordinates* we have

$$
\left.\begin{aligned}
\Theta_k &= \sum_{l=1}^{\nu} \left(a_{k,l} - \tfrac{1}{2}b_l\right) F_l, \\
F_k &= \mathrm{d}\phi^{-1}(\Theta_k, A_k, p), \\
A_k &= hA(t_n + c_k h, \phi(\Theta_k)\phi(C_{h,n})^{-1}Y_n),
\end{aligned}\right\} \quad k = 1, \ldots, \nu,
$$
$$
\begin{aligned}
C_{h,n} &= -\tfrac{1}{2}\sum_{l=1}^{\nu} b_l F_l, \\
\Theta &= \tfrac{1}{2}\sum_{l=1}^{\nu} b_l F_l \quad (= -C_{h,n}), \\
Y_{n+1} &= \phi(\Theta)\phi(C_{h,n})^{-1}Y_n,
\end{aligned}
\tag{A.2}
$$

while *flow-symmetric coordinates* yield

$$
\left.\begin{aligned}
\Theta_k &= \sum_{l=1}^{\nu} (a_{k,l} - w_l) F_l, \\
F_k &= \mathrm{d}\phi^{-1}(\Theta_k, A_k, p), \\
A_k &= hA(t_n + c_k h, \phi(\Theta_k)\phi(C_{h,n})^{-1}Y_n),
\end{aligned}\right\} \quad k = 1, 2, \ldots, \nu,
$$
$$
\begin{aligned}
C_{h,n} &= -\sum_{l=1}^{\nu} w_l F_l, \\
\Theta &= \sum_{l=1}^{\nu} (b_l - w_l) F_l, \\
Y_{n+1} &= \phi(\Theta)\phi(C_{h,n})^{-1}Y_n,
\end{aligned}
\tag{A.3}
$$

where the weights $w_1, \ldots, w_\nu$ are obtained by integrating the Lagrangian cardinal polynomials, as in (7.7).

Herewith a number of important examples, originating in familiar classical RK methods.

- Order-four Gauss–Legendre scheme:

$$
\begin{array}{c|cc}
\frac{1}{2} - \frac{\sqrt{3}}{6} & \frac{1}{4} & \frac{1}{4} - \frac{\sqrt{3}}{6} \\
\frac{1}{2} + \frac{\sqrt{3}}{6} & \frac{1}{4} + \frac{\sqrt{3}}{6} & \frac{1}{4} \\
\hline
\boldsymbol{b}^{\mathrm{T}} & \frac{1}{2} & \frac{1}{2} \\
\boldsymbol{w}^{\mathrm{T}} & \frac{1}{4} + \frac{\sqrt{3}}{8} & \frac{1}{4} - \frac{\sqrt{3}}{8}
\end{array}.
\tag{A.4}
$$

- Order-six Gauss–Legendre scheme:

$$
\begin{array}{c|ccc}
\frac{1}{2} - \frac{\sqrt{15}}{10} & \frac{5}{36} & \frac{2}{9} - \frac{\sqrt{15}}{15} & \frac{5}{36} - \frac{\sqrt{15}}{30} \\
\frac{1}{2} & \frac{5}{36} + \frac{\sqrt{15}}{24} & \frac{2}{9} & \frac{5}{36} - \frac{\sqrt{15}}{24} \\
\frac{1}{2} + \frac{\sqrt{15}}{10} & \frac{5}{36} + \frac{\sqrt{15}}{30} & \frac{2}{9} + \frac{\sqrt{15}}{15} & \frac{5}{36} \\
\hline
\boldsymbol{b}^{\mathrm{T}} & \frac{5}{18} & \frac{4}{9} & \frac{5}{18} \\
\boldsymbol{w}^{\mathrm{T}} & \frac{5}{36} + \frac{\sqrt{15}}{24} & \frac{2}{9} & -\frac{1}{24}
\end{array}
\qquad (A.5)
$$

- Order-four Gauss–Lobatto scheme:

$$
\begin{array}{c|ccc}
0 & 0 & 0 & 0 \\
\frac{1}{2} & \frac{5}{24} & \frac{1}{3} & -\frac{1}{24} \\
1 & \frac{1}{6} & \frac{2}{3} & \frac{1}{6} \\
\hline
\boldsymbol{b}^{\mathrm{T}} & \frac{1}{6} & \frac{2}{3} & \frac{1}{6} \\
\boldsymbol{w}^{\mathrm{T}} & \frac{5}{24} & \frac{1}{3} & -\frac{1}{24}
\end{array}
\qquad (A.6)
$$

Note that, both in the case of the Gauss–Legendre of order six and Gauss–Lobatto of order four, it is true that $C_{h,n} = -F_2$, hence $C_{h,n}$ need not be explicitly evaluated in (A.3).

Algorithms (A.1)–(A.3) are practical for methods of order $p \leq 4$, since for higher-order methods the computation of $\mathrm{d}\phi^{-1}(A, B, p)$ requires the evaluation of a large number of commutators. For higher-order schemes it is recommended instead to use schemes based on graded Lie algebras, along the lines of Section 5.3.

Graded Lie algebras can also be effectively used to optimize existing RK-MK methods. The scheme

$$
\begin{aligned}
A_1 &= hA(t_n, Y_n), \\
B_1 &= A_1, \\
A_2 &= hA\big(t_n + c_2 h, \ \mathrm{expm}(\tfrac{1}{2}B_1)Y_n\big), \\
B_2 &= A_2 - A_1, \\
A_3 &= hA\big(t_n + c_3 h, \ \mathrm{expm}(\tfrac{1}{2}B_1 + \tfrac{1}{2}B_2 - \tfrac{1}{8}[B_1, B_2])Y_n\big), \\
B_3 &= A_3 - A_2, \\
A_4 &= hA\big(t_n + c_4 h, \ \mathrm{expm}(B_1 + B_2 + B_3)Y_n\big), \\
B_4 &= A_4 - 2A_2 + A_1, \\
\Theta &= B_1 + B_2 + \tfrac{1}{3}B_3 + \tfrac{1}{6}B_4 - \tfrac{1}{6}[B_1, B_2] - \tfrac{1}{12}[B_1, B_4], \\
Y_{n+1} &= \mathrm{expm}(\Theta)Y_n,
\end{aligned}
$$

based on the Runge–Kutta tableau

$$
\begin{array}{c|cccc}
0 & & & & \\
\frac{1}{2} & \frac{1}{2} & & & \\
\frac{1}{2} & 0 & \frac{1}{2} & & \\
1 & 0 & 0 & 1 & \\
\hline
 & \frac{1}{6} & \frac{1}{3} & \frac{1}{3} & \frac{1}{6}
\end{array}
\tag{A.7}
$$

requires only two commutators instead of six if implemented using (A.1).

A.2. Higher-order methods for linear equations using graded Lie algebras

In this subsection we present methods for linear equations $Y' = A(t)Y$, whereby the number of commutators is reduced using graded algebras and a further technique due to Blanes et al. (1999) and described briefly in Section 5.4. We do not distinguish between RK–MK and Magnus-type methods because, subject to these reductions, the two methods produce very similar results.

We consider collocation-type schemes. Denote by

$$
\mathrm{VDM}(\boldsymbol{d}) = (d_i^{\,j-1})_{i,j=1}^{\nu}
$$

the Vandermonde matrix generated by the vector $\boldsymbol{d}$. In particular,

- set $V = \mathrm{VDM}(\boldsymbol{c})$, $\boldsymbol{c}^{\mathrm{T}} = (c_1, c_2, \ldots, c_\nu)$ for non-symmetric collocation schemes;

- for symmetric collocation schemes, we take full advantage of symmetry, setting $V = \mathrm{VDM}(\boldsymbol{c} - \frac{1}{2})$.

Then,

$$
\left.
\begin{aligned}
A_k &= hA(t_n + c_k h), \\
B_k &= \sum_{l=1}^{\nu} (V^{-1})_{k,l} A_l, \\
\Theta &= \mathrm{d}\varphi^{-1}(B_1, B_2, \ldots, B_\nu),
\end{aligned}
\right\}
\quad k = 1, 2, \ldots, \nu,
\tag{A.8}
$$

$$
Y_{n+1} = \phi(\Theta)Y_n,
$$

where $\mathrm{d}\varphi^{-1}$ is an order-p truncation to the $\mathrm{d}\phi^{-1}$-equation in the graded basis $B_1, \ldots, B_\nu$. For symmetric collocation schemes, the function $\mathrm{d}\varphi^{-1}$ in (A.8) depends on terms of odd grade only. Specifically, for $\phi = \mathrm{expm}$, we obtain the following.

- For the order-six Gauss–Legendre scheme

$$
\begin{aligned}
\mathrm{d}\varphi^{-1}(B_1, B_2, B_3) = {} & B_1 + \tfrac{1}{12}B_3 - \tfrac{1}{12}[B_1, B_2] + \tfrac{1}{240}[B_2, B_3] \\
& + \tfrac{1}{360}[B_1, [B_1, B_3]] - \tfrac{1}{240}[B_2, [B_1, B_2]] \qquad \text{(A.9)} \\
& + \tfrac{1}{720}[B_1, [B_1, [B_1, B_2]]]
\end{aligned}
$$

(for order two and order four it suffices to consider the first term and the first and third term, respectively, and the nodes of the corresponding quadrature), which can be evaluated with just four commutators using the technique of Blanes et al. (1999),

$$
\begin{aligned}
\mathrm{d}\varphi^{-1}(B_1, B_2, B_3) &\approx C_1 + C_2 + C_3, \\
C_1 &= B_1 + \tfrac{1}{12}B_3 \\
C_2 &= -\tfrac{1}{12}[B_1 + \tfrac{1}{20}B_3, B_2], \\
C_3 &= [B_1, [B_1, \tfrac{1}{360}B_3 - \tfrac{1}{60}C_2]] - \tfrac{1}{20}[B_2, C_2],
\end{aligned}
$$

producing an order-six time-symmetric truncation of $\mathrm{d}\varphi^{-1}(B_1, B_2, B_3)$.

- For the order-four Gauss–Lobatto scheme (Ehle III$_\mathrm{A}$),

$$
\mathrm{d}\varphi^{-1}(B_1, B_2, B_3) = B_1 + \tfrac{1}{12}B_3 - \tfrac{1}{12}[B_1, B_2].
$$

- For the order-six Gauss–Lobatto scheme, with nodes $c_1 = 0$, $c_2 = \tfrac{1}{2} - \tfrac{\sqrt{5}}{10}$, $c_3 = \tfrac{1}{2} + \tfrac{\sqrt{5}}{10}$ and $c_4 = 1$,

$$
\begin{aligned}
\mathrm{d}\varphi^{-1}(B_1, B_2, B_3, B_4) = {} & B_1 + \tfrac{1}{12}B_3 - \tfrac{1}{12}[B_1, B_2] - \tfrac{1}{8}[B_1, B_4] \\
& + \tfrac{1}{240}[B_2, B_3] + \tfrac{1}{360}[B_1, [B_1, B_3]] \\
& - \tfrac{1}{240}[B_2, [B_1, B_2]] + \tfrac{1}{720}[B_1, [B_1, [B_1, B_2]]],
\end{aligned}
$$

which can be evaluated to correct order as

$$
\begin{aligned}
\mathrm{d}\varphi^{-1}(B_1, B_2, B_3, B_4) &\approx C_1 + C_2 + C_3, \\
C_1 &= B_1 + \tfrac{1}{12}B_3 \\
C_2 &= -\tfrac{1}{12}[B_1 + \tfrac{1}{20}B_3, B_2 + \tfrac{3}{2}B_4], \\
C_3 &= [B_1, [B_1, \tfrac{1}{360}B_3 - \tfrac{1}{60}C_2]] - \tfrac{1}{20}[B_2, C_2],
\end{aligned}
$$

with just four commutators.

It may strike the reader that, after the transformation into the self-adjoint basis $\{B_i\}$, the expression of Θ always has the same type of expansion, regardless of the choice of collocation nodes. This is no surprise, the information about the nodes being hidden in the basis elements B_i. Changing into the self-adjoint basis implies that we integrate a Taylor-type expansion, in combination with a truncation of dexp^{-1}, which is independent of the nodes, depending only on the order of the method.

A.3. Methods using the Magnus expansion

Magnus-type methods are well suited to collocation-type techniques. This results in implicit schemes for nonlinear Lie-group equations.

- Order-four Gauss–Legendre:

$$F_1 = \tfrac{1}{4}A_1 + (\tfrac{1}{4} - \tfrac{\sqrt{3}}{6})A_2 + (\tfrac{5}{144} - \tfrac{\sqrt{3}}{48})[A_1, A_2],$$
$$A_1 = hA(t_n + c_1 h, \operatorname{expm}(F_1)Y_n),$$
$$F_2 = (\tfrac{1}{4} + \tfrac{\sqrt{3}}{6})A_1 + \tfrac{1}{4}A_2 - (\tfrac{5}{144} + \tfrac{\sqrt{3}}{48})[A_1, A_2],$$
$$A_2 = hA(t_n + c_2 h, \operatorname{expm}(F_2)Y_n),$$
$$\Theta = \tfrac{1}{2}(A_1 + A_2) - \tfrac{\sqrt{3}}{12}[A_1, A_2],$$
$$Y_{n+1} = \operatorname{expm}(\Theta)Y_n,$$

where $c_i = \tfrac{1}{2} \pm \tfrac{\sqrt{3}}{6}$, $i = 1, 2$.

- Order-four Gauss–Lobatto:

$$F_1 = O,$$
$$A_1 = hA(t_n, Y_n),$$
$$F_2 = \tfrac{5}{24}A_1 + \tfrac{1}{3}A_2 - \tfrac{1}{24}A_3$$
$$\quad - (\tfrac{11}{480}[A_1, A_2] + \tfrac{5}{1152}[A_1, A_3] + \tfrac{1}{144}[A_2, A_3]),$$
$$A_2 = hA(t_n + \tfrac{1}{2}h, \operatorname{expm}(F_2)Y_n),$$
$$F_3 = \tfrac{1}{6}A_1 + \tfrac{2}{3}A_2 + \tfrac{1}{6}A_3 - (\tfrac{1}{15}[A_1, A_2] + \tfrac{1}{60}[A_1, A_3] + \tfrac{1}{15}[A_2, A_3]),$$
$$A_3 = hA(t_n + h, \operatorname{expm}(F_3)Y_n),$$
$$\Theta = \tfrac{1}{6}A_1 + \tfrac{2}{3}A_2 + \tfrac{1}{6}A_3 - (\tfrac{1}{15}[A_1, A_2] + \tfrac{1}{60}[A_1, A_3] + \tfrac{1}{15}[A_2, A_3]),$$
$$Y_{n+1} = \operatorname{expm}(\Theta)Y_n.$$

Relaxing the collocation conditions, it is possible to obtain explicit methods.

- An explicit order-three scheme:

$$A_1 = hA(t_n, Y_n),$$
$$A_2 = hA(t_n + \tfrac{1}{2}h, \operatorname{expm}(A_1)Y_n),$$
$$A_3 = hA(t_n + h, \operatorname{expm}(-A_1 + 2A_2)Y_n),$$
$$\Theta = \tfrac{1}{6}A_1 + \tfrac{2}{3}A_2 + \tfrac{1}{6}A_3 - [A_1 - A_3, \tfrac{1}{15}A_2 + \tfrac{1}{60}A_3],$$
$$Y_{n+1} = \operatorname{expm}(\Theta)Y_n.$$

A.4. Magnus-type methods with geodesic/flow coordinates

Magnus-type methods introduced in Appendix A.3 employ coordinates at Y_n, hence they cease to be self-adjoint for nonlinear problems. Below we describe their self-adjoint modification.

- Order-four Gauss–Legendre method based on geodesic coordinates:

$$F_1 = \tfrac{1}{4}A_1 + \left(\tfrac{1}{4} - \tfrac{\sqrt{3}}{6}\right)A_2 + \left(\tfrac{5}{144} - \tfrac{\sqrt{3}}{24}\right)[A_1, A_2],$$

$$A_1 = hA\left(t_n + c_1 h,\ \mathrm{expm}(F_1 - C_{h,n})\,\mathrm{expm}(C_{h,n})Y_n\right),$$

$$F_2 = \left(\tfrac{1}{4} + \tfrac{\sqrt{3}}{6}\right)A_1 + \tfrac{1}{4}A_2 - \left(\tfrac{5}{144} + \tfrac{\sqrt{3}}{24}\right)[A_1, A_2],$$

$$A_2 = hA\left(t_n + c_2 h,\ \mathrm{expm}(F_2 - C_{h,n})\,\mathrm{expm}(C_{h,n})Y_n\right),$$

$$C_{h,n} = \tfrac{1}{4}(A_1 + A_2) - \tfrac{\sqrt{3}}{24}[A_1, A_2],$$

$$\Theta = 2C_{h,n},$$

$$Y_{n+1} = \mathrm{expm}(\Theta)Y_n.$$

- Order-four Gauss–Legendre methods with flow coordinates:

$$F_1 = \tfrac{1}{4}A_1 + \left(\tfrac{1}{4} - \tfrac{\sqrt{3}}{6}\right)A_2 + \left(\tfrac{1}{288} - \tfrac{\sqrt{3}}{96}\right)[A_1, A_2],$$

$$A_1 = hA\left(t_n + c_1 h,\ \mathrm{expm}(F_1 - C_{h,n})\,\mathrm{expm}(C_{h,n})Y_n\right),$$

$$F_2 = \left(\tfrac{1}{4} + \tfrac{\sqrt{3}}{6}\right)A_1 + \tfrac{1}{4}A_2 - \left(\tfrac{1}{288} + \tfrac{\sqrt{3}}{96}\right)[A_1, A_2],$$

$$A_2 = hA\left(t_n + c_2 h,\ \mathrm{expm}(F_2 - C_{h,n})\,\mathrm{expm}(C_{h,n})Y_n\right),$$

$$C_{h,n} = \left(\tfrac{1}{4} + \tfrac{\sqrt{3}}{8}\right)A_1 + \left(\tfrac{1}{4} - \tfrac{\sqrt{3}}{8}\right)A_2 - \tfrac{\sqrt{3}}{96}[A_1, A_2],$$

$$\Theta = \tfrac{1}{2}(A_1 + A_2) - \tfrac{\sqrt{3}}{48}[A_1, A_2],$$

$$Y_{n+1} = \mathrm{expm}(\Theta - C_{h,n})\,\mathrm{expm}(C_{h,n})Y_n.$$

Although Magnus-type methods with geodesic/flow coordinates exist for every order p, devising such schemes for nonlinear problems and order greater than four is hard. For this reason we shall restrict our attention to order-four methods. We let $c_1, c_2, \ldots, c_\nu$ be the collocation nodes.

- Collocation order-four Magnus method with geodesic coordinates:

$$\left.\begin{aligned} F_k &= \sum_{l=1}^{\nu} a_{k,l} A_l + \tfrac{1}{2}\sum_{l,j=1}^{\nu} \left(a_{k;l,j} + \tfrac{1}{2}b_l a_{k,l}\right)[A_l, A_j], \\ A_k &= hA(t_n + c_k h,\ \mathrm{expm}(F_k - C_{h,n})\,\mathrm{expm}(C_{h,n})Y_n), \end{aligned}\right\} \quad k = 1, \ldots, \nu,$$

$$C_{h,n} = \tfrac{1}{2}\sum_{l=1}^{\nu} b_l A_l + \tfrac{1}{4}\sum_{l,j=1}^{\nu} b_{l,j}[A_l, A_j],$$

$$\Theta = 2C_{h,n},$$

$$Y_{n+1} = \mathrm{expm}(\Theta)Y_n.$$

- Collocation order-four Magnus method with flow coordinates:

$$
\left.
\begin{aligned}
F_k &= \sum_{l=1}^{\nu} a_{k,l} A_l + \tfrac{1}{2} \sum_{l,j=1}^{\nu} (a_{k;l,j} + w_l a_{k,l})[A_l, A_j], \\
A_k &= hA(t_n + c_k h, \ \mathrm{expm}(F_k - C_{h,n}) \ \mathrm{expm}(C_{h,n}) Y_n),
\end{aligned}
\right\} \quad k = 1, \dots, \nu,
$$

$$
C_{h,n} = \sum_{l=1}^{\nu} w_l A_l + \tfrac{1}{2} \sum_{l,j=1}^{\nu} w_{l,j}[A_l, A_j],
$$

$$
\Theta = \sum_{l=1}^{\nu} b_l A_l + \tfrac{1}{2} \sum_{l,j=1}^{\nu} (b_{l,j} + w_l b_j)[A_l, A_j],
$$

$$
Y_{n+1} = \mathrm{expm}(\Theta - C_{h,n}) \ \mathrm{expm}(C_{h,n}) Y_n.
$$

In both cases the $a_{k;l,j}$s are evaluated according to (5.15) and the $b_{l,j}$s and $w_{l,j}$s are evaluated from (5.15) for $\theta = 1$ and $\theta = \tfrac{1}{2}$ respectively. The weights w_l are given by (7.7).

To conclude, it should be noted that all Lie-group methods for nonlinear problems require a number of exponential evaluations in the internal stages. This is a consequence of the fact that the function $A(t, Z)$ may fail to be an element of $\mathfrak{g}$ for arguments $Z \notin \mathcal{G}$. If $A(t, Z) \in \mathfrak{g}$ for all matrices Z then exponentiations (or, with greater generality, evaluations of the map ϕ) in the internal stages may be disregarded, at the cost of a minor increase of local truncation error. If $A(t, Z) \in \mathfrak{g}$ only when $Z \in \mathcal{G}$, however, disregarding the evaluation of the map ϕ in the internal stages would compromise the assurance that the numerical approximation Y_{n+1} resides in $\mathcal{G}$. However, as observed by Liu (1998), in some cases it is possible to devise simplified versions of the methods, whereby $A(t, Z)$ is projected on $\mathfrak{g}$ according to need. Specifically, setting $A = A(t, Z)$, we note that

$$
\boldsymbol{P}(A) = \tfrac{1}{2}(A - A^{\mathrm{T}})
$$

is a projector onto $\mathfrak{so}(N)$, the algebra of skew-symmetric matrices,

$$
\boldsymbol{P}(A) = A - \delta \boldsymbol{I}, \qquad \delta = \tfrac{1}{N} \mathrm{tr} A,
$$

is a projector onto $\mathfrak{sl}(N)$, the algebra of matrices with zero trace, and

$$
\boldsymbol{P}(A) = \tfrac{1}{2}(A + J A^{\mathrm{T}} J), \qquad J = \begin{bmatrix} \boldsymbol{O} & \boldsymbol{I} \\ -\boldsymbol{I} & \boldsymbol{O} \end{bmatrix},
$$

is a projector onto $\mathfrak{sp}(N)$, the algebra of symplectic matrices. Using these projectors in the internal stages of Lie-group methods may significantly reduce the number of evaluations of the map ϕ.

B. Fast computation of 3D rotations

Rotations in three dimensions are ubiquitous in computational mechanics, hence it is important to have fast algorithms for their computation. The Lie algebra $\mathfrak{so}(3)$ can be realized either as the set of all skew-symmetric 3×3 matrices with the matrix commutator as the bracket, or as the Euclidean space $\mathbb{R}^3$ with the vector product as the bracket. As we will see, some formulae are most easily expressed by representing $\mathfrak{so}(3)$ as 3-vectors, while other formulae appear more naturally in terms of skew-symmetric 3×3 matrices. It is convenient to switch back and forth between these forms. The Lie-algebra isomorphism between these two representations is given by the *hat map*,

$$\widehat{x} = \begin{bmatrix} 0 & -x_3 & x_2 \\ x_3 & 0 & -x_1 \\ -x_2 & x_1 & 0 \end{bmatrix}, \tag{B.1}$$

mapping $x \in \mathbb{R}^3$ into a 3×3 skew-symmetric matrix $\widehat{x}$ such that $\widehat{x}y = x \times y$. (Note that, for clarity's sake, we have abandoned our convention of denoting matrices with upper-case letters.) In particular, the identities

$$[\widehat{x}, \widehat{y}] = \widehat{x}\widehat{y} - \widehat{y}\widehat{x} = \widehat{x \times y}, \tag{B.2}$$
$$\widehat{x}\widehat{y}\widehat{x} = -(x^T y)\widehat{x}, \tag{B.3}$$
$$\widehat{x}^2\widehat{y} + \widehat{y}\widehat{x}^2 = -(x^T x)\widehat{y} - (x^T y)\widehat{x}, \tag{B.4}$$
$$\widehat{x}^2\widehat{y}^2 - \widehat{y}^2\widehat{x}^2 = -(x^T y)\widehat{x \times y} \tag{B.5}$$

are obeyed. For future convenience, we let

$$\varphi = \|x\| = (x^T x)^{1/2}.$$

Note that, since $\widehat{x}^3 = -\varphi^2\widehat{x}$, we deduce that for any real analytic function $f(z)$ we can easily obtain real functions $c_0(z)$, $c_1(z)$ and $c_2(z)$ such that

$$f(\widehat{x}) = c_0(\varphi)I + c_1(\varphi)\widehat{x} + c_2(\varphi)\widehat{x}^2.$$

Interesting examples include

$$\mathrm{expm}(\widehat{x}) = I + \frac{\sin\varphi}{\varphi}\widehat{x} + \frac{1}{2}\frac{\sin^2(\varphi/2)}{(\varphi/2)^2}\widehat{x}^2, \tag{B.6}$$

$$(I - \widehat{x})^{-1} = I + \frac{1}{1 + \varphi^2}(\widehat{x} + \widehat{x}^2), \tag{B.7}$$

$$\mathrm{cay}(\widehat{x}) = \left(I + \tfrac{1}{2}\widehat{x}\right)\left(I - \tfrac{1}{2}\widehat{x}\right)^{-1} \tag{B.8}$$
$$= I + c\widehat{x} + \frac{c}{2}\widehat{x}^2,$$

where

$$c \equiv c(\varphi) = \frac{4}{4 + \varphi^2}. \tag{B.9}$$

The first of these, (B.6), is well known in literature as the *Rodrigues formula* for the exponential mapping (Marsden and Ratiu 1994).

In many instances we need to compute repeatedly expressions in the general form $f(\mathrm{ad}_{\boldsymbol{x}})(\boldsymbol{y})$ for some function $f(x)$. These expressions in $\mathfrak{so}(3)$ can be computed rapidly in a very similar manner. Since $\mathrm{ad}_{\boldsymbol{x}}(\boldsymbol{y}) = \widehat{\boldsymbol{x}}\boldsymbol{y}$, it is true that

$$f(\mathrm{ad}_{\boldsymbol{x}})(\boldsymbol{y}) = f(\widehat{\boldsymbol{x}})\boldsymbol{y},$$

and we may use the same technique as above to simplify $f(\widehat{\boldsymbol{x}})$:

$$\mathrm{dexp}_{\boldsymbol{x}} = \frac{\exp\widehat{\boldsymbol{x}} - \boldsymbol{I}}{\widehat{\boldsymbol{x}}} = \boldsymbol{I} + \frac{\sin^2(\varphi/2)}{\varphi^2/2}\widehat{\boldsymbol{x}} + \frac{\varphi - \sin\varphi}{\varphi^3}\widehat{\boldsymbol{x}}^2, \qquad \text{(B.10)}$$

$$\mathrm{dexp}_{\boldsymbol{x}}^{-1} = \frac{\widehat{\boldsymbol{x}}}{\exp\widehat{\boldsymbol{x}} - \boldsymbol{I}} = \boldsymbol{I} - \frac{1}{2}\widehat{\boldsymbol{x}} - \frac{\varphi\cot(\varphi/2) - 2}{2\varphi^2}\widehat{\boldsymbol{x}}^2, \qquad \text{(B.11)}$$

$$\mathrm{dcay}_{\boldsymbol{x}} = c\left(\boldsymbol{I} + \tfrac{1}{2}\widehat{\boldsymbol{x}}\right), \qquad \text{(B.12)}$$

$$\mathrm{dcay}_{\boldsymbol{x}}^{-1} = \boldsymbol{I} - \tfrac{1}{2}\widehat{\boldsymbol{x}} + \tfrac{1}{4}\boldsymbol{x}\boldsymbol{x}^{\mathrm{T}}, \qquad \text{(B.13)}$$

with c as in (B.9). These formulae are based on the representation of $\mathfrak{so}(3)$ as 3-vectors, so (B.12), for instance, should read $\mathrm{dcay}_{\boldsymbol{x}}(\boldsymbol{y}) = c(\boldsymbol{y} + \tfrac{1}{2}\widehat{\boldsymbol{x}}\boldsymbol{y}) = c(\boldsymbol{y} + \tfrac{1}{2}\boldsymbol{x}\times\boldsymbol{y})$.

If $U = \mathrm{expm}(\widehat{\boldsymbol{x}})$ is an orthogonal matrix, the matrix $\widehat{\boldsymbol{x}}$ can be obtained by means of the matrix logarithm as

$$\widehat{\boldsymbol{x}} = \mathrm{logm}(U) = \frac{\sin^{-1}\|\boldsymbol{y}\|}{\|\boldsymbol{y}\|}\widehat{\boldsymbol{y}}, \qquad \widehat{\boldsymbol{y}} = \tfrac{1}{2}(U - U^{\mathrm{T}}). \qquad \text{(B.14)}$$

Similarly, we may invert the Cayley map as

$$\widehat{\boldsymbol{x}} = \mathrm{cay}^{-1}(U) = 2\left(\frac{1 - \sqrt{1 - \|\boldsymbol{y}\|^2}}{\|\boldsymbol{y}\|^2}\right)\widehat{\boldsymbol{y}}, \qquad \widehat{\boldsymbol{y}} = \tfrac{1}{2}(U - U^{\mathrm{T}}). \qquad \text{(B.15)}$$

Also, canonical coordinates of the second kind (6.15) can be evaluated explicitly. Letting

$$C_1 = \begin{bmatrix} 0 & 1 & 0 \\ -1 & 0 & 0 \\ 0 & 0 & 0 \end{bmatrix}, \qquad C_2 = \begin{bmatrix} 0 & 0 & 1 \\ 0 & 0 & 0 \\ -1 & 0 & 0 \end{bmatrix}, \qquad C_3 = \begin{bmatrix} 0 & 0 & 0 \\ 0 & 0 & 1 \\ 0 & -1 & 0 \end{bmatrix}$$

be a Chevalley basis, we have

$$\mathrm{expm}(\widehat{\boldsymbol{x}}) = \mathrm{expm}(\alpha_1 C_1)\,\mathrm{expm}(\alpha_2 C_2)\,\mathrm{expm}(\alpha_3 C_3),$$